PRACTICING PHYSICS

CONCEPTUAL Physics

PRACTICING PHYSICS

CONCEPTUAL Physics

ninth edition

Paul G. Hewitt

City College of San Francisco

San Francisco Boston New York
Capetown Hong Kong London Madrid Mexico City
Montreal Munich Paris Singapore Sydney Tokyo Toronto

Cover Credit: G. Brad Lewis/Stone

ISBN 0-321-05153-X

---

5 6 7 8 9—BTP—04 03

www.aw.com/physics

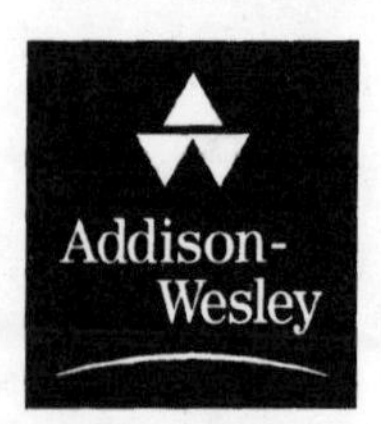

Welcome to the Conceptual Physics Practice Book

These practice pages supplement *Conceptual Physics, Ninth Edition.* Their purpose is as the name implies — practice — not testing. You'll find it is easier to learn physics by *doing* it — by *practicing*. AFTER you've worked through a page, check your responses with the reduced pages with answers beginning on page 116.

Pages 167 – 224 show solutions to the odd-numbered exercises and problems in the textbook.

At the end of this book are sample multiple-choice questions for all eight parts of the textbook.

Table of Contents

Name ______________________________ Date ____________

CONCEPTUAL **Physics** PRACTICE PAGE

Chapter 1 About Science
Making Hypotheses

The word science comes for Latin, meaning "to know." The word *hypothesis* comes from Greek, "under an idea." A hypothesis (an educated guess) often leads to new knowledge and may help to establish a theory.

Examples:

1. It is well known that objects generally expand when heated. An iron plate gets slightly bigger, for example, when put in a hot oven. But what of a hole in the middle of the plate? Will the hole get bigger or smaller when expansion occurs? One friend may say the size of the hole will increase, and another says it will decrease.

   a. What is your hypothesis about hole size, and if you are wrong, is there a test for finding out?

   ____

   ____

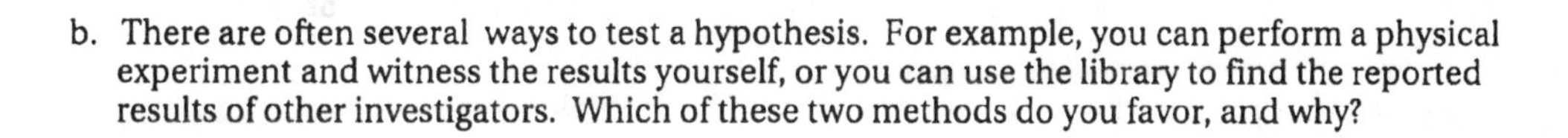

   b. There are often several ways to test a hypothesis. For example, you can perform a physical experiment and witness the results yourself, or you can use the library to find the reported results of other investigators. Which of these two methods do you favor, and why?

   ____

   ____

2. Before the time of the printing press, books were hand-copied by scribes, many of whom were monks in monasteries. There is the story of the scribe who was frustrated to find a smudge on an important page he was copying. The smudge blotted out part of the sentence that reported the number of teeth in the head of a donkey. The scribe was very upset and didn't know what to do. He consulted with other scribes to see if any of their books stated the number of teeth in the head of a donkey. After many hours of fruitless searching through the library, it was agreed that the best thing to do was to send a messenger by donkey to the next monastery and continue the search there. What would be your advice?

____

Making Distinctions

Many people don't seem to see the difference between a thing and the *abuse* of the thing. For example, a city council that bans skateboarding may not distinguish between skateboarding and reckless skateboarding. A person who advocates that a particular technology be banned may not distinguish between that technology and the abuses of that technology. There's a difference between a thing and the abuse of the thing.

On a separate sheet of paper, list other examples where use and abuse are often not distinguished. Compare your list with others in your class.

Hewitt Drew it!

CONCEPTUAL **Physics** PRACTICE PAGE

Pinhole Image Formation

Look carefully at the round spots of light on the shady ground beneath trees. These are ***sunballs***, which are images of the sun. They are cast by openings between leaves in the trees that act as pinholes. (Did you make a pinhole "camera" back in middle school?) Large sunballs, several centimeters in diameter or so, are cast by openings that are relatively high above the ground, while small ones are produced by closer "pinholes." The interesting point is that the ratio of the diameter of the sunball to its distance from the pinhole is the same as the ratio of the sun's diameter to its distance from the pinhole. We know the sun is approximately 150,000,000 km from the pinhole, so careful measurements of the ratio of diameter/distance for a sunball leads you to the diameter of the sun. That's what this page is about. Instead of measuring sunballs under the shade of trees on a sunny day, make your own easier-to-measure sunball.

1. Poke a small hole in a piece of card. Perhaps an index card will do, and poke the hole with a sharp pencil or pen. Hold the card in the sunlight and note the circular image that is cast. This is an image of the sun. Note that its size doesn't depend on the size of the hole in the card, but only on its distance. The image is a circle when cast on a surface perpendicular to the rays — otherwise it's "stretched out" as an ellipse.

2. Try holes of various shapes; say a square hole, or a triangular hole. What is the shape of the image when its distance from the card is large compared with the size of the hole? Does the shape of the pinhole make a difference?

____

____

3. Measure the diameter of a small coin. Then place the coin on a viewing area that is perpendicular to the sun's rays. Position the card so the image of the sunball exactly covers the coin. Carefully measure the distance between the coin and the small hole in the card. Complete the following:

$$\frac{\text{Diameter of sunball}}{\text{Distance to pinhole}} = \underline{\qquad\qquad}$$

With this ratio, estimate the diameter of the sun. Show your work on a separate piece of paper.

4. If you did this on a day when the sun is partially eclipsed, what shape of image would you expect to see?

____

WHAT SHAPE DO SUNBALLS HAVE DURING A PARTIAL ECLIPSE OF THE SUN?

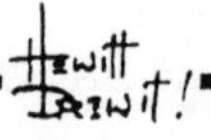

Name ______________________________ Date ______________

CONCEPTUAL Physics PRACTICE PAGE

Chapter 2 Newton's First Law of Motion—Inertia

Static Equilibrium

1. Little Nellie Newton wishes to be a gymnast and hangs from a variety of positions as shown. Since she is not accelerating, the net force on her is zero. This means the upward pull of the rope(s) equals the downward pull of gravity. She weighs 300 N. Show the scale reading for each case.

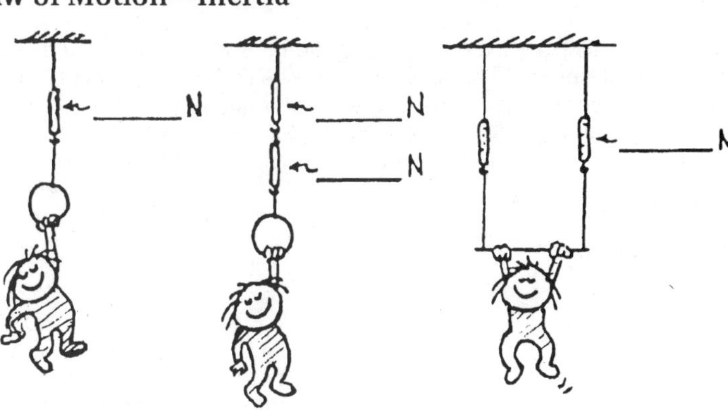

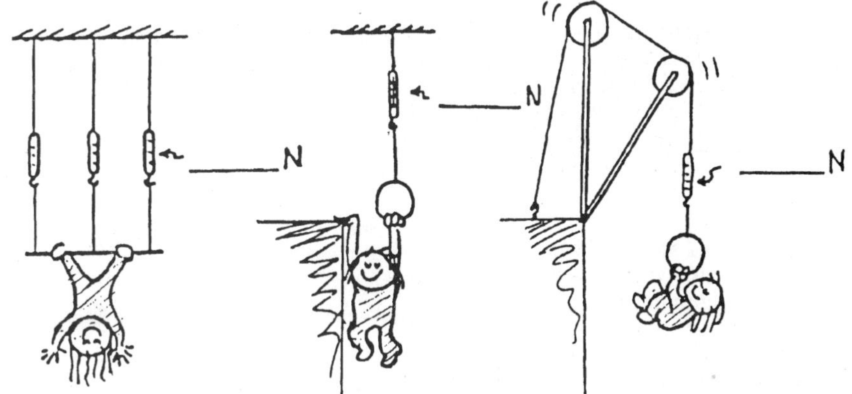

2. When Burl the painter stands in the exact middle of his staging, the left scale reads 600 N. Fill in the reading on the right scale. The total weight of Burl and staging must be __________ N.

3. Burl stands farther from the left. Fill in the reading on the right scale.

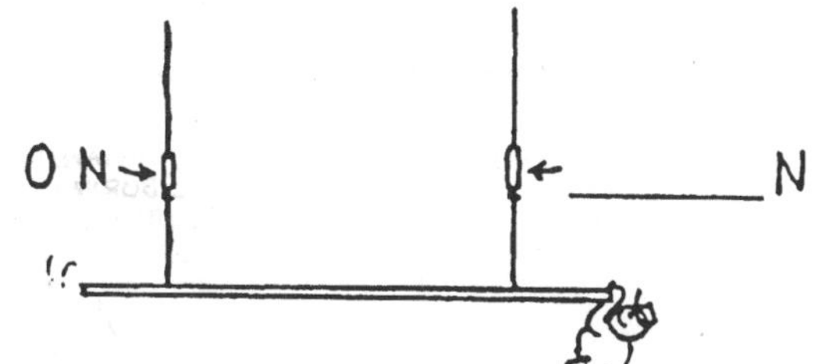

4. In a silly mood, Burl dangles from the right end. Fill in the reading on the right scale.

Hewitt Drew it!

Chapter 2 Newton's First Law of Motion—Inertia

The Equilibrium Rule: $\Sigma F = 0$

1. Manuel weighs 1000 N, and stands in the middle of a board that weighs 200 N. The ends of the board rest on bathroom scales. (We can assume the weight of the board acts at its center). Fill in the correct weight reading on each scale.

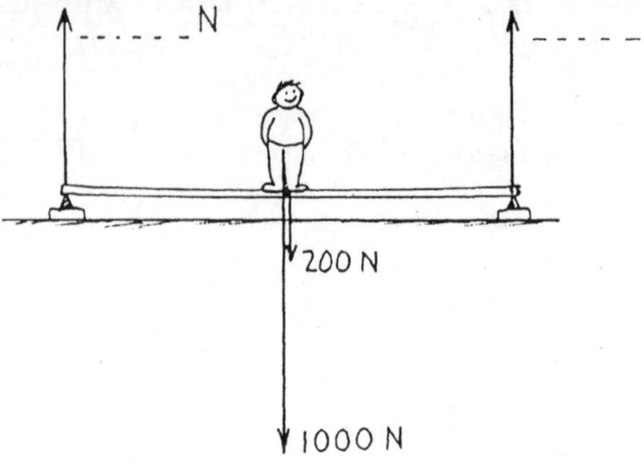

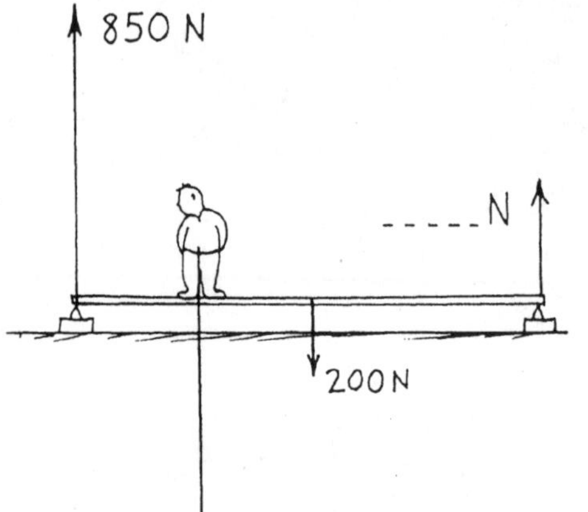

2. When Manuel moves to the left as shown, the scale closest to him reads 850 N. Fill in the weight reading for the far scale.

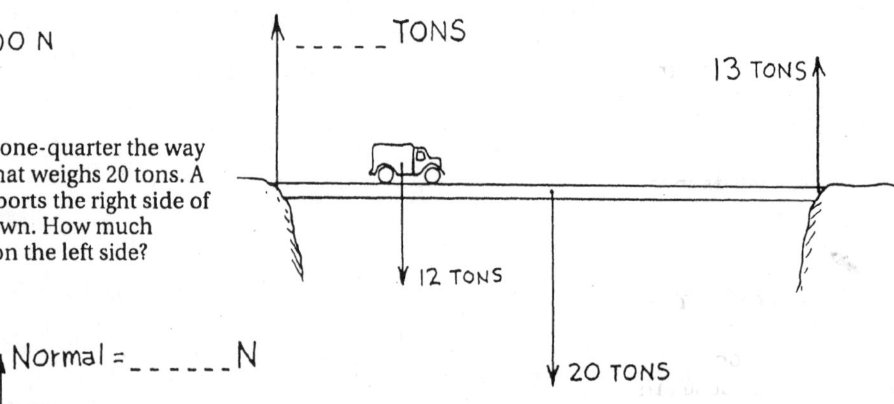

3. A 12-ton truck is one-quarter the way across a bridge that weighs 20 tons. A 13-ton force supports the right side of the bridge as shown. How much support force is on the left side?

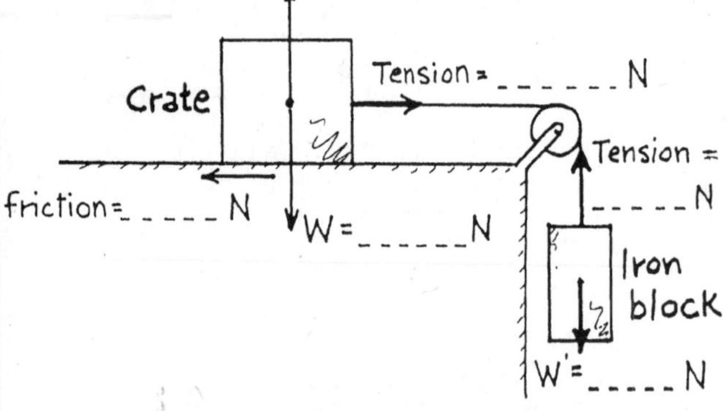

4. A 1000-N crate resting on a horizontal surface is connected to a 500-N iron block through a frictionless pulley as shown. Friction between the crate and surface is enough to keep the system at rest. The arrows show the forces that act on the crate and the block. Fill in the magnitude of each force.

5. If the crate and block in the preceding question move at constant speed, the tension in the rope (is the same) (increases) (decreases.)

The sliding system is then in (static equilibrium) (dynamic equilibrium).

Hewitt Drew it!

Name Kevin Hewston Date 9/26/05

CONCEPTUAL Physics PRACTICE PAGE

Chapter 3 Linear Motion

Free Fall Speed

1. Aunt Minnie gives you \$10 per second for 4 seconds. How much money do you have after 4 seconds? \$40

2. A ball dropped from rest picks up speed at 10 m/s per second. After it falls for 4 seconds, how fast is it going? 40 m/s

3. You have \$20, and Uncle Harry gives you \$10 each second for 3 seconds. How much money do you have after 3 seconds? \$50

4. A ball is thrown straight down with an initial speed of 20 m/s. After 3 seconds, how fast is it going? 60 m/s

5. You have \$50 and you pay Aunt Minnie \$10/second. When will your money run out? 5 s

6. You shoot an arrow straight up at 50 m/s. When will it run out of speed? 5 s

7. So what will be the arrow's speed 5 seconds after you shoot it? 0 m/s

8. What will its speed be 6 seconds after you shoot it? 7 seconds? 10 m/s 20 m/s

$V = V_0 + at$

$d = V_0 t + \frac{1}{2} a t^2$

Free Fall Distance

$d = 50(5) + \frac{1}{2}(-10)(5)^2$

$250 + 125 = 125$

1. Speed is one thing; distance another. *Where* is the arrow you shoot up at 50 m/s when it runs out of speed? 125 m

$V = 0$, $V_0 = 50\ m/s$

$0 = 50 - 10t$, $t = 5$

2. How high will the arrow be 7 seconds after being shot up at 50 m/s? 105 m

3 a. Aunt Minnie drops a penny into a wishing well and and it falls for 3 seconds before hitting the water. How fast is it going when it hits? 30 m/s

b. What is the penny's average speed during its 3-second drop? 15 m/s

c. How far down is the water surface? 45 m

FROM REST,
$v = 10t$
$d = 5t^2$

4. Aunt Minnie didn't get her wish, so she goes to a deeper wishing well and throws a penny straight down into it at 10 m/s. How far does this penny go in 3 seconds? 75 m/s

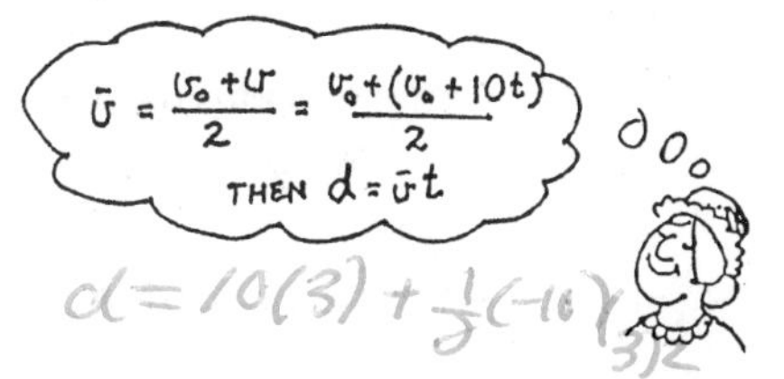

Distinguish between "how fast," "how far," and "how long"!

$d = 10(3) + \frac{1}{2}(-10)(3)^2$

Hewitt Drew it!

CONCEPTUAL **Physics** PRACTICE PAGE

Acceleration of Free Fall

A rock dropped from the top of a cliff picks up speed as it falls. Pretend that a speedometer and odometer are attached to the rock to show readings of speed and distance at 1-second intervals. Both speed and distance are zero at time = zero (see sketch). Note that after falling 1 second the speed reading is 10 m/s and the distance fallen is 5 m. The readings for succeeding seconds of fall are not shown and are left for you to complete. So draw the position of the speedometer pointer and write in the correct odometer reading for each time. Use $g = 10\ \text{m/s}^2$ and neglect air resistance.

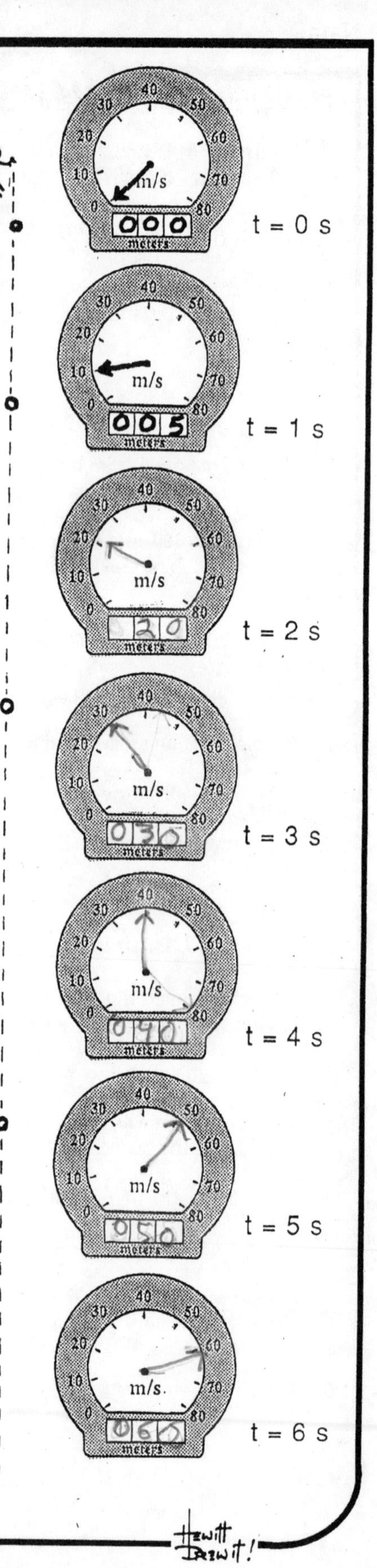

YOU NEED TO KNOW:

Instantaneous speed of fall from rest:

$$v = gt$$

Distance fallen from rest:

$$d = \frac{1}{2}gt^2$$

1. The speedometer reading increases by the same amount, 10 m/s, each second. This increase in speed per second is called acceleration.

2. The distance fallen increases as the square of the time.

3. If it takes 7 seconds to reach the ground, then its speed at impact is 70 m/s, the total distance fallen is 245 m, and its acceleration of fall just before impact is 10 m/s^2.

Name ______________________ Date __________

CONCEPTUAL **Physics** PRACTICE PAGE

Chapter 4 Newton's Second Law of Motion

Force and Acceleration

1. Skelly the skater, total mass 25 kg, is propelled by rocket power.

   a. Complete Table I (neglect resistance)

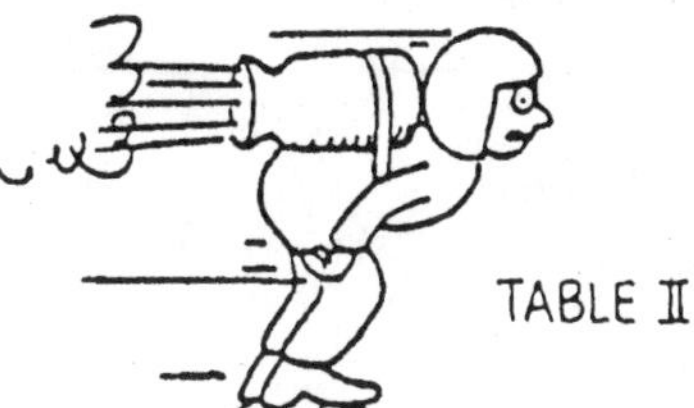

TABLE I

FORCE	ACCELERATION
100 N	
200 N	
	10 m/s^2

   b. Complete Table II for a constant 50-N resistance.

TABLE II

FORCE	ACCELERATION
50 N	0 m/s^2
100 N	
200 N	

2. Block A on a horizontal friction-free table is accelerated by a force from a string attached to Block B. B falls vertically and drags A horizontally. Both blocks have the same mass m. (Neglect the string's mass.)

   (Circle the correct answers)

   a. The *mass* of the system [A + B] is (m) (2 m).

   b. The *force* that accelerates [A + B] is the weight of (A) (B) (A + B).

   c. The weight of B is ($mg/2$) (mg) (2 mg).

   d. Acceleration of [A + B] is (less than g) (g) (more than g).

   e. Use $a =$ to show the acceleration of [A + B] as a fraction of g. ______________

If B were allowed to fall by itself, not dragging A, then wouldn't its acceleration be g?

Yes, because the force that accelerates it would only be acting on its own mass – not twice the mass!

To better understand this, consider 3 and 4 on the other side!

Force and Acceleration continued

3. Suppose A is still a 1-kg block, but B is a low-mass feather (or a coin).

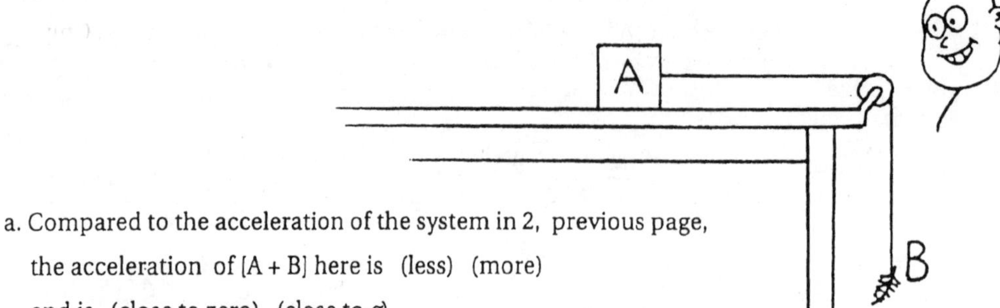

a. Compared to the acceleration of the system in 2, previous page, the acceleration of [A + B] here is (less) (more) and is (close to zero) (close to *g*).

b. In this case the acceleration of B is (practically that of free fall) (constrained).

4. Suppose A is a feather or coin, and B has a mass of 1 kg.

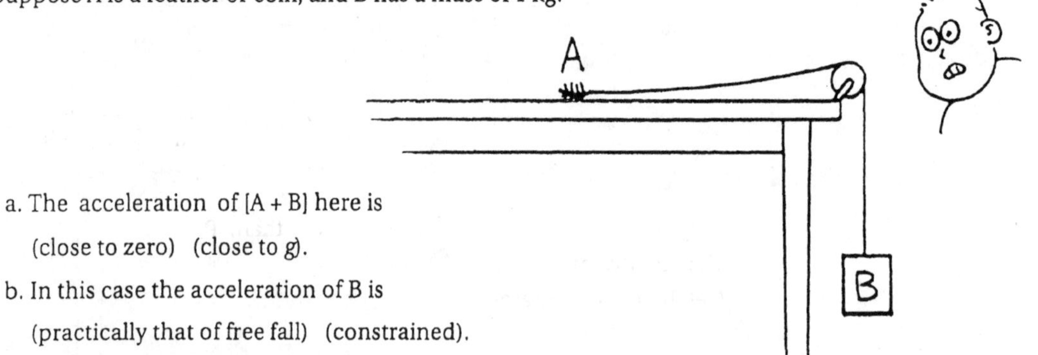

a. The acceleration of [A + B] here is (close to zero) (close to *g*).

b. In this case the acceleration of B is (practically that of free fall) (constrained).

5. Summarizing 2, 3, and 4, where the weight of one object causes the acceleration of two objects, we see the range of possible accelerations is

(between zero and *g*) (between zero and infinity) (between *g* and infinity).

6. A ball rolls down a uniform-slope ramp.

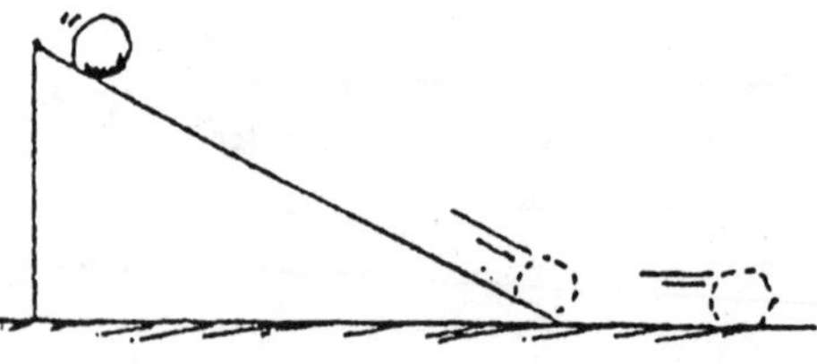

a. Acceleration is (decreasing) (constant) (increasing).

b. If the ramp were steeper, acceleration would be (more) (the same) (less).

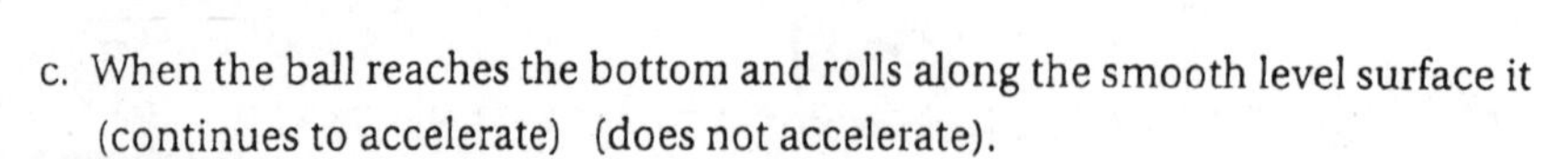

c. When the ball reaches the bottom and rolls along the smooth level surface it (continues to accelerate) (does not accelerate).

Hewitt Drew it!

Name ______________________________ Date __________

CONCEPTUAL **Physics** PRACTICE PAGE

Chapter 4 Newton's Second Law of Motion
Friction

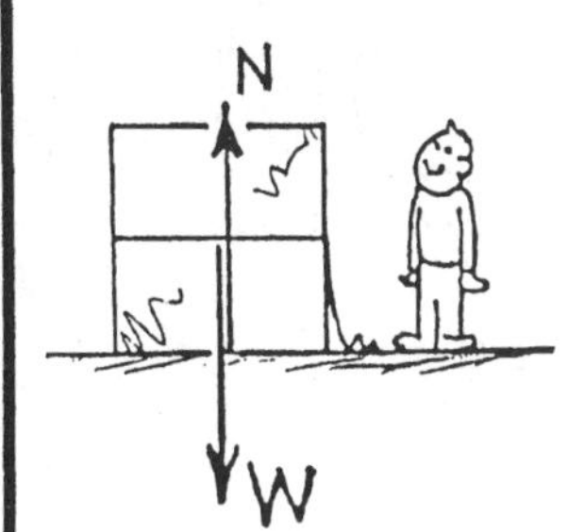

1. A crate filled with delicious junk food rests on a horizontal floor. Only gravity and the support force of the floor act on it, as shown by the vectors for weight W and normal force N.

   a. The net force on the crate is (zero) (greater than zero).

   b. Evidence for this is ______________________________.

2. A slight pull P is exerted on the crate, not enough to move it. A force of friction f now acts,

   a. which is (less than) (equal to) (greater than) P.

   b. Net force on the crate is (zero) (greater than zero).

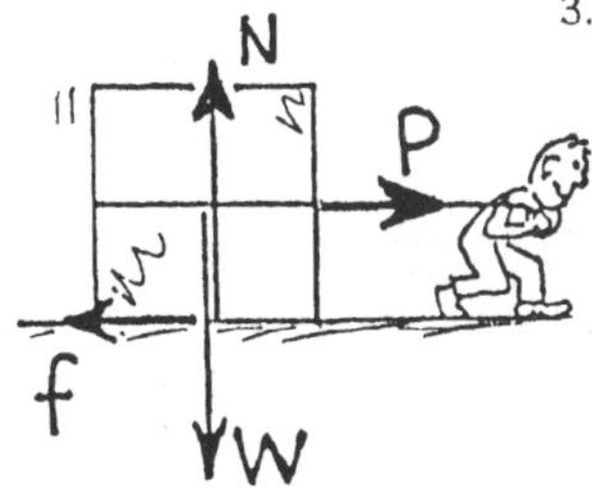

3. Pull P is increased until the crate begins to move. It is pulled so that it moves with constant velocity across the floor.

   a. Friction f is (less than) (equal to) (greater than) P.

   b. Constant velocity means acceration is (zero) (greater than zero).

   c. Net force on the crate is (less than) (equal to) (greater than) zero.

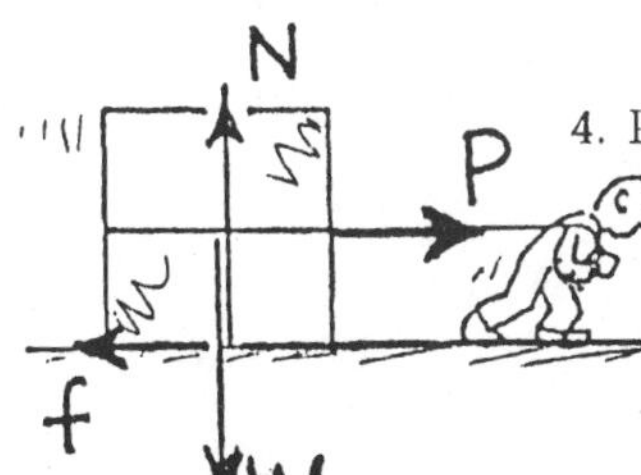

4. Pull P is further increased and is now greater than friction f.

   a. Net force on the crate is (less than) (equal to) (greater than) zero.

   b. The net force acts toward the right, so acceleration acts toward the (left) (right).

5. If the pulling force P is 150 N and the crate doesn't move, what is the magnitude of f? __________

6. If the pulling force P is 200 N and the crate doesn't move, what is the magnitude of f? __________

7. If the force of sliding friction is 250 N, what force is necessary to keep the crate sliding at constant velocity? __________

8. If the mass of the crate is 50 kg and sliding friction is 250 N, what is the acceleration of the crate when the pulling force is 250 N? __________ 300 N? __________ 500 N? __________

Hewitt Drew it!

Fallling and Air Resistance

Bronco skydives and parachutes from a stationary helicopter. Various stages of fall are shown in positions *a* through *f*. Using Newton's 2nd law,

$$a = \frac{F_{NET}}{m} = \frac{W-R}{m}$$

find Bronco's acceleration at each position (answer in the blanks to the right). You need to know that Bronco's mass *m* is 100 kg so his weight is a constant 1000 N. Air resistance *R* varies with speed and cross-sectional area as shown.

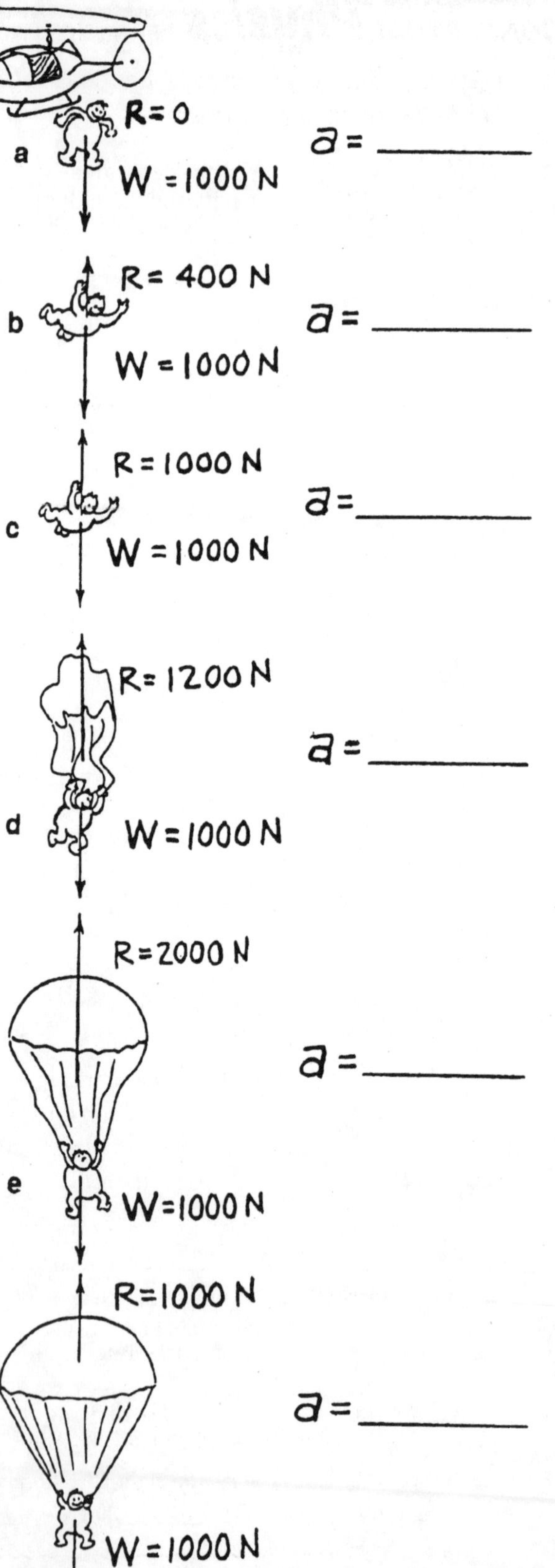

Circle the correct answers.

1. When Bronco's speed is least, his acceleration is

   (least) (most).

2. In which position(s) does Bronco experience a downward acceleration?

   (a) (b) (c) (d) (e) (f)

3. In which position(s) does Bronco experience an upward acceleration?

   (a) (b) (c) (d) (e) (f)

4. When Bronco experiences an upward acceleration, his velocity is

   (still downward) (upward also).

5. In which position(s) is Bronco's velocity constant?

   (a) (b) (c) (d) (e) (f)

6. In which position(s) does Bronco experience terminal velocity?

   (a) (b) (c) (d) (e) (f)

7. In which position(s) is terminal velocity greatest?

   (a) (b) (c) (d) (e) (f)

8. If Bronco were heavier, his terminal velocity would be

   (greater) (less) (the same).

Hewitt Drew it!

Name ______________________ Date __________

CONCEPTUAL Physics PRACTICE PAGE

Chapter 5 Newton's Third Law of Motion

Action and Reaction Pairs

1. In the example below, the action-reaction pair is shown by the arrows (vectors), and the action-reaction described in words. In (*a*) through (*g*) draw the other arrow (vector) and state the reaction to the given action. Then make up your own example in (*h*).

Example:

Fist hits wall.

Wall hits fist.

Head bumps ball.

(*a*) ______________________

Windshield hits bug.

(*b*) ______________________

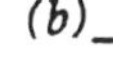

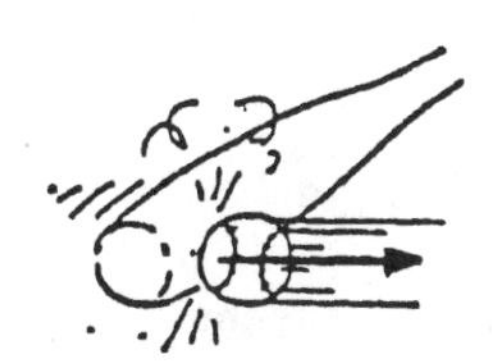

Bat hits ball.

(*c*) ______________________

Hand touches nose.

(*d*) ______________________

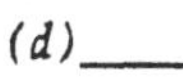

Hand pulls on flower.

(*e*) ______________________

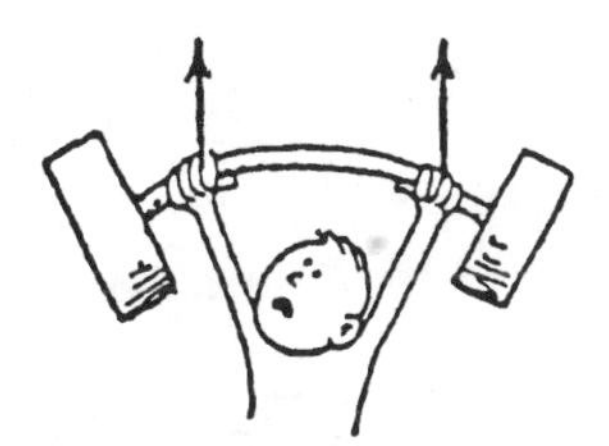

Athlete pushes bar upward.

(*f*) ______________________

____

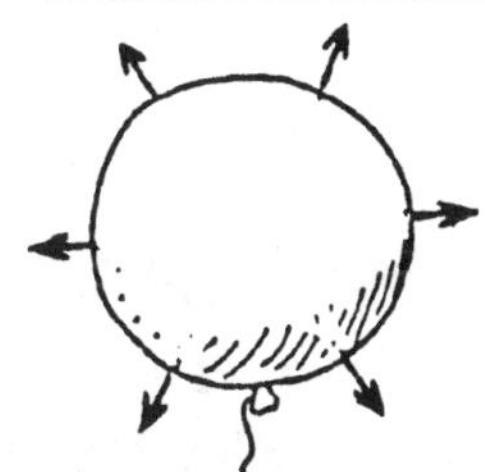

Compressed air pushes balloon surface outward.

(*g*) ______________________

____

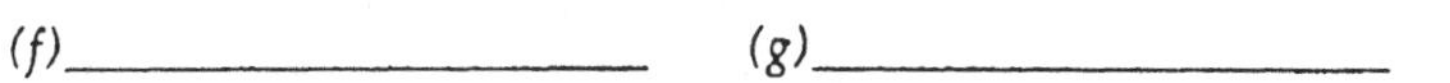

(*h*) ______________________

____

____

____

2. Draw arrows to show the chain of at least six pairs of action-reaction forces below.

YOU CAN'T TOUCH WITHOUT BEING TOUCHED— *NEWTON'S THIRD LAW*

Hewitt Drew it!

Interactions

3. Nellie Newton holds an apple weighing 1 newton at rest on the palm of her hand. The force vectors shown are the forces that act on the apple.

N
W

a. To say the weight of the apple is 1 N is to say that a downward gravitational force of 1 N is exerted on the apple by
(the earth) (her hand).

b. Nellie's hand supports the apple with normal force N, which acts in a direction opposite to W. We can say N
(equals W) (has the same magnitude as W).

c. Since the apple is at rest, the net force on the apple is (zero) (nonzero).

d. Since N is equal and opposite to W, we (can) (cannot) say that N and W comprise an action-reaction pair. The reason is because action and reaction always
(act on the same object) (act on different objects),
and here we see N and W
(both acting on the apple) (acting on different objects).

e. In accord with the rule, "If ACTION is A acting on B, then REACTION is B acting on A," if we say *action* is the earth pulling down on the apple, *reaction* is
(the apple pulling up on the earth) (N, Nellie's hand pushing up on the apple).

f. To repeat for emphasis, we see that N and W are equal and opposite to each other
(and comprise an action-reaction pair) (but do *not* comprise an action-reaction pair).

To identify a pair of action-reaction forces in any situation, first identify the pair of interacting objects involved. Something is interacting with something else. In this case the whole earth is interacting (gravitationally) with the apple. So the earth pulls downward on the apple (call it action), while the apple pulls upward on the earth (reaction).
Simply put, earth pulls on apple (action); apple pulls on earth (reaction).
Better put, apple and earth *pull on each other* with equal and opposite forces that comprise a *single* interaction.

g. Another pair of forces is N [shown] and the downward force of the apple against Nellie's hand [not shown]. This force pair (is) (isn't) an action-reaction pair.

h. Suppose Nellie now pushes upward on the apple with a force of 2 N. The apple
(is still in equilibrium) (accelerates upward), and compared to W, the magnitude of N is
(the same) (twice) (not the same, and not twice).

i. Once the apple leaves Nellie's hand, N is (zero) (still twice the magnitude of W),
and the net force on the apple is (zero) (only W) (still W - N, a negative force).

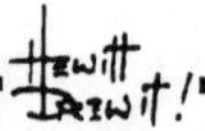
Hewitt Drew it!

Name ______________________ Date __________

CONCEPTUAL **Physics** PRACTICE PAGE

Chapter 5 Newton's Third Law of Motion
Vectors and the Parallelogram Rule

1. When vectors A and B are at an angle to each other, they add to produce the resultant C by the *parallelogram rule.* Note that C is the diagonal of a parallelogram where A and B are adjacent sides. Resultant C is shown in the first two diagrams, *a* and *b.* Construct the resultant C in diagrams *c* and *d.* Note that in diagram *d* you form a rectangle (a special case of a parallelogram).

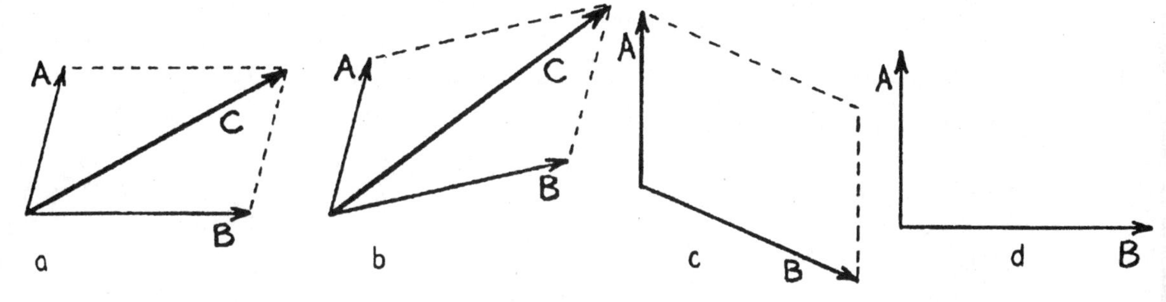

2. Below we see a top view of an airplane being blown offcourse by wind in various directions. Use the parallelogram rule to show the resulting speed and direction of travel for each case. In which case does the airplane travel fastest across the ground? _____ Slowest? ______

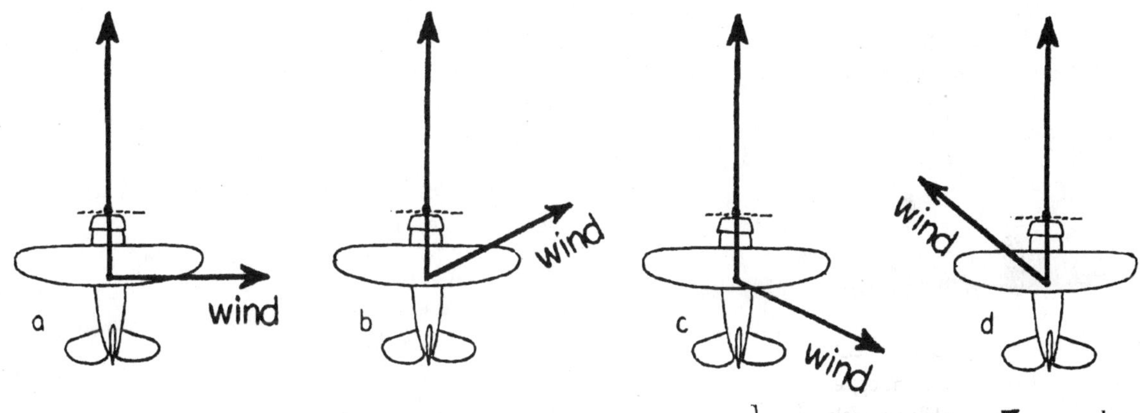

3. To the right we see top views of 3 motorboats crossing a river. All have the same speed relative to the water, and all experience the same water flow.

Construct resultant vectors showing the speed and direction of the boats.

a. Which boat takes the shortest path to the opposite shore?______

b. Which boat reaches the opposite shore first? ________

c. Which boat provides the fastest ride? ________

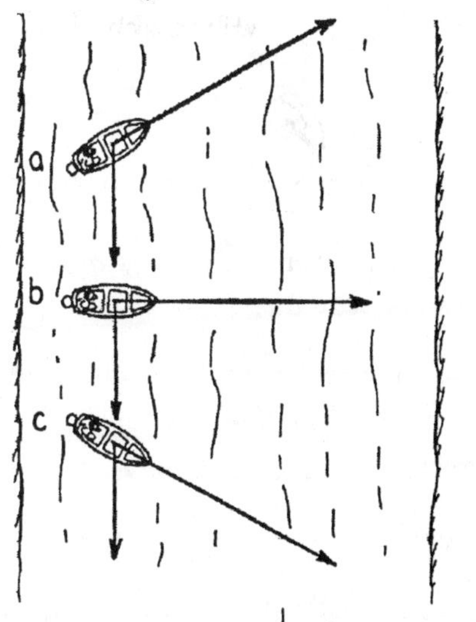

Hewitt Drew it!

CONCEPTUAL **Physics** PRACTICE PAGE

Velocity Vectors and Components

1. Draw the resultants of the four sets of vectors below.

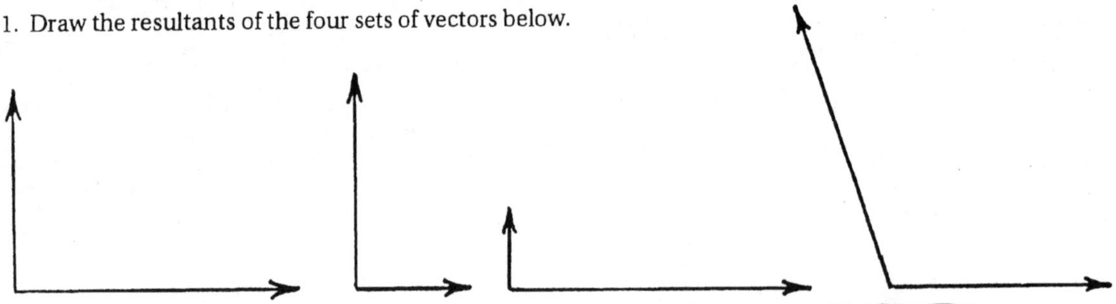

2. Draw the horizontal and vertical components of the four vectors below.

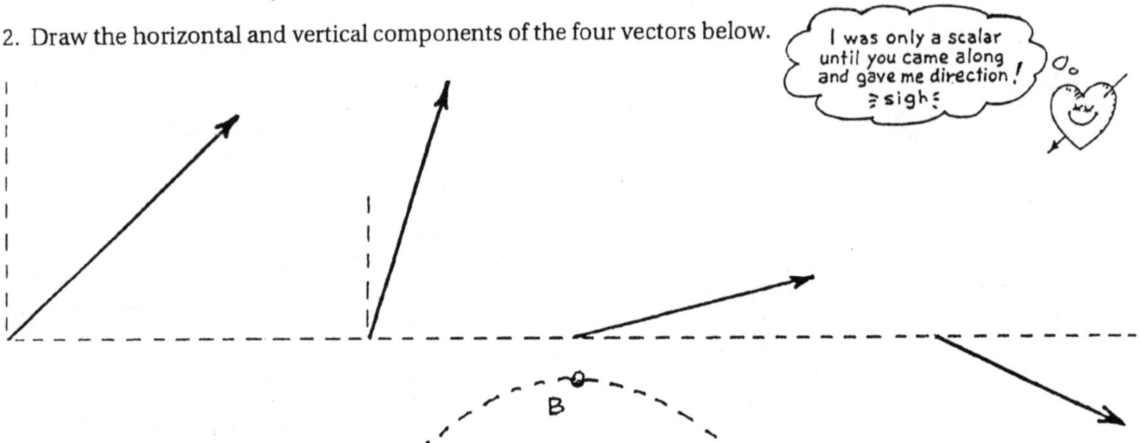

I was only a scalar until you came along and gave me direction! ≈sigh≈

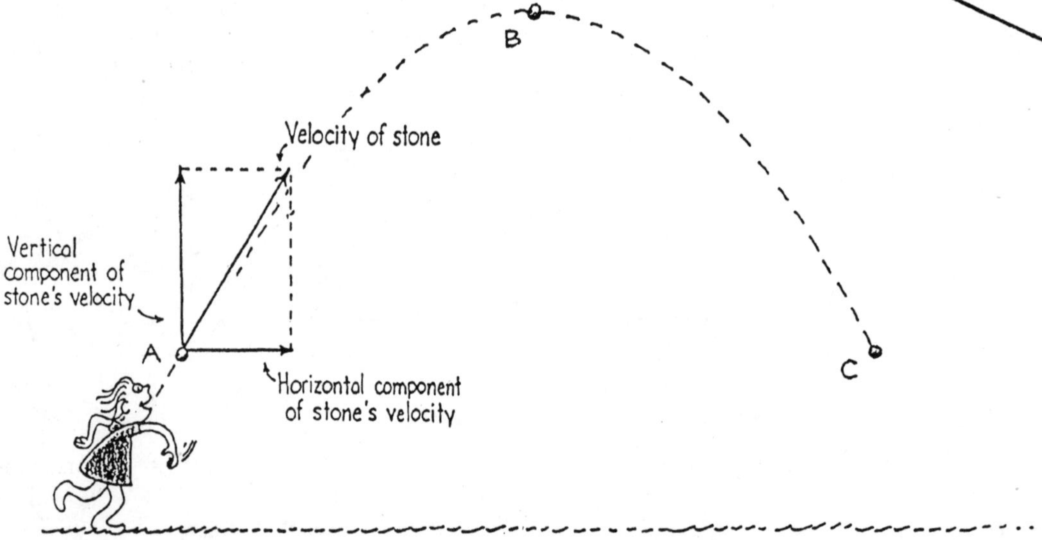

B

Velocity of stone

Vertical component of stone's velocity

A

Horizontal component of stone's velocity

C

3. She tosses the ball along the dashed path. The velocity vector, complete with its horizontal and vertical components is shown at position A. Carefully sketch the appropriate velocity vectors with appropriate components for positions B and C.

a. Since there is no acceleration in the horizontal direction, how does the horizontal component of velocity compare for positions A, B, and C? ______________________

b. What is the value of the vertical component of velocity at position B? ______________________

c. How does the vertical component of velocity at position C compare with that of position A? __________

____

Hewitt Drew it!

Name ______________________ Date __________

CONCEPTUAL Physics PRACTICE PAGE

Chapter 5 Newton's Third Law of Motion

Force and Velocity Vectors

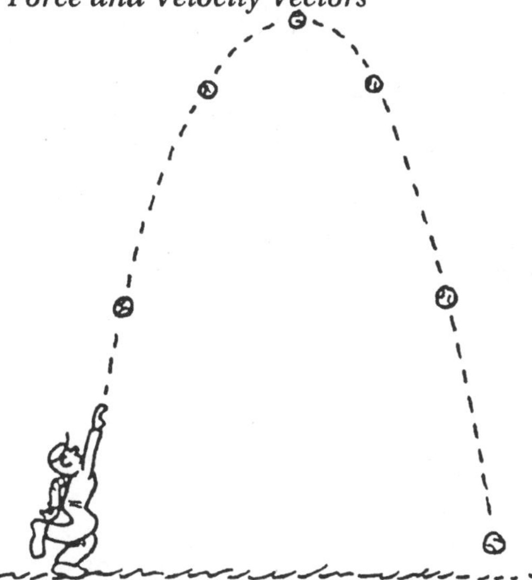

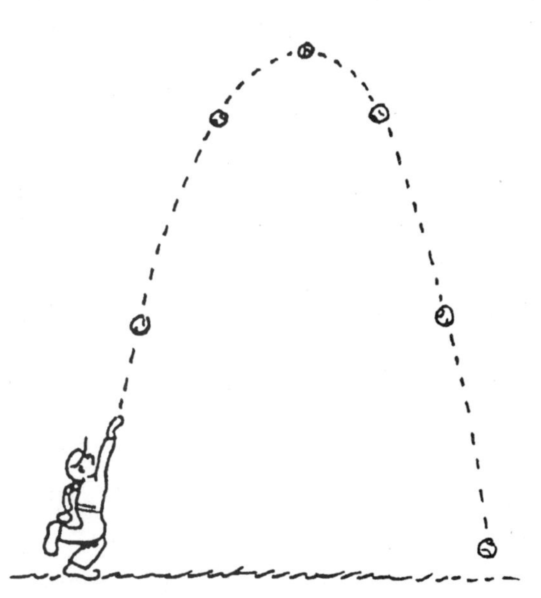

1. Draw sample vectors to represent the force of gravity on the ball in the positions shown above (after it leaves the thrower's hand). Neglect air drag.

2. Draw sample bold vectors to represent the velocity of the ball in the positions shown above. With lighter vectors, show the horizontal and vertical components of velocity for each position.

3. (a) Which velocity component in the previous question remains constant ? Why?

   ____

   (b) Which velocity component changes along the path? Why?

   ____

4. It is important to distinguish between force and velocity vectors. Force vectors combine with other force vectors, and velocity vectors combine with other velocity vectors. Do velocity vectors combine with force vectors? __________

5. All forces on the bowling ball, weight down and support of alley up, are shown by vectors at its center before it strikes the pin (*a*). Draw vectors of all the forces that act on the ball (*b*) when it strikes the pin, and (*c*) after it strikes the pin.

(a)

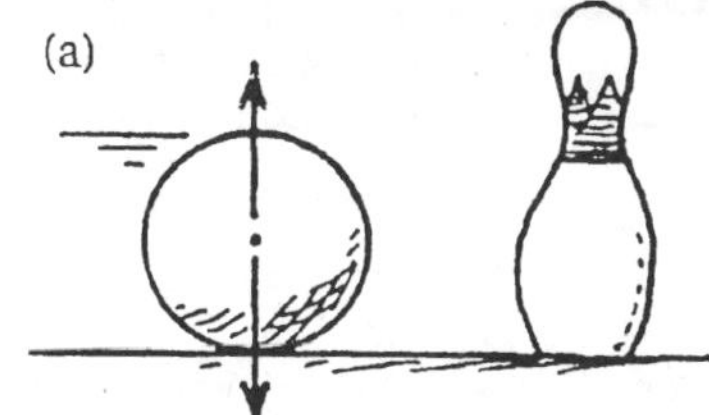

(b)

(c)

thanx to Howard Brand

Hewitt Drew it!

Force Vectors and the Parallelogram Rule

1. The heavy ball is supported in each case by two strands of rope. The tension in each strand is shown by the vectors. Use the parallelogram rule to find the resultant of each vector pair.

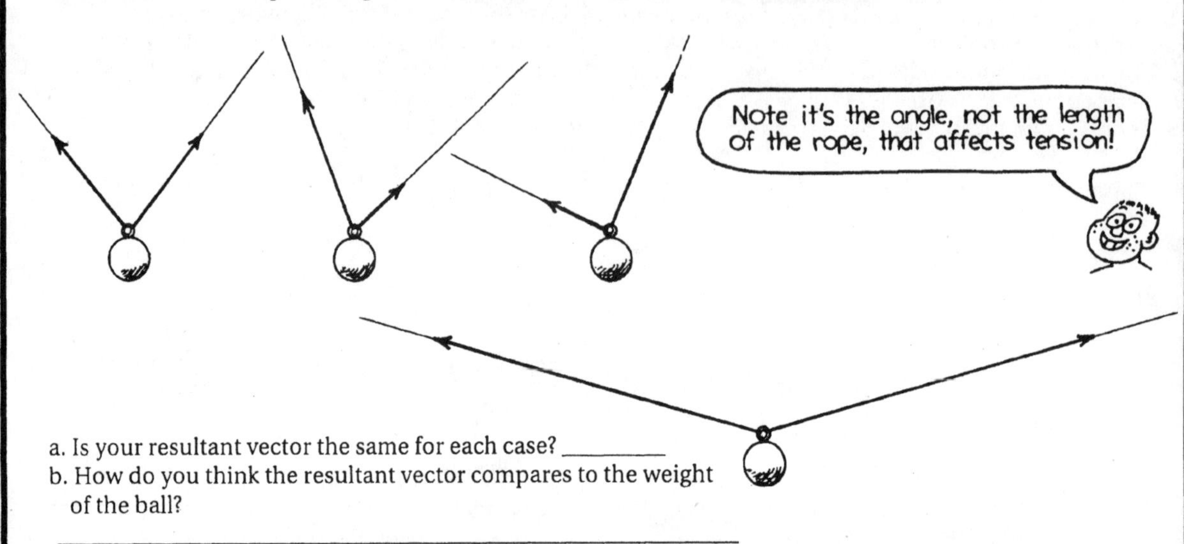

a. Is your resultant vector the same for each case? ________

b. How do you think the resultant vector compares to the weight of the ball?

____

2. Now let's do the opposite of what we've done above. More often, we know the weight of the suspended object, but we don't know the rope tensions. In each case below, the weight of the ball is shown by the vector W. Each dashed vector represents the resultant of the pair of rope tensions. Note that each is equal and opposite to vectors W (they must be; otherwise the ball wouldn't be at rest).

a. Construct parallelograms where the ropes define adjacent sides and the dashed vectors are the diagonals.

b. How do the relative lengths of the sides of each parallelogram compare to rope tensions?

c. Draw rope-tension vectors, clearly showing their relative magnitudes.

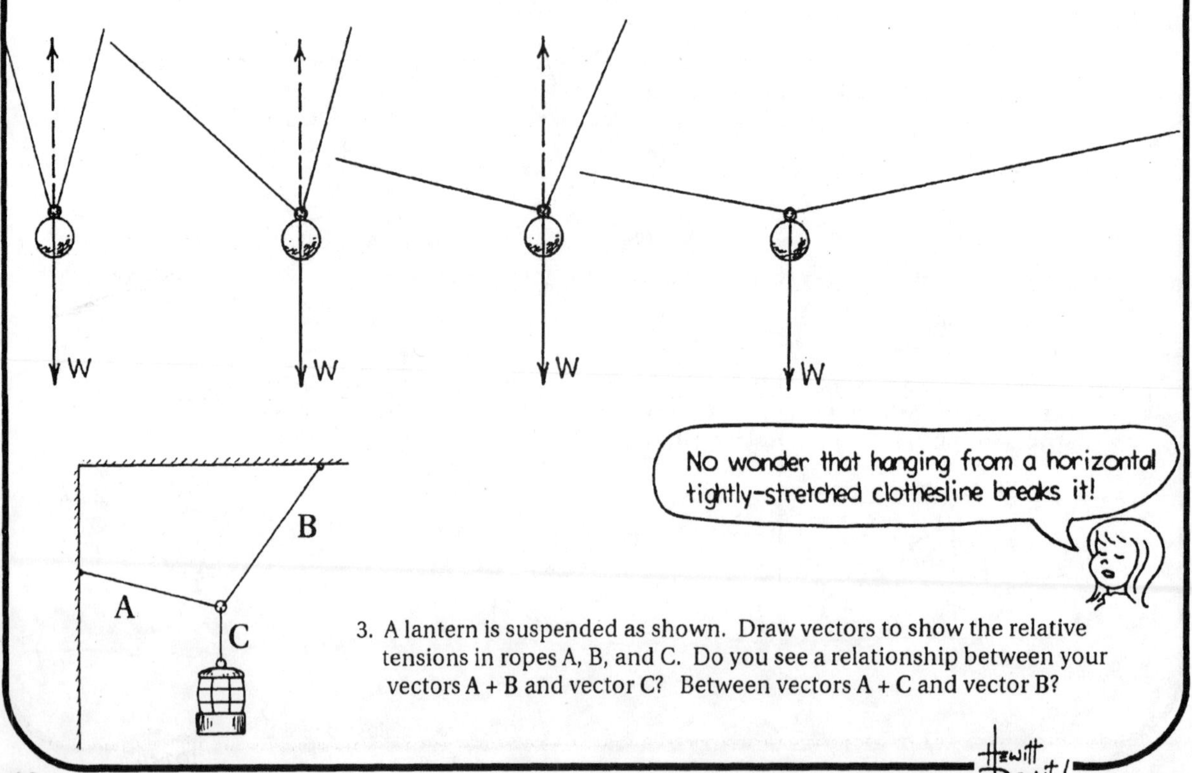

W
W
W
W
No wonder that hanging from a horizontal tightly-stretched clothesline breaks it!
B
A
C

3. A lantern is suspended as shown. Draw vectors to show the relative tensions in ropes A, B, and C. Do you see a relationship between your vectors A + B and vector C? Between vectors A + C and vector B?

Hewitt Drew it!

Name ________________________________ Date __________

CONCEPTUAL **Physics** PRACTICE PAGE

Chapter 5 Newton's Third Law of Motion

Force-Vector Diagrams

In each case, a rock is acted on by one or more forces. Draw an accurate vector diagram showing all forces acting on the rock, and no other forces. Use a ruler, and do it in pencil so you can correct mistakes. The first two are done as examples. Show by the parallelogram rule in 2 that the vector sum of **A** + **B** is equal and opposite to **W** (that is, **A** + **B** = -**W**). Do the same for 3 and 4. Draw and label vectors for the weight and normal support forces in 5 to 10, and for the appropriate forces in 11 and 12.

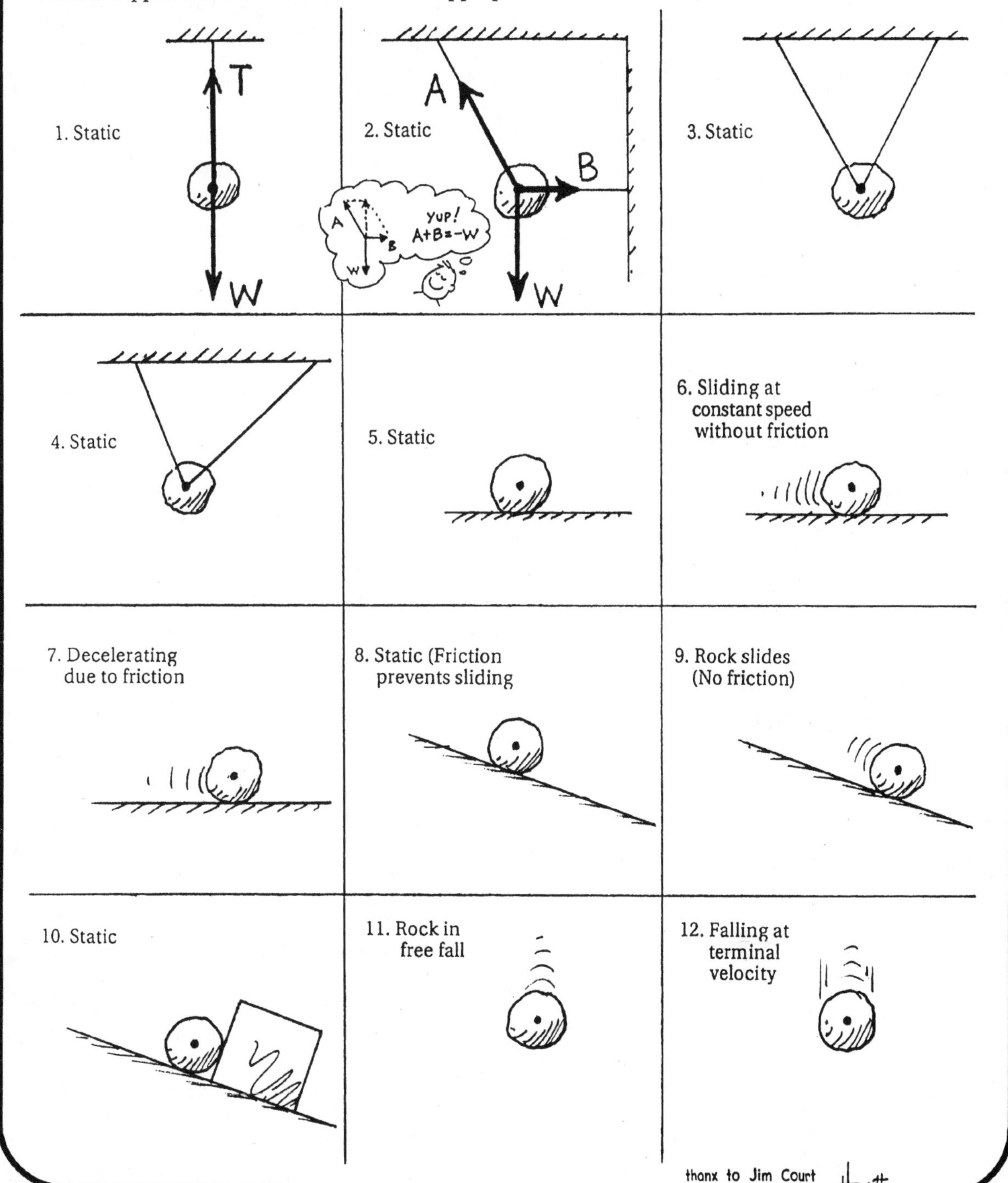

thanx to Jim Court

Name ______________________ Date __________

CONCEPTUAL Physics PRACTICE PAGE

Appendix D More About Vectors
Vectors and Sailboats

(Do not attempt this until you have studied Appendix D!)

1. The sketch shows a top view of a small railroad car pulled by a rope. The force *F* that the rope exerts on the car has one component along the track, and another component perpendicular to the track.

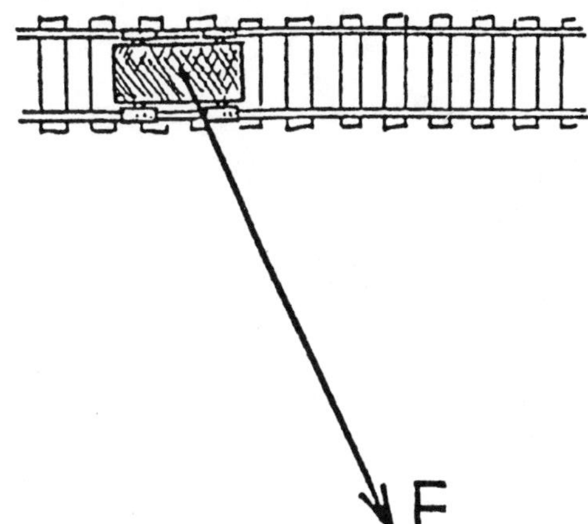

   a. Draw these components on the sketch. Which component is larger?

   ____

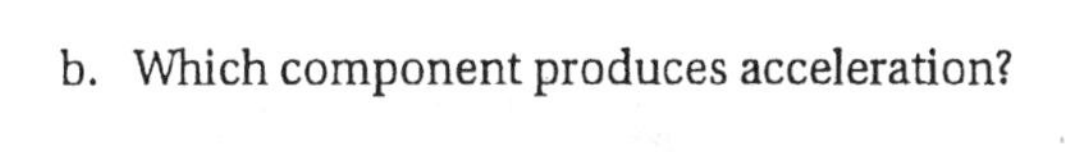

   b. Which component produces acceleration?

   ____

   c. What would be the effect of pulling on the rope if it were perpendicular to the track?

   ____

2. The sketches below represent simplified top views of sailboats in a cross-wind direction. The impact of the wind produces a FORCE vector on each as shown. (We do NOT consider *velocity* vectors here!)

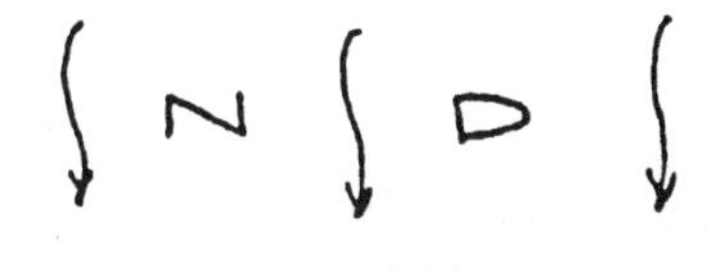

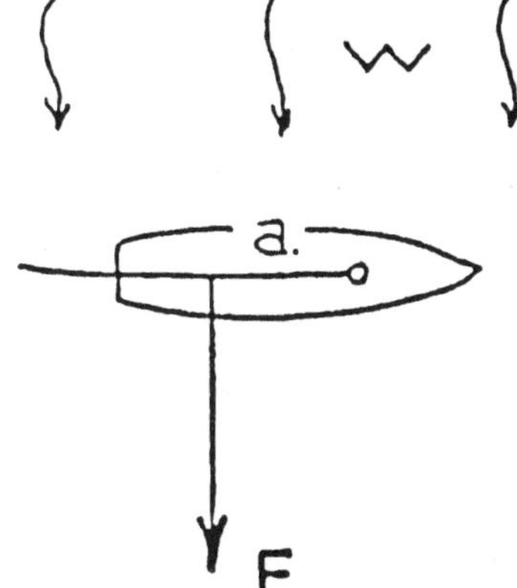

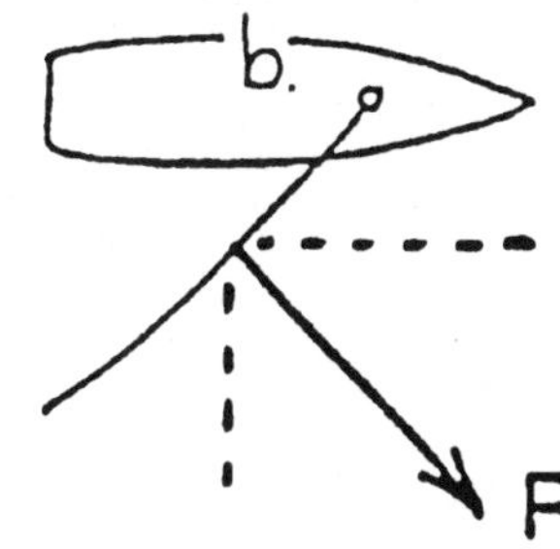

   a. Why is the position of the sail above useless for propelling the boat along its forward direction? (Relate this to Question 1c. above. Where the train is constrained by tracks to move in one direction, the boat is similarly constrained to move along one direction by its deep vertical fin — the *keel.*)

   ____

   ____

   b. Sketch the component of force parallel to the direction of the boat's motion (along its keel), and the component perpendicular to its motion. Will the boat move in a forward direction? (Relate this to Question 1b. above.)

   ____

   ____

Hewitt Drew it!

3. The boat to the right is oriented at an angle into the wind. Draw the force vector and its forward and perpendicular components.

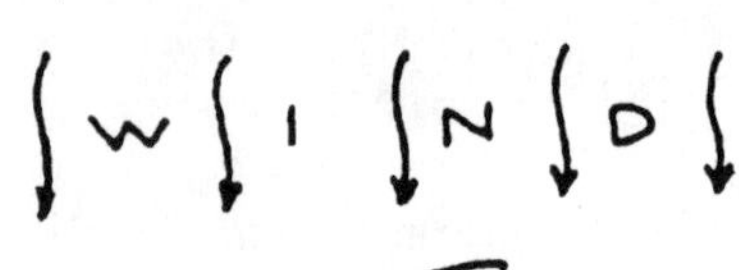

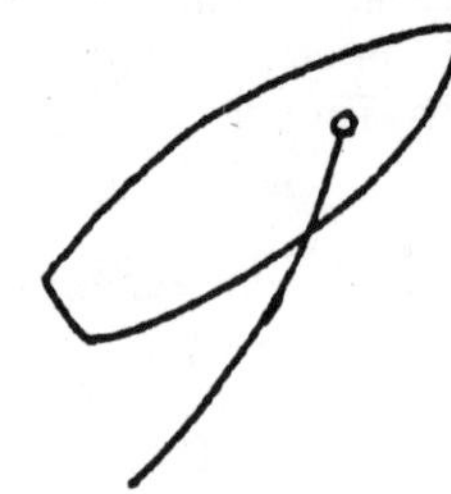

   a. Will the boat move in a forward direction and tack into the wind? Why or why not?

   ____

   ____

4. The sketch below is a top view of five identical sailboats. Where they exist, draw force vectors to represent wind impact on the sails. Then draw components parallel and perpendicular to the keels of each boat.

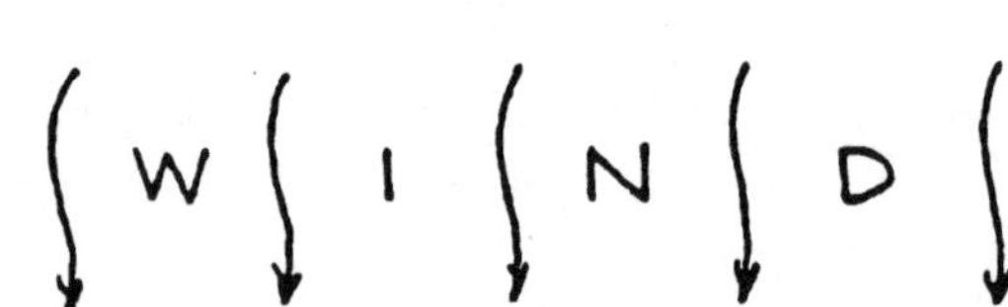

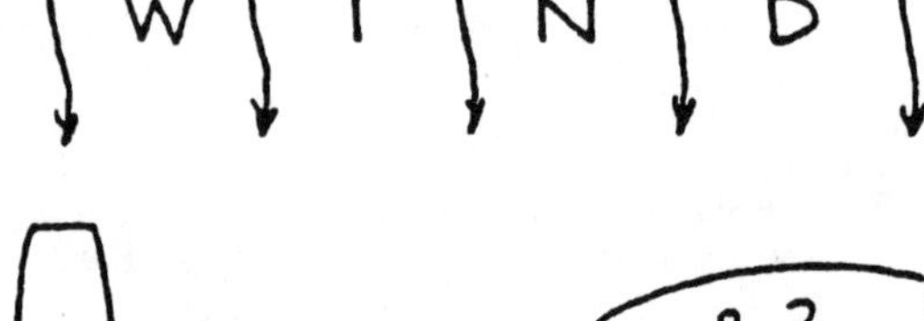

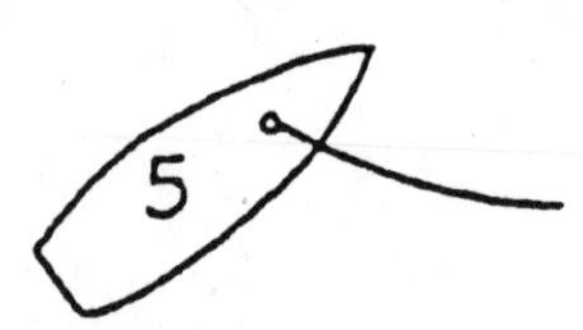

   a. Which boat will sail the fastest in a forward direction?

   ____

   b. Which will respond least to the wind?

   ____

   c. Which will move in a backward direction?

   ____

   d. Which will experience less and less wind impact with increasing speed?

   ____

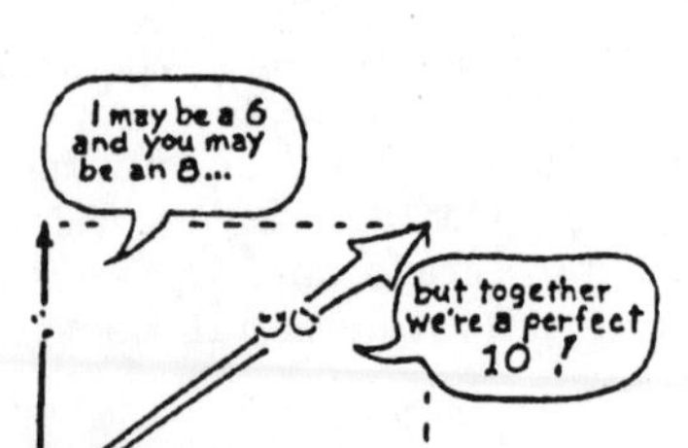

Hewitt Drew it!

Name ______________________________ Date __________

CONCEPTUAL **Physics** PRACTICE PAGE

Chapter 6 Momentum
Impulse and Momentum

1. A moving car has momentum. If it moves twice as fast, its momentum is ____________ as much.

2. Two cars, one twice as heavy as the other, move down a hill at the same speed. Compared to the lighter car, the momentum of the heavier car is __________ as much.

3. The recoil momentum of a gun that kicks is (more than) (less than) (the same as) the momentum of the bullet it fires.

4. If a man firmly holds a gun when fired, then the momentum of the bullet is equal to the recoil momentum of the
(gun alone) (gun-man system) (man alone)

5. Suppose you are traveling in a bus at highway speed on a nice summer day and the momentum of an unlucky bug is suddenly changed as it splatters onto the front window.

a. Compared to the force that acts on the bug, how much force acts on the bus?
(more) (the same) (less)

b. The time of impact is the same for both the bug and the bus. Compared to the impulse on the bug, this means the impulse on the bus is
(more) (the same) (less)

c. Although the momentum of the bus is very large compared to the momentum of the bug, the change in momentum of the bus, compared to the *change* of momentum of the bug is
(more) (the same) (less)

d. Which undergoes the greater acceleration?
(bus) (both the same) (bug)

e. Which therefore, suffers the greater damage?
(bus) (both the same) (the bug of course!)

Hewitt Drew it!

Systems

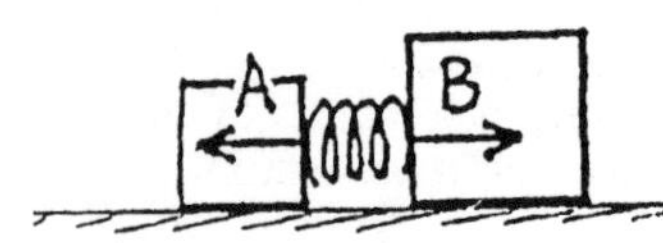

1. When the compressed spring is released, Blocks A and B will slide apart. There are 3 systems to consider here, indicated by the closed dashed lines below — System A, System B, and System A+B. Ignore the vertical forces of gravity and the support force of the table.

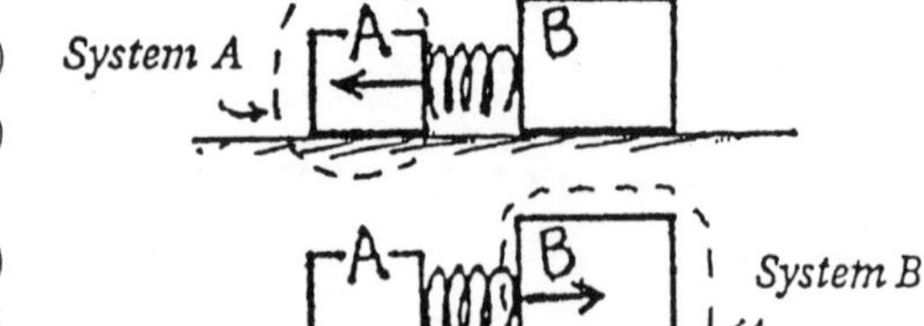

a. Does an external force act on System A? (yes) (no)

Will the momentum of System A change? (yes) (no)

b. Does an external force act on System B? (yes) (no)

Will the momentum of System B change? (yes) (no)

c. Does an external force act on System A+B? (yes) (no)

Will the momentum of System A+B change? (yes) (no)

2. Billiard ball A collides with billiard ball B at rest. Isolate each system with a closed dashed line. Draw only the external force vectors that act on each system.

System A

System B

System A+B

a. Upon collision, the momentum of System A (increases) (decreases) (remains unchanged).

b. Upon collision, the momentum of System B (increases) (decreases) (remains unchanged).

c. Upon collision, the momentum of System A+B (increases) (decreases) (remains unchanged).

3. A girl jumps upward. In the sketch to the left, draw a closed dashed line to indicate the system of the girl.

a. Is there an external force acting on her? (yes) (no)

Does her momentum change? (yes) (no)

Is the girl's momentum conserved? (yes) (no)

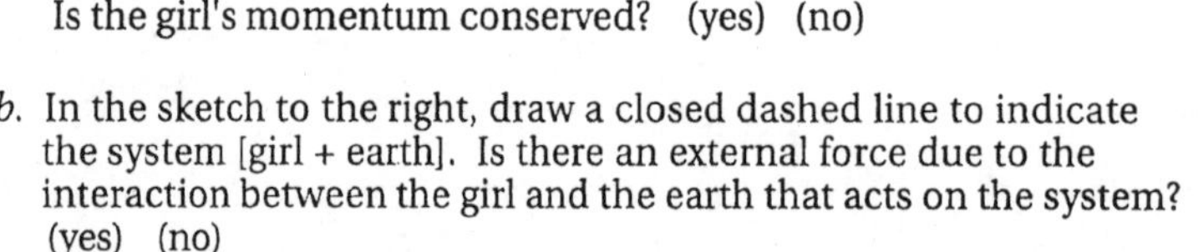

b. In the sketch to the right, draw a closed dashed line to indicate the system [girl + earth]. Is there an external force due to the interaction between the girl and the earth that acts on the system? (yes) (no)

Is the momentum of the system conserved? (yes) (no)

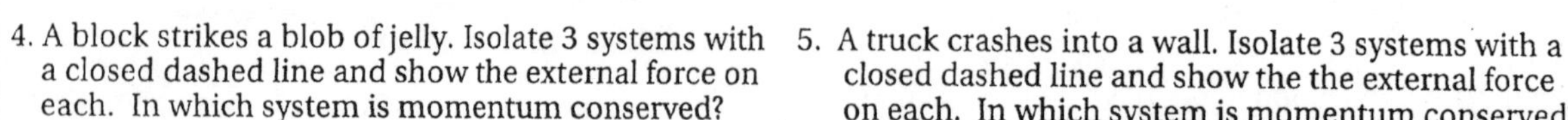

4. A block strikes a blob of jelly. Isolate 3 systems with a closed dashed line and show the external force on each. In which system is momentum conserved?

5. A truck crashes into a wall. Isolate 3 systems with a closed dashed line and show the the external force on each. In which system is momentum conserved?

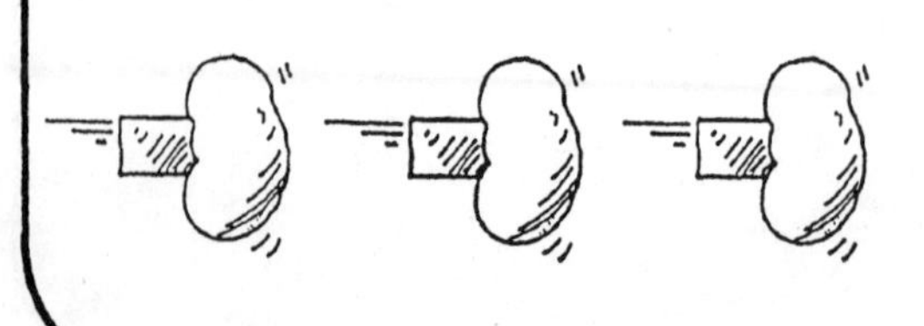

thanx to Cedric Linder

Hewitt Drew it!

Chapter 6 Momentum
Momentum Conservation

Momentum conservation for colliding balls, freight cars, and fish are worked out in the textbook. Here we consider more collisions. In the table below, fill in the numerical values for total momentum before and after the collisions of the two-body systems. Also fill in the blanks for velocity.

1. Bumper cars are fun. Assume each car with its occupant has a mass of 200 kg.

Momentum of Two-Car System	
BEFORE	AFTER

2. Granny whizzes around the rink and is suddenly confronted with Ambrose at rest directly in her path. Rather than knock him over, she picks him up and continues in motion without "braking."

DATA

Granny's mass; 50 kg

Granny's initial speed; 3 m/s

Ambrose's mass; 25 kg

Ambrose's initial speed; 0 m/s

Momentum of Granny-Ambrose System	
BEFORE	AFTER

Name ______________________ Date ________

CONCEPTUAL Physics PRACTICE PAGE

Chapter 7 Energy
Work and Energy

1. How much work (energy) is needed to lift an object that weighs 200 N to a height of 4 m?

____

2. How much power is needed to lift the 200-N object to a height of 4 m in 4 s?

____

3. What is the power output of an engine that does 60 000 J of work in 10 s?

____

4. The block of ice weighs 500 newtons.

   a. How much force is needed to push it up the incline (neglect friction)?

   ____

   b. How much work is required to push it up the incline compared with lifting the block vertically 3 m?

   ____

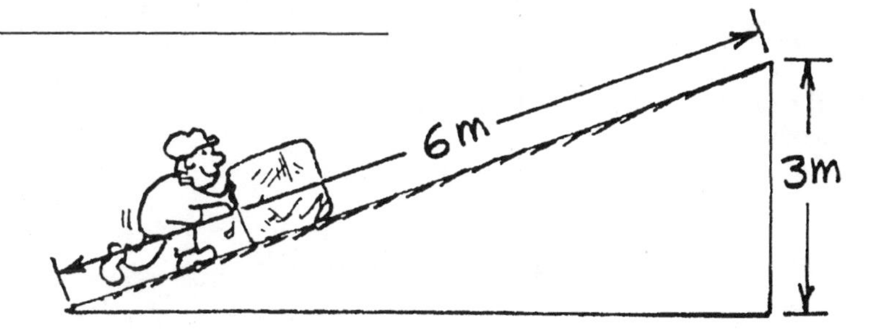

5. All the ramps are 5 m high. We know that the KE of the block at the bottom of the ramp will be equal to the loss of PE (conservation of energy). Find the speed of the block at ground level in each case. [Hint: Do you recall from earlier chapters how long it takes something to fall a vertical distance of 5 m from a positon of rest (assume $g = 10\ m/s^2$)? And how much speed a falling object acquires in this time? This gives you the answer to Case 1. Discuss with your classmates how energy conservation gives you the answers to Cases 2 and 3.]

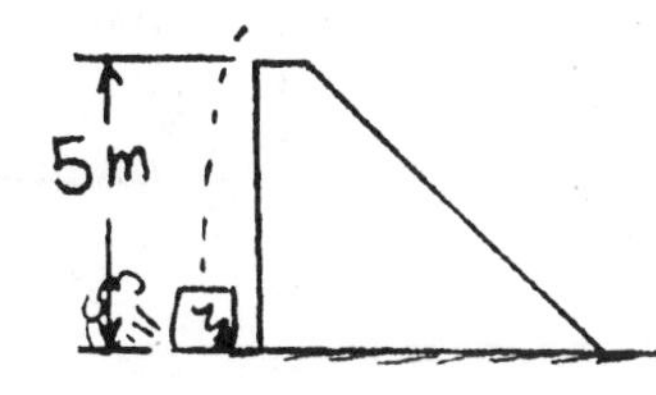

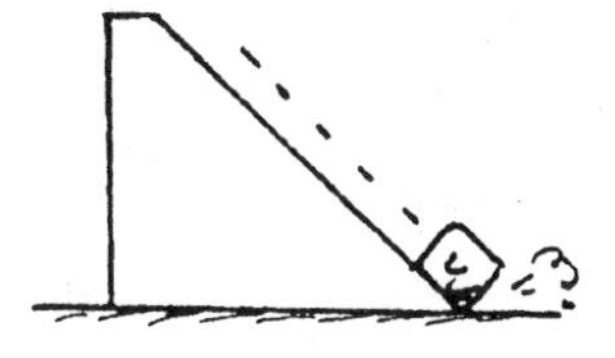

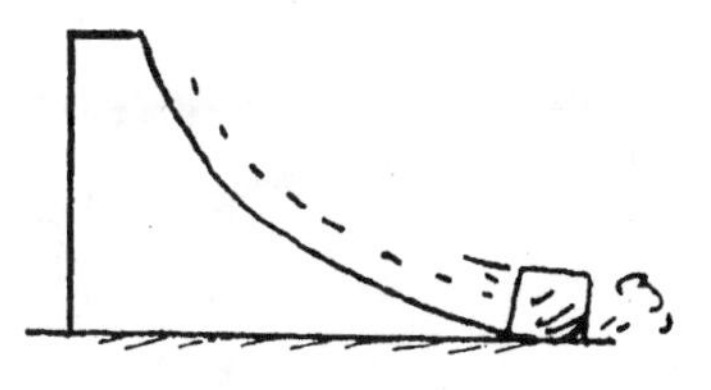

Case 1: Speed = ________ m/s   Case 2: Speed = ________ m/s   Case 3: Speed = ________ m/s.

Hewitt Drew it!

6. Which block gets to the bottom of the incline first? Assume no friction. (Be careful!) Explain your answer.

____

____

____

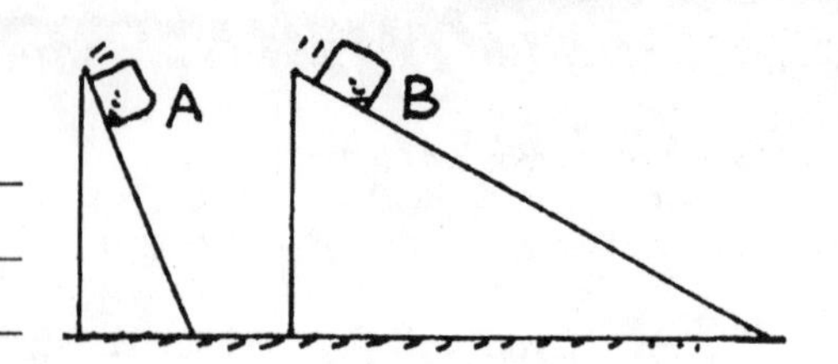

7. The KE and PE of a block freely sliding down a ramp are shown in only one place in the sketch. Fill in the missing values.

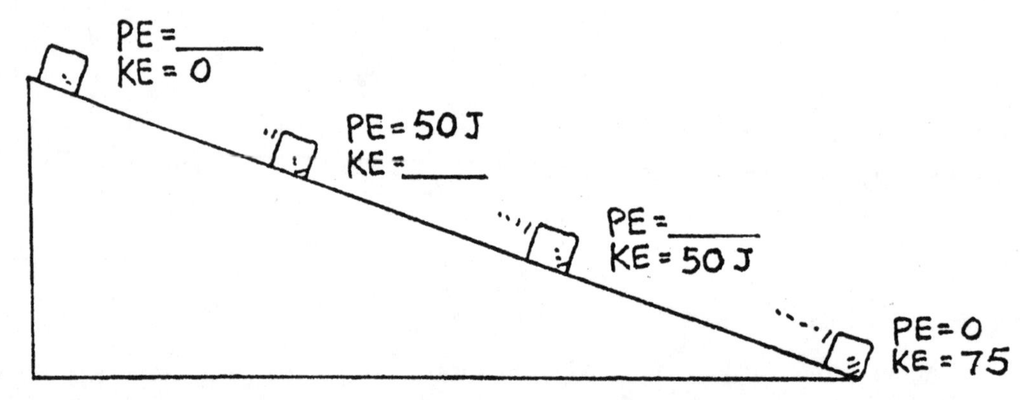

8. A big metal bead slides due to gravity along an upright friction-free wire. It starts from rest at the top of the wire as shown in the sketch. How fast is it traveling as it passes

Point B?________

Point D?________

Point E?________

At what point does it have the maximum speed?________

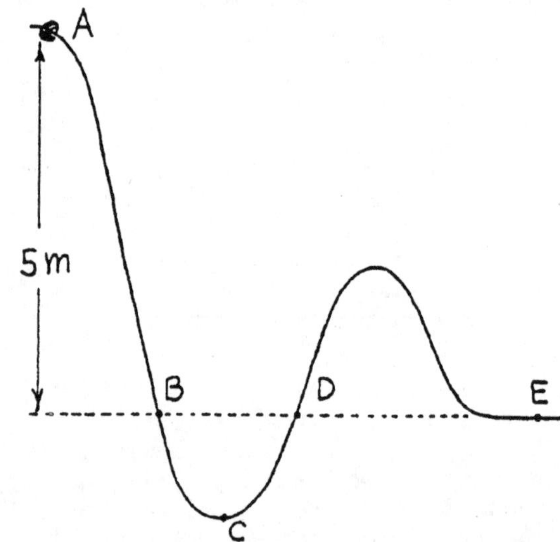

9. Rows of wind-powered generators are used in various windy locations to generate electric power. Does the power generated affect the speed of the wind? Would locations behind the 'windmills' be windier if they weren't there? Discuss this in terms of energy conservation with your classmates.

____

____

____

Hewitt Drew it!

Name ______________________ Date ________

CONCEPTUAL Physics PRACTICE PAGE

Chapter 7 Energy

Conservation of Energy

1. Fill in the blanks for the six systems shown.

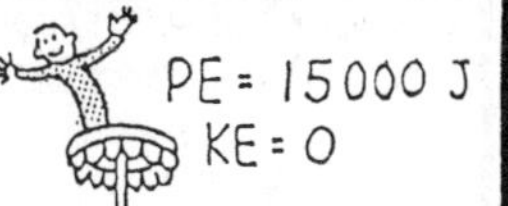

$v = 30$ km/h
KE = 10^6 J

$v = 60$ km/h
KE = ______

$v = 90$ km/h
KE = ______

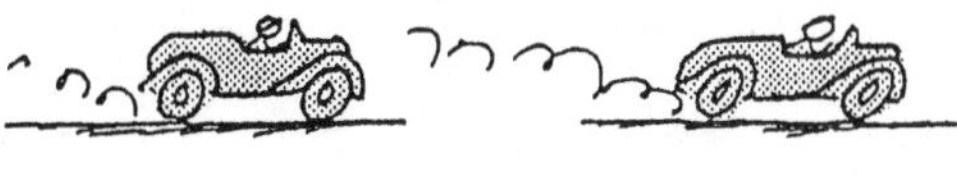

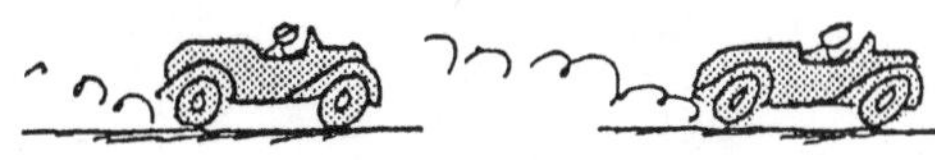

PE = 11250 J
KE = ______

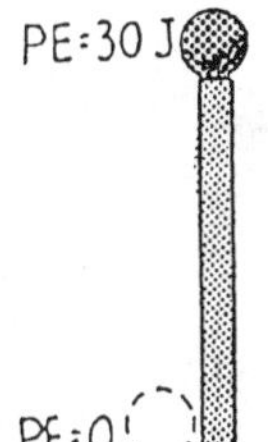

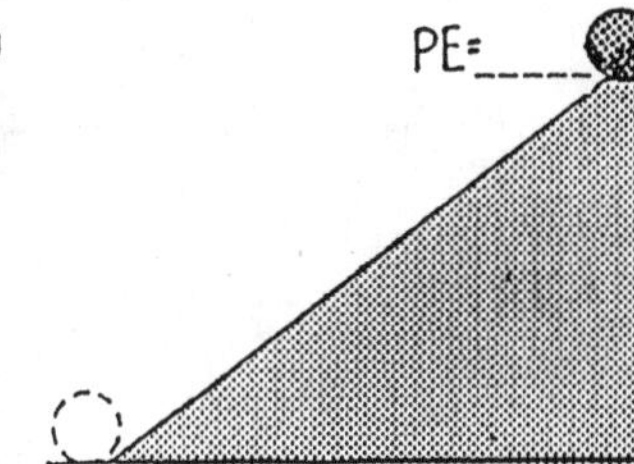

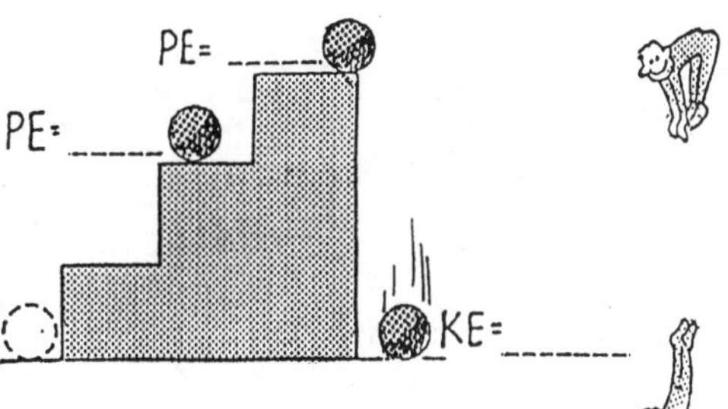

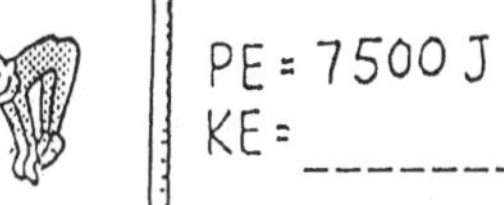

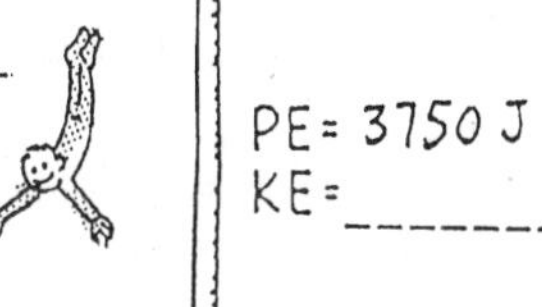

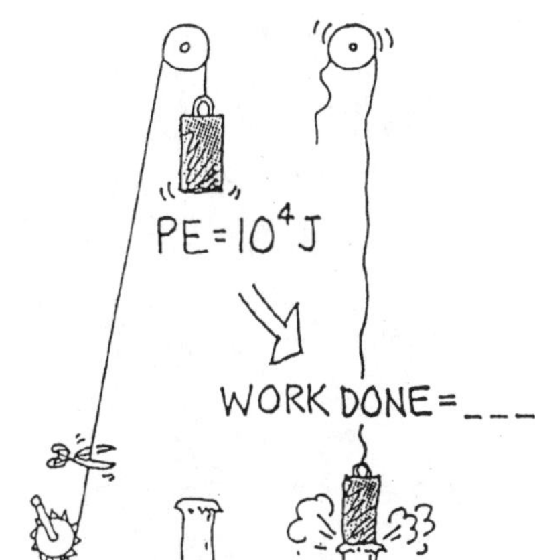

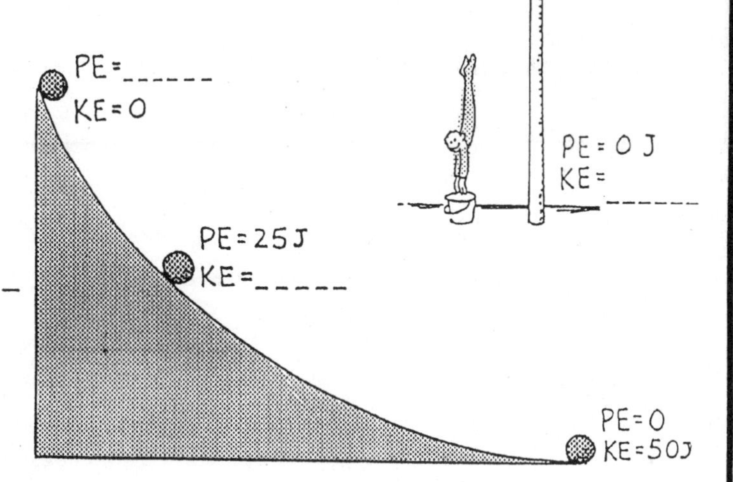

PE = 0 J
KE = ______

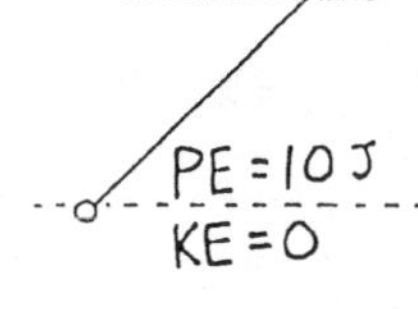

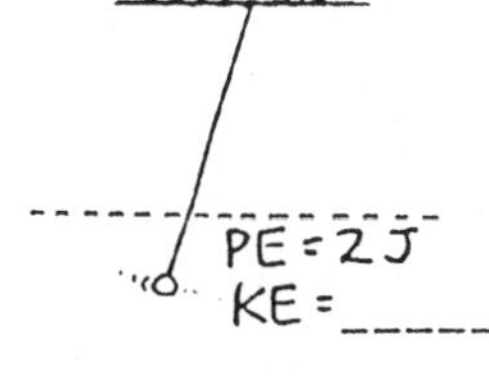

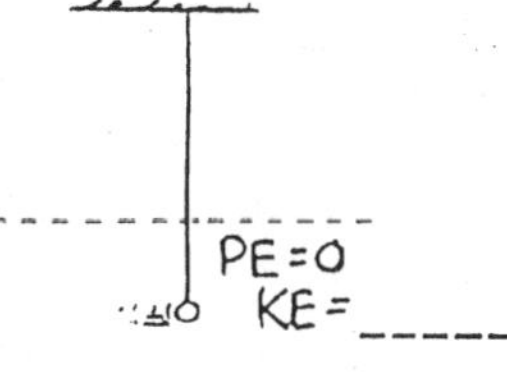

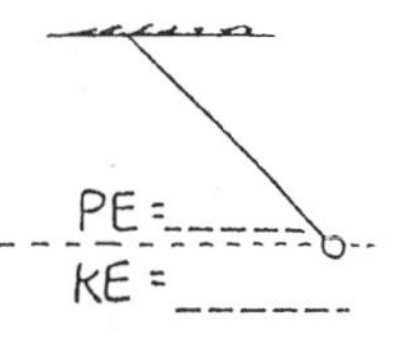

Hewitt Drew it!

2. The woman supports a 100-N load with the friction-free pulley systems shown below. Fill in the spring-scale readings that show how much force she must exert.

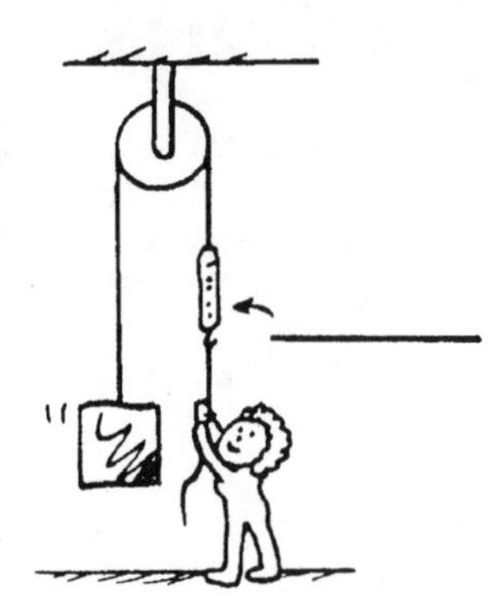

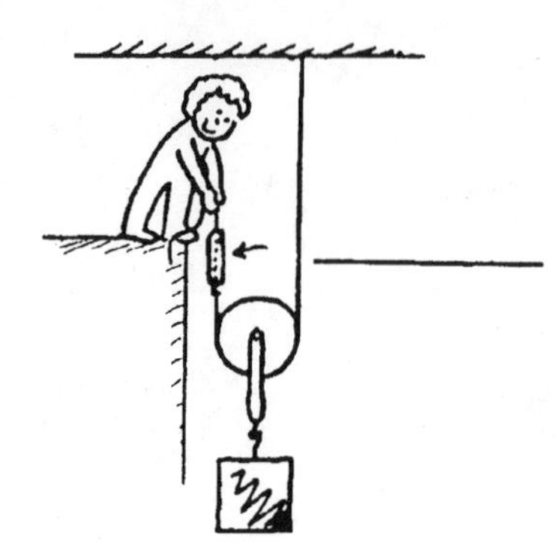

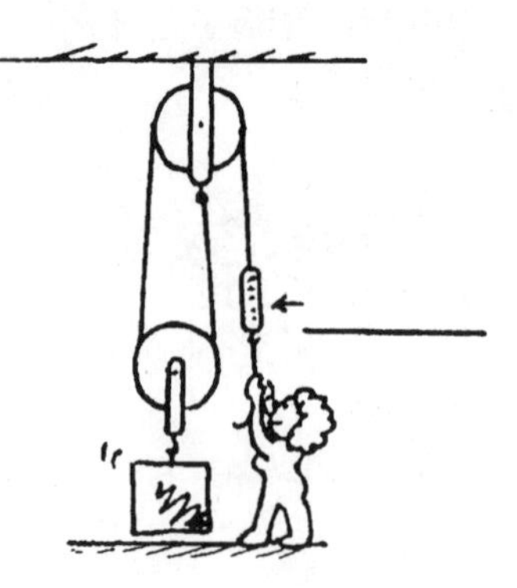

3. A 600-N block is lifted by the friction-free pulley system shown.

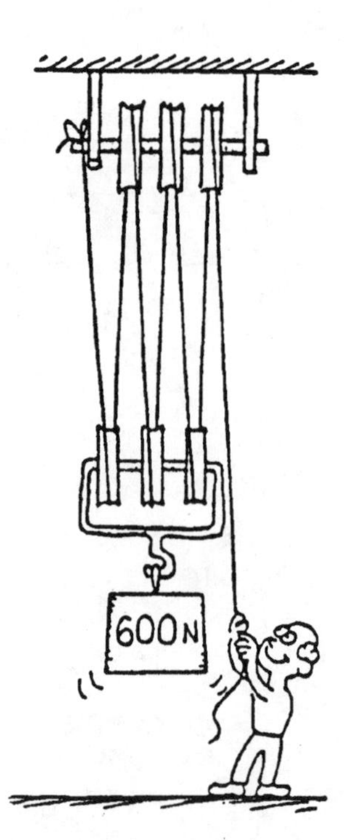

a. How many strands of rope support the 600-N weight?

______

b. What is the tension in each strand?

______

c. What is the tension in the end held by the man?

______

d. If the man pulls his end down 60 cm, how many cm will the weight rise?

______

e. If the man does 60 joules of work, what will be the increase of PE of the 600-N weight?

______

4. Why don't balls bounce as high during the second bounce as they do in the first?

______

______

______

Can you see how the conservation of energy applies to all changes in nature?

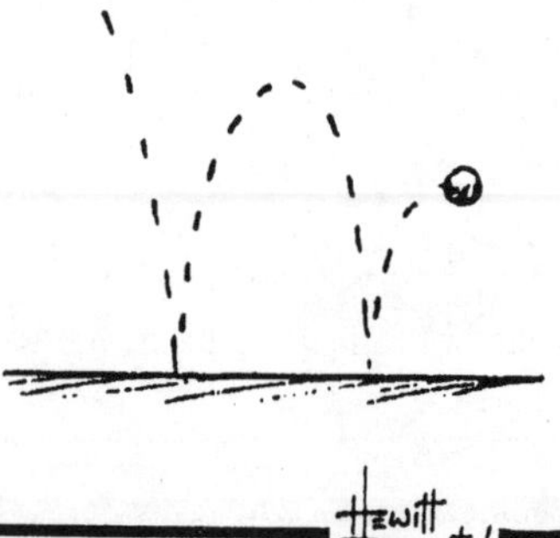

Hewitt Drew it!

Name ______________________________ Date ____________

CONCEPTUAL **Physics** PRACTICE PAGE

Chapter 7 Energy

Momentum and Energy

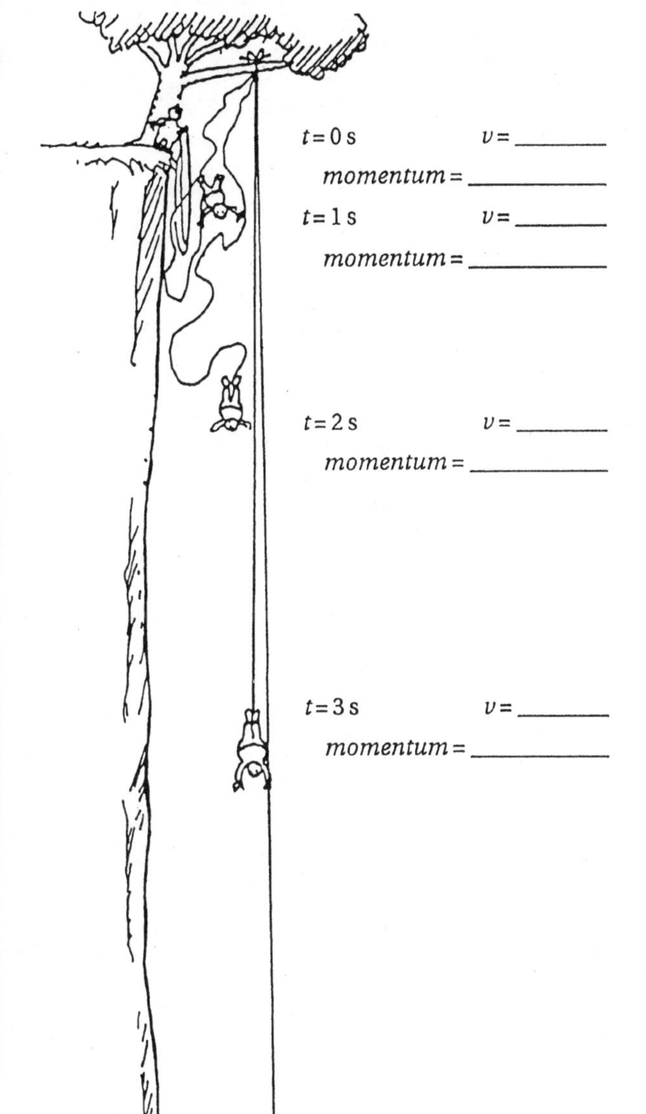

Bronco Brown wants to put $Ft = \Delta mv$ to the test and try bungee jumping. Bronco leaps from a high cliff and experiences free fall for 3 seconds. Then the bungee cord begins to stretch, reducing his speed to zero in 2 seconds. Fortunately, the cord stretches to its maximum length just short of the ground below.

Fill in the blanks. Bronco's mass is 100 kg. Acceleration of free fall is 10 m/s².

Express values in SI units (*distance* in m, *velocity* in m/s, *momentum* in kg-m/s, *impulse* in N-s, and *deceleration* in m/s²).

The 3-s free-fall distance of Bronco just before the bungee cord begins to stretch

= ____________

Δmv during the 3-s interval of free fall

= ____________

Δmv during the 2-s interval of slowing down

= ____________

Impulse during the 2-s interval of slowing down

= ____________

Average force exerted by the cord during the 2-s interval of slowing down

= ____________

How about *work* and *energy*? How much KE does Bonco have 3 s after his jump?

____

How much does gravitational PE decrease during this 3 s? ____________

What two kinds of PE are changing during the slowing-down interval?

____

Energy and Momentum

A Honda Civic and a Lincoln Town Car are initially at rest on a horizontal parking lot at the edge of a steep cliff. For simplicity, we assume that the Town Car has twice as much mass as the Civic. Equal constant forces are applied to each car and they accelerate across equal distances (we ignore the effects of friction). When they reach the far end of the lot the force is suddenly removed, whereupon they sail through the air and crash to the ground below. (The cars are beat up to begin with, and this is a scientific experiment!)

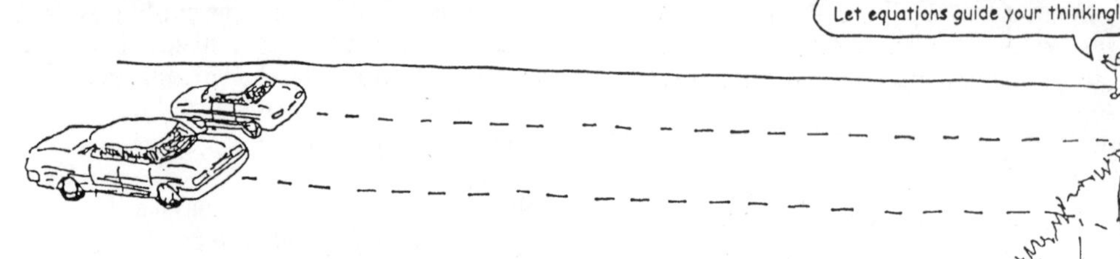
Let equations guide your thinking!

1. Which car has the greater acceleration ? (Think $a = F/m$)

   ____

2. Which car spends more time along the surface of the lot? (The faster or slower one?)

   ____

3. Which car has the larger impulse imparted to it by the applied force? (Think Impulse = Ft)

   Defend your answer.

   ____

4. Which car has the greater momentum at the cliff's edge? (Think $Ft = \Delta mv$) Defend your answer.

   ____

Impulse = Δmomentum
$Ft = \Delta mv$

5. Which car has the greater work done on it by the applied force? (Think $W = Fd$) Defend your answer in terms of the distance traveled.

   ____

Work = $Fd = \Delta KE = \Delta \frac{1}{2}mv^2$

6. Which car has the greater kinetic energy at the edge of the cliff? (Think $W = \Delta KE$) Does your answer follow from your explanation of 5? Does it contradict your answer to 3? Why or why not?

   ____

Making the distinction between momentum and kinetic energy is high-level physics.

7. Which car spends more time in the air, from the edge of the cliff to the ground below?

   ____

8. Which car lands farthest horizontally from the edge of the cliff onto the ground below?

   ____

Challenge: Suppose the slower car crashes a horizontal distance of 10 m from the ledge. Then at what horizontal distance does the faster car hit?

Hewitt Drew it!

Name ______________________________ Date __________

CONCEPTUAL Physics PRACTICE PAGE

Chapter 8 Rotational Motion

Torques

1. Apply what you know about torques by making a mobile. Shown below are five horizontal arms with fixed 1- and 2-kg masses attached, and four hangers with ends that fit in the loops of the arms, lettered A through R. You are to figure where the loops should be attached so that when the whole system is suspended from the spring scale at the top, it will hang as a proper mobile, with its arms suspended horizontally. This is best done by working from the bottom upward. Circle the loops where the hangers should be attached. When the mobile is complete, how many kilograms will be indicated on the scale? (Assume the horizontal struts and connecting hooks are practically massless compared to the 1- and 2-kg masses.) On a separate sheet of paper, make a sketch of your completed mobile.

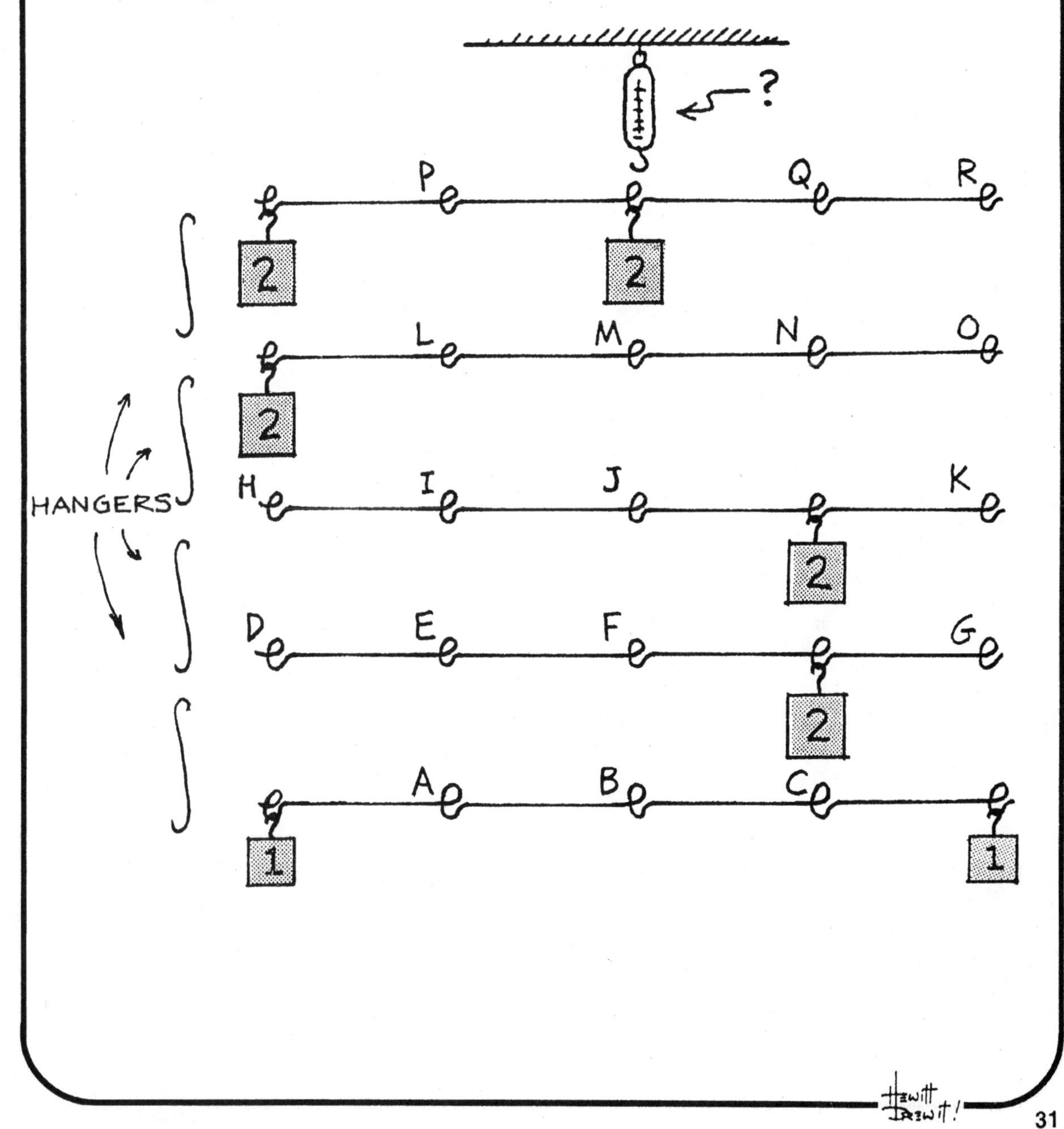

Hewitt Drew it!

2. Complete the data for the three seesaws in equilibrium.

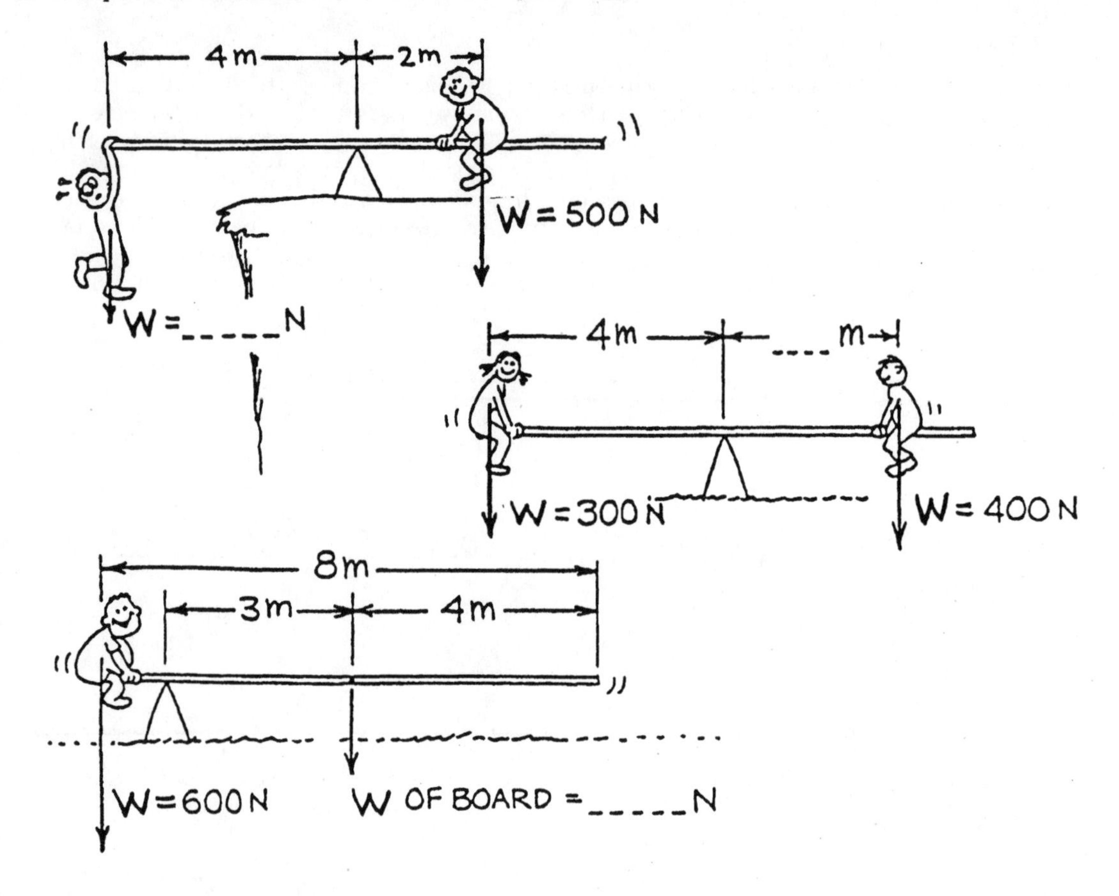

3. The broom balances at its CG. If you cut the broom in half at the CG and weigh each part of the broom, which end would weigh more?

____

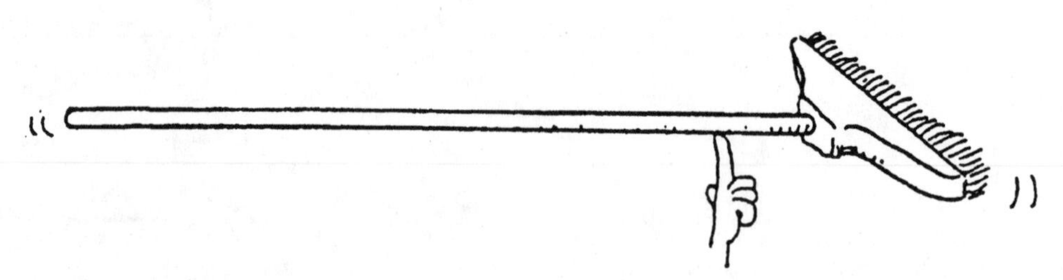

Explain why each end has or does not have the same weight? (Hint: Compare this to one of the seesaw systems above.)

____

____

Name ______________________________ Date __________

CONCEPTUAL Physics PRACTICE PAGE

Chapter 8 Rotational Motion
Torques and Rotation

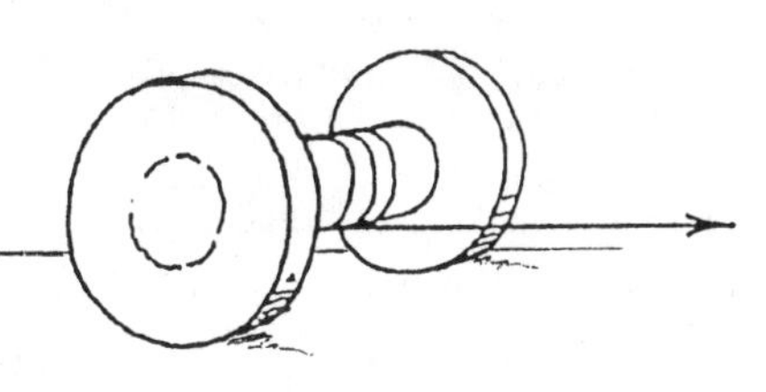

1. Pull the string gently and the spool rolls. The direction of roll depends on the way the torque is applied.

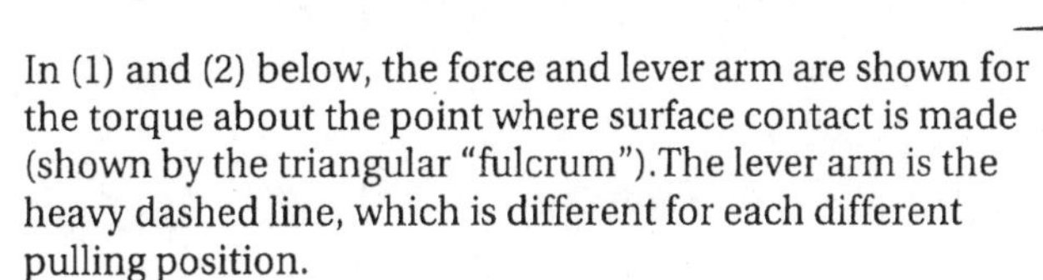

In (1) and (2) below, the force and lever arm are shown for the torque about the point where surface contact is made (shown by the triangular "fulcrum"). The lever arm is the heavy dashed line, which is different for each different pulling position.

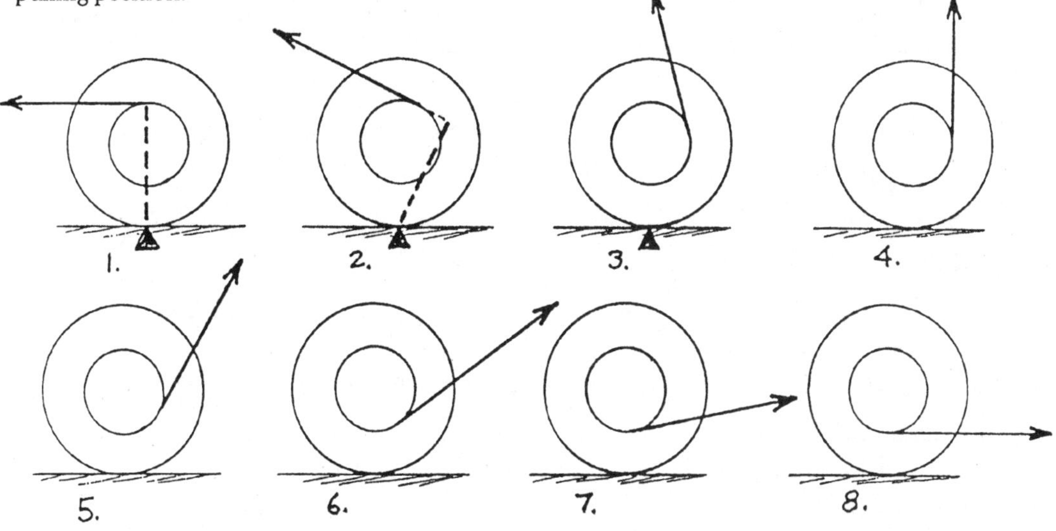

a. Construct the lever arm for the other positions.

b. Lever arm is longer when the string is on the (top) (bottom) of the spool spindle.

c. For a given pull, the torque is greater when the string is on the (top) (bottom).

d. For the same pull, rotational acceleration is greater when the string is on (top) (bottom) (makes no difference).

e. At which positions does the spool roll to the left? __________

f. At which positions does the spool roll to the right? __________

g. At which position does the spool not roll at all? __________

h. Why does the spool slide rather than roll at this position?

Be sure your right angle is between the force's *line of action* and the lever arm.

____

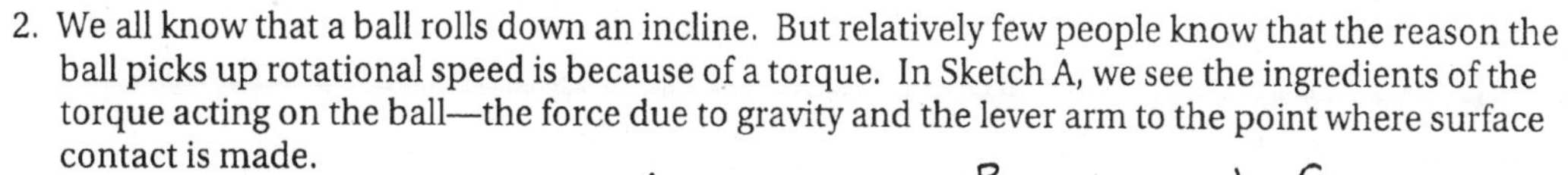

2. We all know that a ball rolls down an incline. But relatively few people know that the reason the ball picks up rotational speed is because of a torque. In Sketch A, we see the ingredients of the torque acting on the ball—the force due to gravity and the lever arm to the point where surface contact is made.

a. Constuct the lever arms for positions B and C.

b. As the incline becomes steeper, the torque (increases) (decreases)

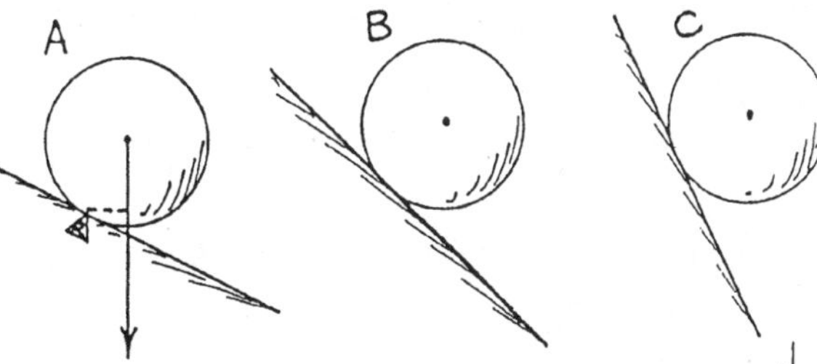

Hewitt Drew it!

Acceleration and Circular Motion

Newton's 2nd law, $a = F/m$, tells us that net force and its corresponding acceleration are always in the same direction. (Both force and acceleration are vector quantities). But force and acceleration are not always in the direction of velocity (another vector).

1. You're in a car at a traffic light. The light turns green and the driver "steps on the gas."
   a. Your body lurches (forward) (not at all) (backward).
   b. The car accelerates (forward) (not at all) (backward).
   c. The force on the car acts (forward) (not at all) (backward).

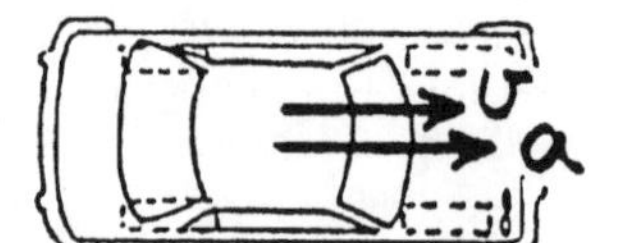

The sketch shows the top view of the car. Note the directions of the velocity and acceleration vectors.

2. You're driving along and approach a stop sign. The driver steps on the brakes.
   a. Your body lurches (forward) (not at all) (backward).
   b. The car accelerates (forward) (not at all) (backward).
   c. The force on the car acts (forward) (not at all) (backward).

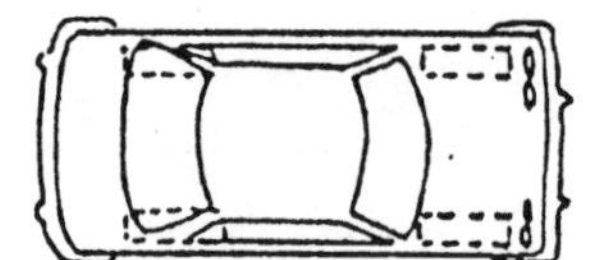

The sketch shows the top view of the car. Draw vectors for velocity and acceleration.

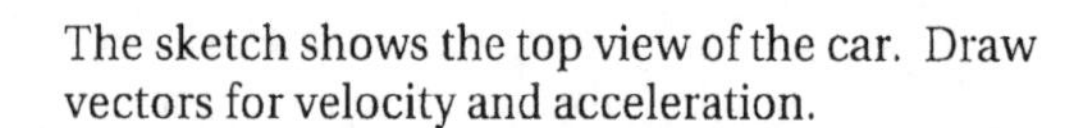

3. You continue driving, and round a sharp curve to the left at constant speed.
   a. Your body leans (inward) (not at all) (outward).
   b. The direction of the car's acceleration is (inward) (not at all) (outward).
   c. The force on the car acts (inward) (not at all) (outward).

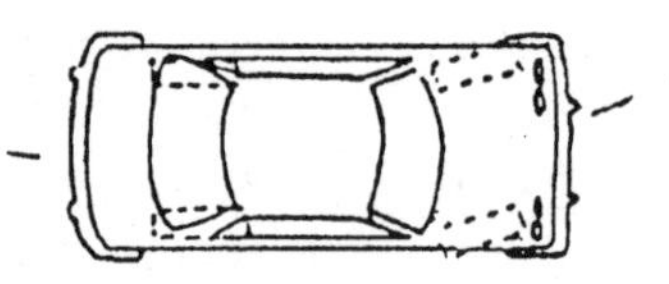

Draw vectors for velocity and acceleration of the car.

4. In general, the directions of lurch and acceleration, and therefore the directions of lurch and force, are (the same) (not related) (opposite).

5. The whirling stone's direction of motion keeps changing.
   a. If it moves faster, its direction changes (faster) (slower).
   b. This indicates that as speed increases, acceleration (increases) (decreases) (stays the same).

6. Consider whirling the stone on a shorter string—that is, of smaller radius.
   a. For a given speed, the rate that the stone changes direction is (less) (more) (the same).
   b. This indicates that as the radius decreases, acceleration (increases) (decreases) (stays the same).

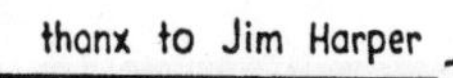

Name ______________________________ Date __________

CONCEPTUAL Physics PRACTICE PAGE

Chapter 8 Rotational Motion
Simulated Gravity and Frames of Reference

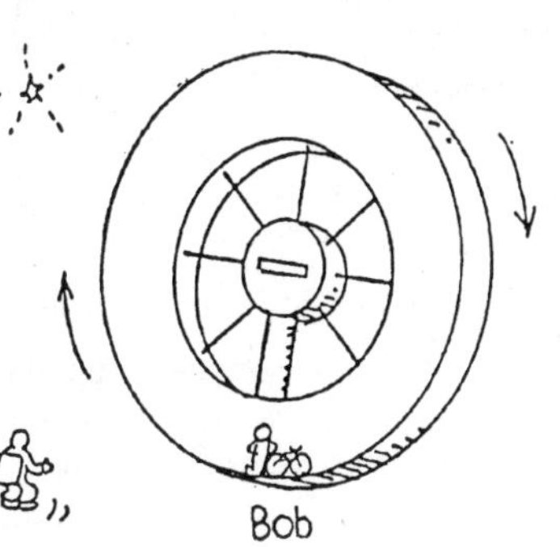

Susie Spacewalker and Bob Biker are in outer space. Bob experiences earth-normal gravity in a rotating habitat, where centripetal force on his feet provides a normal support force that feels like weight. Suzie hovers outside in a weightless condition, motionless relative to the stars and the center of mass of the habitat.

1. Susie sees Bob rotating clockwise in a circular path at a linear speed of 30 km/h. Suzie and Bob are facing each other, and from Bob's point of view, he is at rest and he sees Suzie moving

   (clockwise) (counter clockwise).

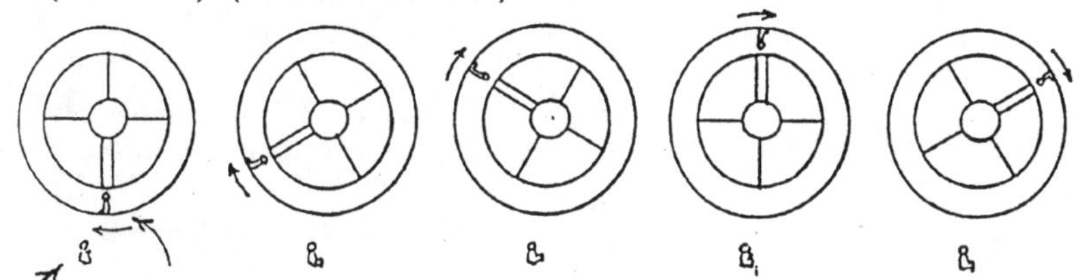

2. The rotating habitat seems like home to Bob—until he rides his bicycle. When he rides in the opposite direction as the habitat rotates, Suzie sees him moving (faster) (slower).

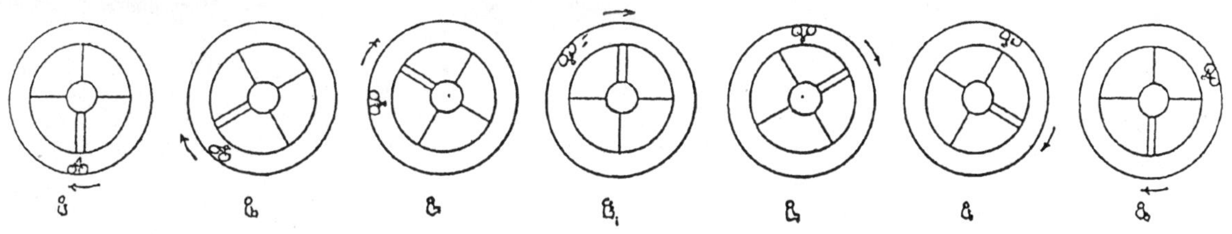

3. As Bob's bicycle speedometer reading increases, his rotational speed

   (decreases) (remains unchanged) (increases) and the normal force that feels like weight

   (decreases) (remains unchanged) (increases). So friction between the tires and the floor

   (decreases) (remains unchanged) (increases).

4. When Bob nevertheless gets his speed up to 30 km/h, as read on his bicycle speedometer, Suzie sees him

   (moving at 30 km/h) (motionless) (moving at 60 km/h)

thanx to Bob Becker Hewitt Drew it!

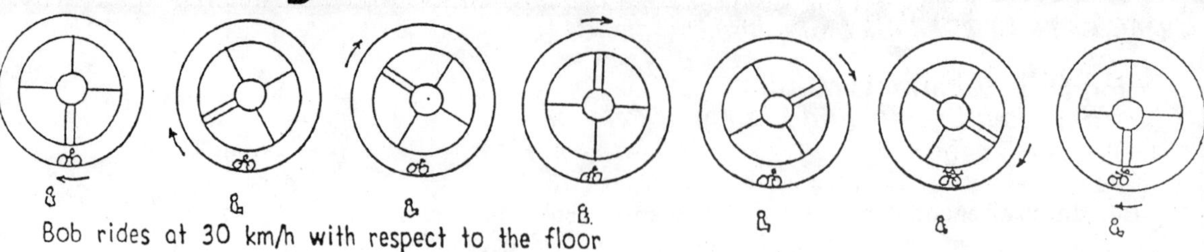

Bob rides at 30 km/h with respect to the floor

5. Bounding off the floor a bit while riding at 30 km/h, and neglecting wind effects, Bob

   (drifts toward the ceiling in midspace as the floor whizzes by him at 30 km/h)
   (falls as he would on earth)
   (slams onto the floor with increased force)

   and he finds himself

   (in the same frame of reference as Suzie)
   (as if he rode at 30 km/h on the earth's surface)
   (pressed harder against the bicycle seat).

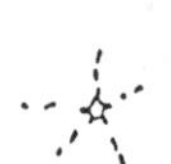

6. Bob manuevers back to his initial condition, whirling at rest with the habitat, standing beside his bicycle. But not for long. Urged by Suzie, he rides in the opposite direction, clockwise with the rotation of the habitat.

   Now Suzie sees him moving (faster) (slower).

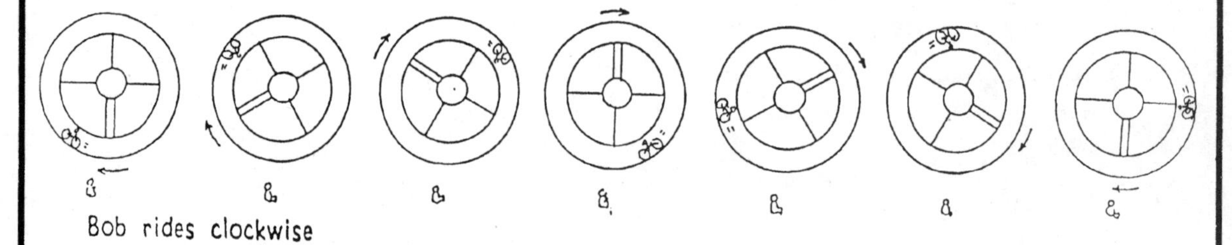

7. As Bob gains speed, the normal support force that feels like weight

   (decreases) (remains unchanged) (increases).

8. When Bob's speedometer reading gets up to 30 km/h, Suzie sees him moving

   (30 km/h) (not at all) (60 km/h)

   and Bob finds himself

   (weightless like Suzie)
   (just as if he rode at 30 km/h on the earth's surface)
   (pressed harder against the bicycle seat).

Next, Bob goes bowling. You decide whether the game depends on which direction the ball is rolled!

Hewitt Drew it!

Name ______________________________ Date __________

CONCEPTUAL **Physics** PRACTICE PAGE

Chapter 9 Gravity
Inverse-Square Law

1. Paint spray travels radially away from the nozzle of the can in straight lines. Like gravity, the strength (intensity) of the spray obeys an inverse-square law. Complete the diagram by filling in the blank spaces.

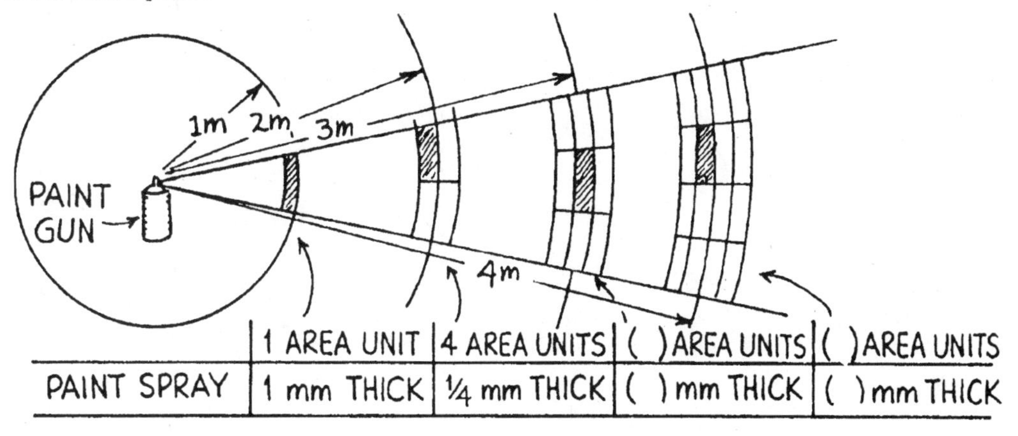

	1 AREA UNIT	4 AREA UNITS	( ) AREA UNITS	( ) AREA UNITS
PAINT SPRAY	1 mm THICK	¼ mm THICK	( ) mm THICK	( ) mm THICK

2. A small light source located 1 m in front of an opening of area 1 m² illuminates a wall behind. If the wall is 1 m behind the opening (2 m from the light source), the illuminated area covers 4 m². How many square meters will be illuminated if the wall is

5 m from the source? ______________

10 m from the source? ______________

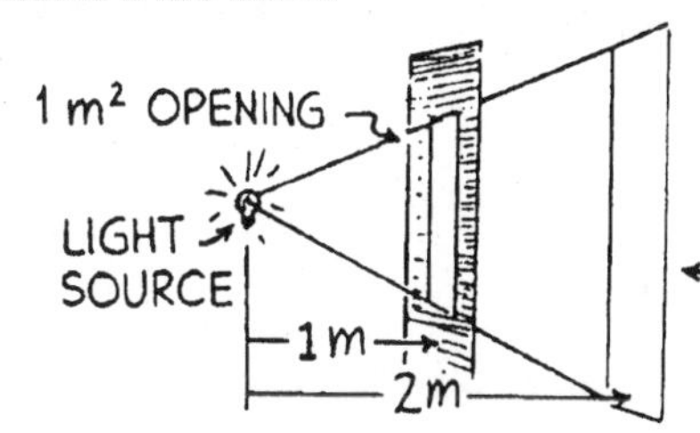

3. If we stand on a weighing scale and find that we are pulled toward the earth with a force of 500 N, then we weigh ________ N. Strictly speaking, we weigh ________ N relative to the earth. How much does the earth weigh? If we tip the scale upside down and repeat the weighing process, we can say that we and the earth are still pulled together with a force of _______ N, and therefore, relative to us, the whole 6 000 000 000 000 000 000 000 000-kg earth weighs ________ N! Weight, unlike mass, is a relative quantity.

We are pulled to the earth with a force of 500 N, so we weigh 500 N.

DO YOU SEE WHY IT MAKES SENSE TO DISCUSS THE EARTH'S MASS, BUT NOT ITS WEIGHT?

The earth is pulled toward us with a force of 500 N, so it weighs 500 N.

Hewitt Drew it!

Name ______________________ Date __________

CONCEPTUAL *Physics* PRACTICE PAGE

Chapter 9 Gravity
Our Ocean Tides

1. Consider two equal-mass blobs of water, A and B, initially at rest in the moon's gravitational field. The vector shows the gravitational force of the moon on A.

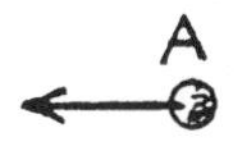

a. Draw a force vector on B due to the moon's gravity.

b. Is the force on B more or less than the force on A? ________

c. Why?____

d. The blobs accelerate toward the moon. Which has the greater acceleration? (A) (B)

e. Because of the different accelerations, with time

(A gets farther ahead of B) (A and B gain identical speeds) and the distance between A and B

(increases) (stays the same) (decreases).

f. If A and B were connected by a rubber band, with time the rubber band would

(stretch) (not stretch).

g. This (stretching) (non-stretching) is due to the (difference) (non-difference) in the moon's gravitational pulls.

h. The two blobs will eventually crash into the moon. To orbit around the moon instead of crashing into it, the blobs should move

(away from the moon) (tangentially). Then their accelerations will consist of changes in

(speed) (direction).

2. Now consider the same two blobs located on opposite sides of the Earth.

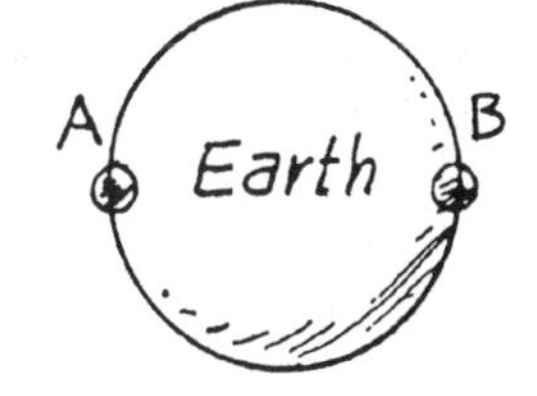

a. Because of differences in the moon's pull on the blobs, they tend to

(spread away from each other) (approach each other).

b. Does this spreading produce ocean tides? (Yes) (No)

c. If Earth and moon were closer, gravitational force between them would be

(more) (the same) (less), and the difference in gravitational forces on the near and far parts of the ocean would be (more) (the same) (less).

d. Because the Earth's orbit about the sun is slightly elliptical, Earth and sun are closer in December than in June. Taking the sun's tidal force into account, on a world average, ocean tides are greater in (December) (June) (no difference).

Hewitt Drew it!

Name ______________________________ Date ____________

CONCEPTUAL **Physics** PRACTICE PAGE

Chapter 10 Projectile and Satellite Motion

Independence of Horizontal and Vertical Components of Motion

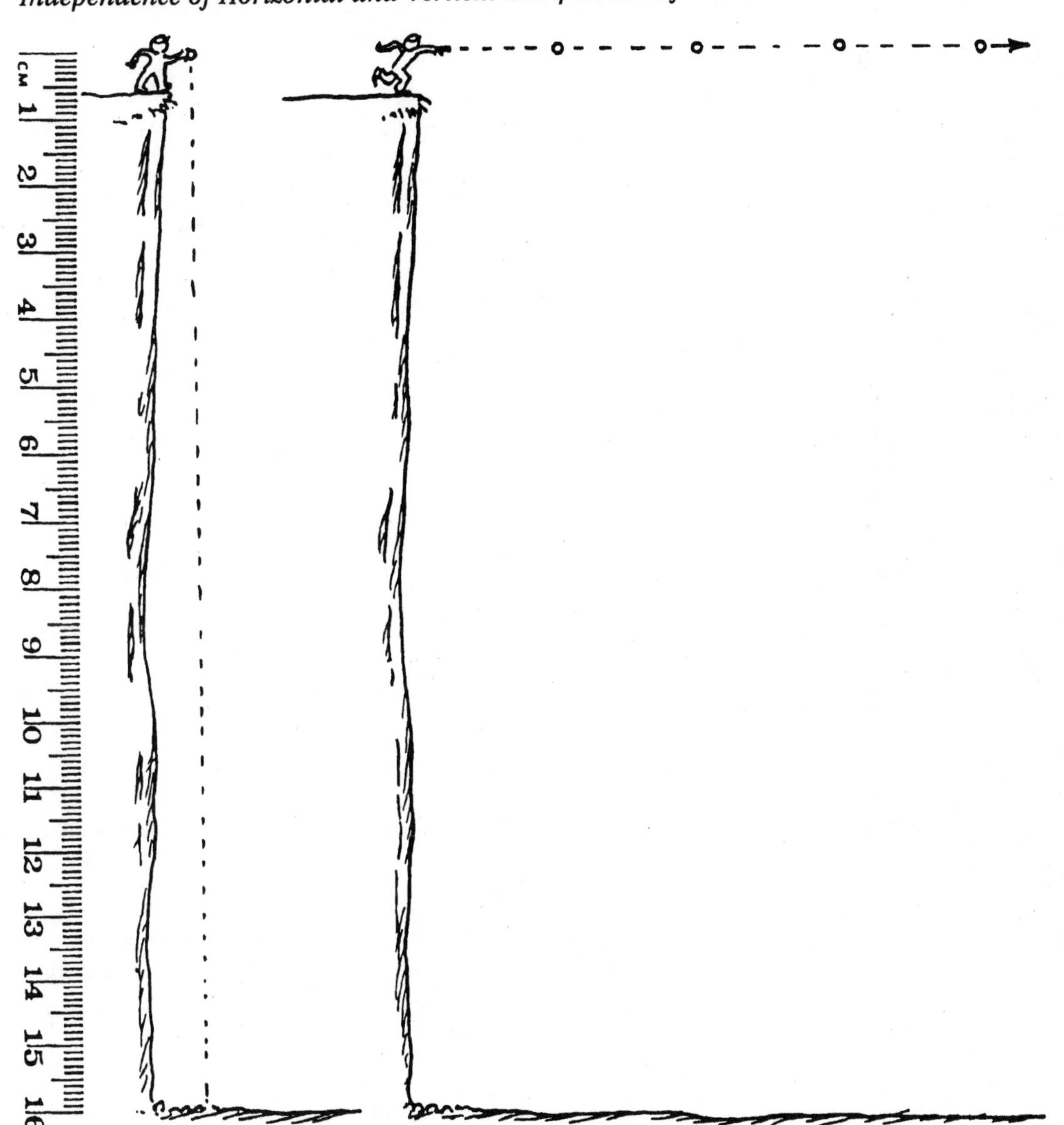

1. Above left: Use the scale 1 cm: 5 m and draw the positions of the dropped ball at 1-second intervals. Neglect air drag and assume $g = 10\ m/s^2$. Estimate the number of seconds the ball is in the air.

   ____________ seconds.

2. Above right: The four positions of the thrown ball with *no gravity* are at 1-second intervals. At 1 cm: 5 m, carefully draw the positions of the ball *with* gravity. Neglect air drag and assume $g = 10\ m/s^2$. Connect your positions with a smooth curve to show the path of the ball. How is the motion in the vertical direction affected by motion in the horizontal direction?

   ____

Hewitt Drew it!

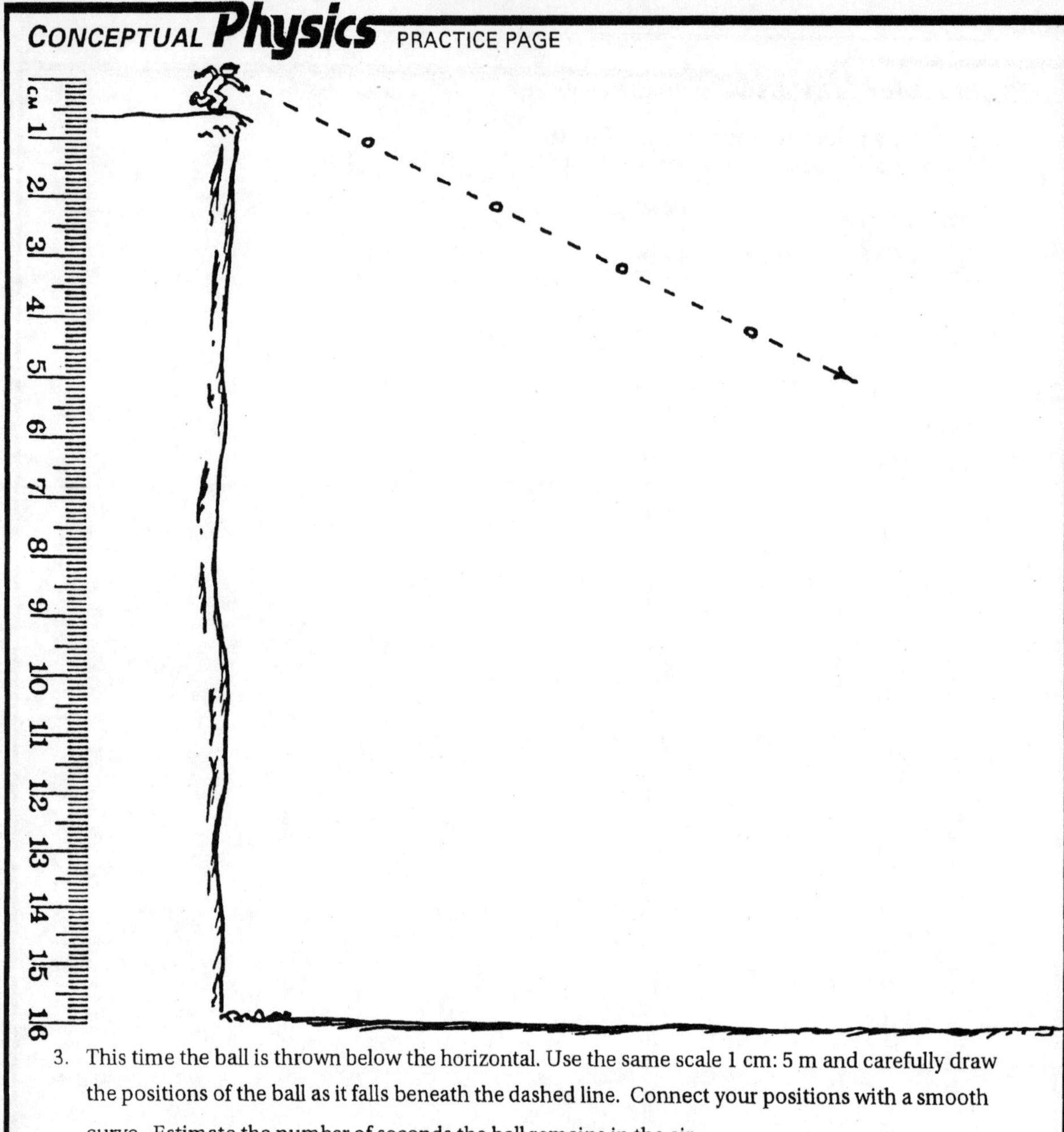

3. This time the ball is thrown below the horizontal. Use the same scale 1 cm: 5 m and carefully draw the positions of the ball as it falls beneath the dashed line. Connect your positions with a smooth curve. Estimate the number of seconds the ball remains in the air.________________ s

4. Suppose that you are an accident investigator and you are asked to figure whether or not the car was speeding before it crashed through the rail of the bridge and into the mudbank as shown. The speed limit on the bridge is 55 mph = 24 m/s. What is your conclusion?

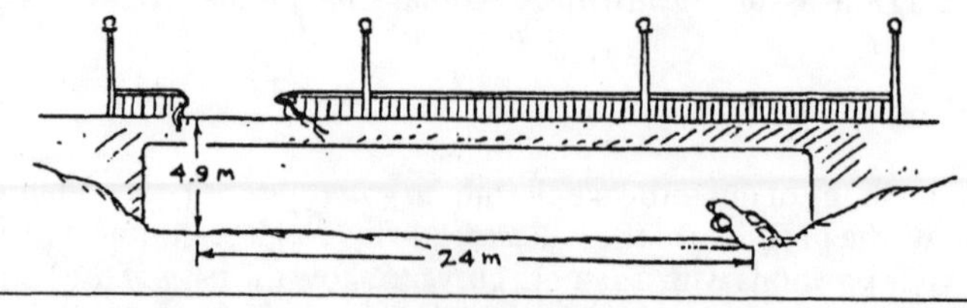

____

____

____

Hewitt Drew it!

Name ______________________ Date __________

CONCEPTUAL **Physics** PRACTICE PAGE

Chapter 10 Projectile and Satellite Motion

Tossed Ball

A ball tossed upward has initial velocity components 30 m/s vertical, and 5 m/s horizontal. The position of the ball is shown at 1-second intervals. Air resistance is negligible, and $g = 10\ \text{m/s}^2$. Fill in the boxes, writing in the values of velocity ***components*** ascending, and your calculated ***resultant velocities*** descending.

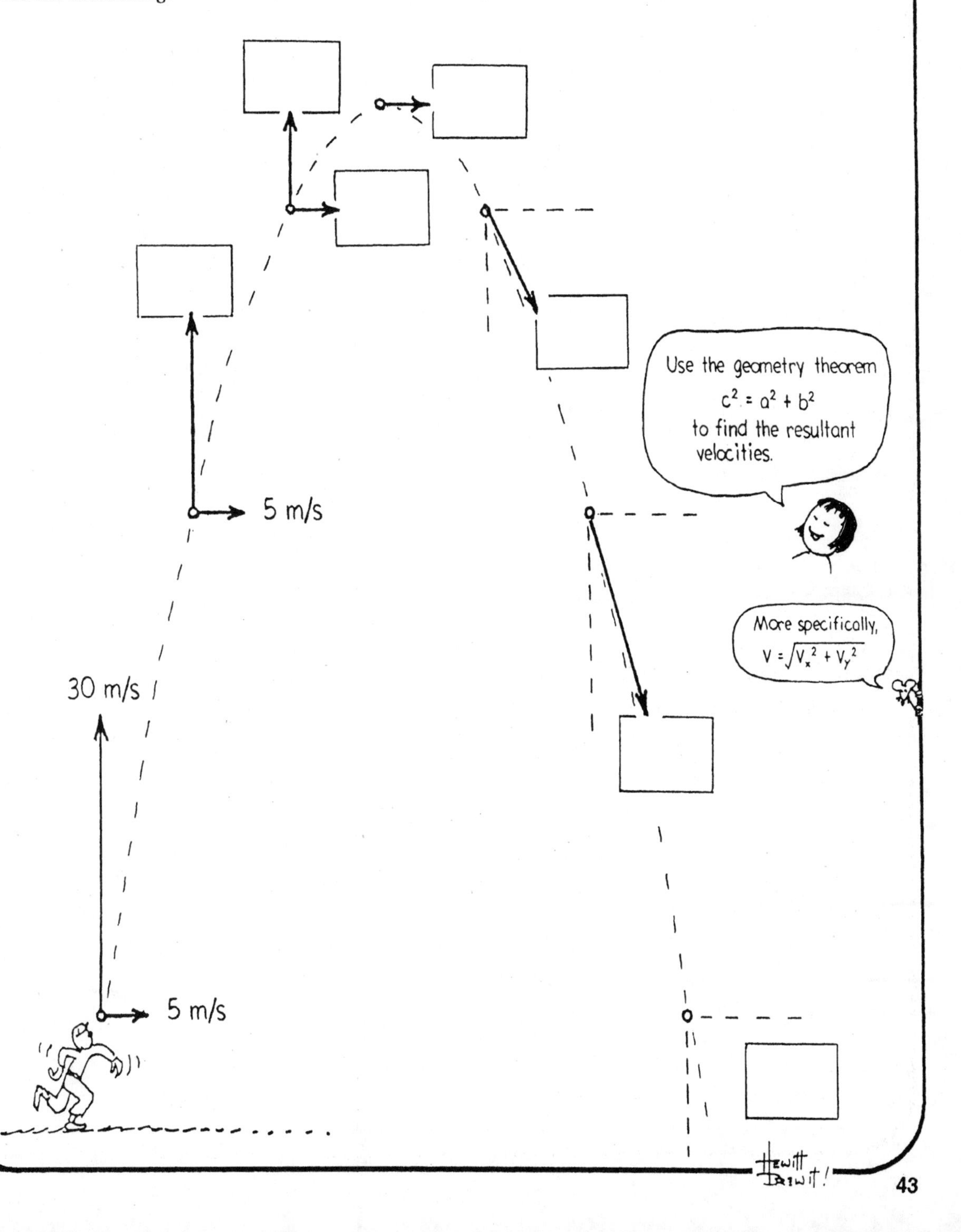

30 m/s
5 m/s
5 m/s
Use the geometry theorem
$c^2 = a^2 + b^2$
to find the resultant velocities.
More specifically,
$V = \sqrt{V_x^2 + V_y^2}$

Name ______________________ Date __________

CONCEPTUAL **Physics** PRACTICE PAGE

Chapter 10 Projectile and Satellite Motion

Satellite in Circular Orbit

1. Figure A shows "Newton's Mountain," so high that its top is above the drag of the atmosphere. The cannonball is fired and hits the ground as shown.

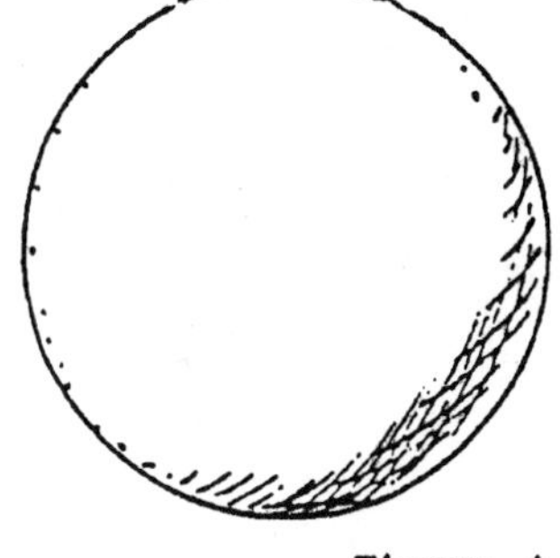

Figure A

   a. You draw the path the cannonball might take if it were fired a little bit faster.

   b. Repeat for a still greater speed, but still less than 8 km/s.

   c. Then draw the orbital path it would take if its speed were 8 km/s.

   d. What is the shape of the 8 km/s curve?

   ____

   e. What would be the shape of the orbital path if the cannonball were fired at a speed of about 9 km/s?

   ____

2. Figure B shows a satellite in circular orbit.

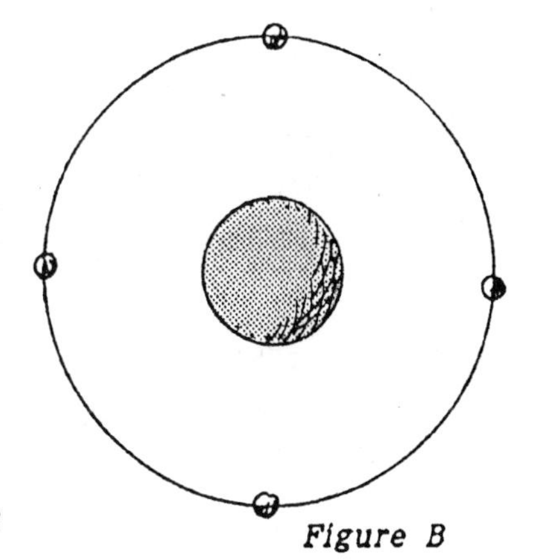

Figure B

   a. At each of the four positions draw a vector that represents the gravitational *force* exerted on the satellite.

   b. Label the force vectors *F*.

   c. Then draw at each position a vector to represent the *velocity* of the satellite at that position, and label it *V*.

   d. Are all four *F* vectors the same length? Why or why not?

   ____

   e. Are all four *V* vectors the same length? Why or why not?

   ____

   f. What is the angle between your *F* and *V* vectors? ______________________

   g. Is there any component of *F* along *V*? ______________________

   h. What does this tell you about the work the force of gravity does on the satellite?

   ____

   i. Does the KE of the satellite in Figure B remain constant, or does it vary? ______________________

   j. Does the PE of the satellite remain constant, or does it vary?

   ____

Hewitt Drew it!

Satellite in Elliptical Orbit

a. Repeat the procedure you used for the circular orbit, drawing vectors *F* and *V* for each position, including proper labeling. Show equal magnitudes with equal lengths, and greater magnitudes with greater lengths, but don't bother making the scale accurate.

b. Are your vectors *F* all the same magnitude? Why or why not?

c. Are your vectors *V* all the same magnitude? Why or why not?

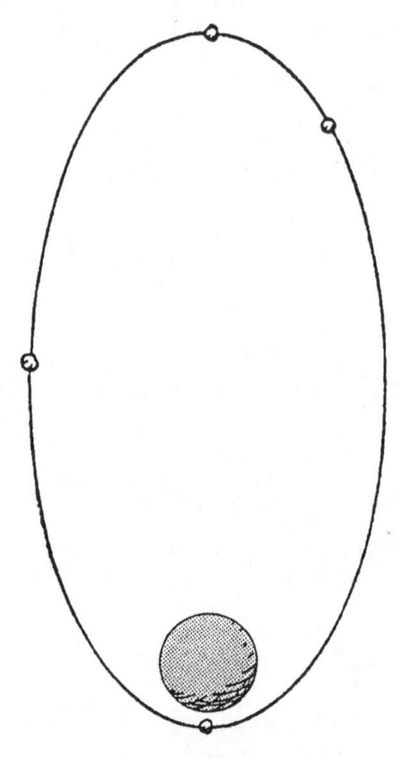

Figure C

d. Is the angle between vectors *F* and *V* everywhere the same, or does it vary?

e. Are there places where there is a component of *F* along *V*?

f. Is work done on the satellite when there is a component of *F* along and in the same direction of *V* and if so, does this increase or decrease the KE of the satellite?

g. When there is a component of *F* along and opposite to the direction of *V*, does this increase or decrease the KE of the satellite?

h. What can you say about the sum KE + PE along the orbit?

Be very very careful when placing both velocity and force vectors on the same diagram. Not a good practice, for one may construct the resultant of the vectors -- ouch!

Hewitt Drew it!

Name ______________________ Date __________

CONCEPTUAL Physics PRACTICE PAGE

Mechanics Overview

1. The sketch shows the elliptical path described by a satellite about the earth. In which of the marked positions, A – D, (put S for "same everywhere") does the satellite experience the maximum . . .

   a. gravitational force?________

   b. speed? ________

   c. momentum?________

   d. kinetic energy?________

   e. gravitational potential energy?________

   f. total energy (KE + PE)?______

   g. acceleration?__________

   h. angular momentum?________

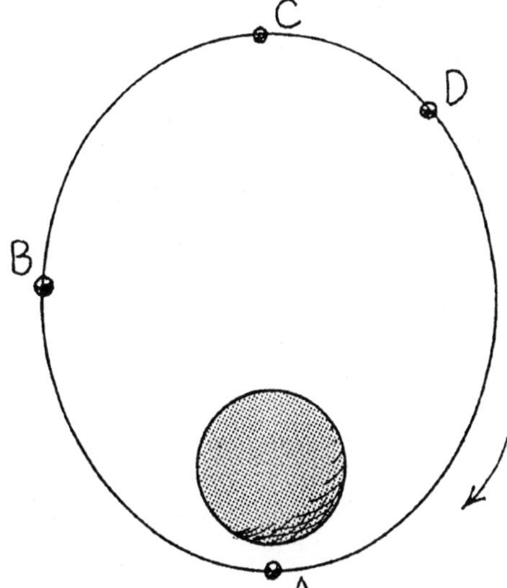

2. Answer the above questions for a satellite in circular orbit.

   a. _______ b._______ c._______ d._______ e._______ f. _______ g._______ h._______

3. In which position(s) is there momentarily no work being done on the satellite by the force of gravity? Why?

   ____

4. Work changes energy. Let the equation for work, $W = Fd$, guide your thinking on these: Defend your answers in terms of $W = Fd$.

   a. In which position will a several-minutes thrust of rocket engines pushing the satellite forward do the most work on the satellite and give it the greatest change in kinetic energy? (Hint: think about where the most distance will be traveled during the application of a several-minutes thrust?)

   ____

   b. In which position will a several-minutes thrust of rocket engines pushing the satellite forward do the least work on the satellite and give it the least boost in kinetic energy?

   ____

   c. In which positon will a several-minutes thrust of a retro-rocket (pushing opposite to the satellite's direction of motion) do the most work on the satellite and change its kinetic energy the most?

   ____

Hewitt Drew it!

Name ______________________ Date __________

CONCEPTUAL **Physics** PRACTICE PAGE

Chapter 11 The Atomic Nature of Matter

Atoms and Atomic Nuclei

ATOMS ARE CLASSIFIED BY THEIR ATOMIC NUMBER, WHICH IS THE SAME AS THE NUMBER OF __________ IN THE NUCLEUS.

TO CHANGE THE ATOMS OF ONE ELEMENT INTO THOSE OF ANOTHER, __________ MUST BE ADDED OR SUBTRACTED!

Use the periodic table in your text to help you answer the following questions.

1. When the atomic nuclei of hydrogen and lithium are squashed together (nuclear fusion) the element that is produced is

   ____

2. When the atomic nuclei of a pair of lithium nuclei are fused, the element produced is

   ____

3. When the atomic nuclei of a pair of aluminum nuclei are fused, the element produced is

   ____

4. When the nucleus of a nitrogen atom absorbs a proton, the resulting element is

   ____

5. What element is produced when a gold nucleus gains a proton?

   ____

6. Which results in the more valuable product – *adding* or *subtracting* protons from gold nuclei?

   ____

   ____

7. What element is produced when a uranium nucleus ejects an elementary particle composed of two protons and two neutrons?

   ____

8. If a uranium nucleus breaks into two pieces (nuclear fission) and one of the pieces is zirconium (atomic number 40), the other piece is the element

   ____

9. Which has more mass, a nitrogen molecule (N_2) or an oxygen molecule (O_2)?

   ____

10. Which has the greater number of atoms, a gram of helium or a gram of neon?

    ____

I LIKE THE WAY YOUR ATOMS ARE PUT TOGETHER!

SIGH

Hewitt Drew it!

Name ______________________________ Date ____________

CONCEPTUAL **Physics** PRACTICE PAGE

Chapter 12 Solids
Scaling

1. Consider a cube 1 cm x 1 cm x 1 cm (about the size of a sugar cube). Its volume is 1 cm^3. The surface area of one of its faces is 1 cm^2. The total surface area of the cube is 6 cm^2 because it has 6 sides. Now consider a second cube, scaled up by a factor of 2 so it is 2 cm x 2 cm x 2 cm.

a. What is the total surface area of each cube?

1st cube ________ cm^2; 2nd cube ________ cm^2

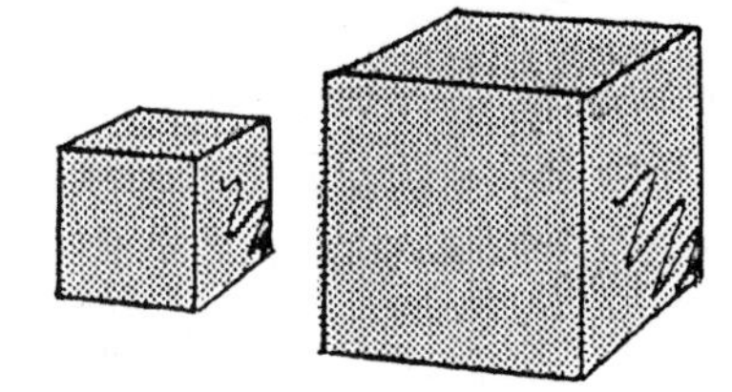

b. What is the volume of each cube?

1st cube ________ cm^3; 2nd cube ________ cm^3

c. Compare the ratio of surface area to volume for each cube.

1st cube: $\frac{\text{surface area}}{\text{volume}}$ = ________; 2nd cube: $\frac{\text{surface area}}{\text{volume}}$ = ________

2. Now consider a third cube, scaled up by a factor of 3 so it is 3 cm x 3 cm x 3 cm.

a. What is its total surface area? ________ cm^2

b. What is its volume? ________ cm^3

c. What is its ratio of surface area to volume?

$\frac{\text{surface area}}{\text{volume}}$ = ________

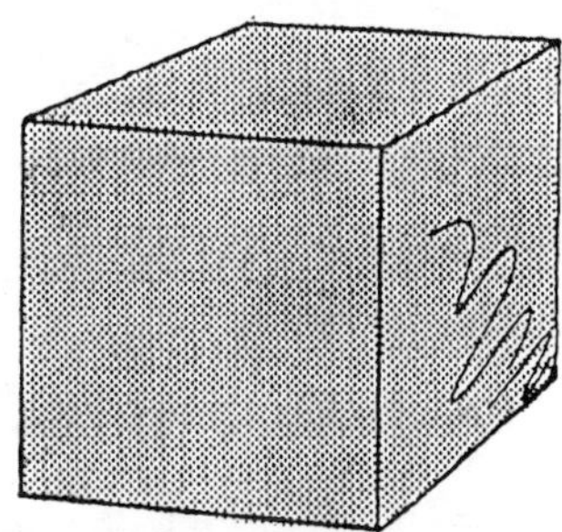

3. When the size of a cube is scaled up by a certain factor (2 and then 3 for the above examples), the area increases as the __________ of the factor, and the volume increases as the __________ of the factor.

4. Does the ratio of surface area to volume increase or decrease as things are scaled up?

____

5. Does the rule for the scaling up of cubes apply also to other shapes? ______________
Would your answers have been different if we started with a sphere of diameter 1 cm and scaled it up to a sphere of diameter 2 cm, and then 3 cm? ____________________

6. The effects of scaling are beneficial to some creatures and detrimental to others. Check either beneficial (B) or detrimental (D) for each of the following:

a. an insect falling from a tree______ b. an elephant falling from the same tree_____
c. a small fish trying to flee a big fish_____ d. a big fish chasing a small fish_______
e. a hungry mouse ________ f. an insect that falls in the water________

Hewitt Drew it!

Scaling Circles

1. Complete the table.

CIRCLES		
RADIUS	CIRCUMFERENCE	AREA
1 cm	$2\pi(1\text{cm}) = 2\pi\ \text{cm}$	$\pi(1\text{cm})^2 = \pi\text{cm}^2$
2 cm		
3 cm		
10 cm		

FOR THE CIRCUMFERENCE OF A CIRCLE, $C = 2\pi r$

AND FOR THE AREA OF A CIRCLE, $A = \pi r^2$

2. From your completed table, when the radius of a circle is doubled, its area increases by a factor of __________. When the radius is increased by a factor of 10, the area increases by a factor of __________.

3. Consider a round pizza that costs $5.00. Another pizza of the same thickness has twice the diameter. How much should the larger pizza cost?

____

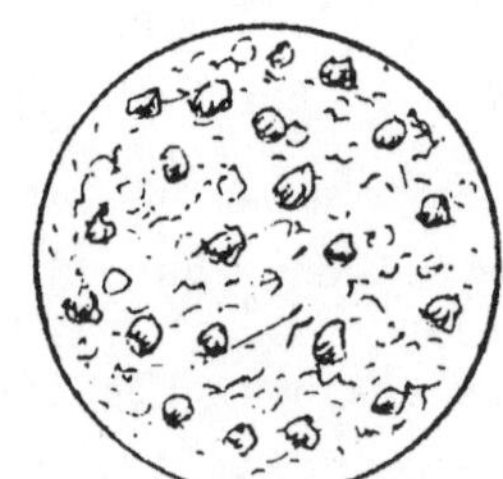

4. *True or false*: If the radius of a circle is increased by a certain factor, say 5, then the area increases by the *square* of the factor, in this case 5^2 or 25. ______________________________

So if you scale up the radius of a circle by a factor of 10, its area will increase by a factor of __________.

5. *(Application:)* Suppose you raise chickens and spend $50 to buy wire for a chicken pen. To hold the most chickens inside, you should make the shape of the pen

(square) (circular) (either, for both provide the same area)

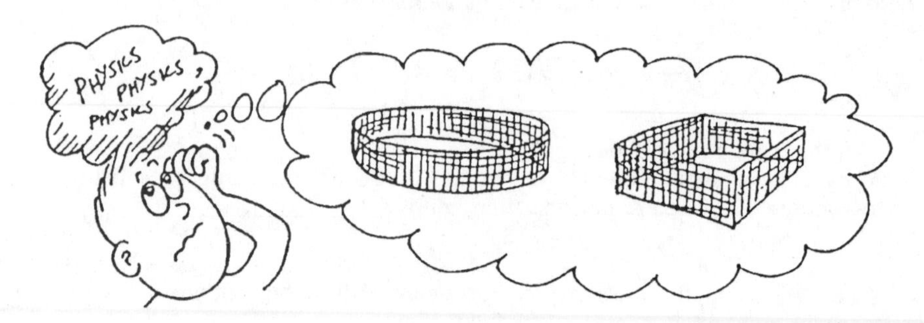

Hewitt Drew it!

Name ______________________ Date ____________

CONCEPTUAL *Physics* PRACTICE PAGE

Chapter 13 Liquids
Archimedes' Principle I

1. Consider a balloon filled with 1 liter of water (1000 cm^3) in equilibrium in a container of water, as shown in Figure 1.

   a. What is the mass of the 1 liter of water?

   ____

   WATER DOES NOT SINK IN WATER!

   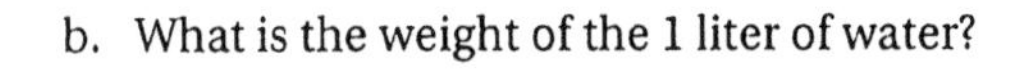

   b. What is the weight of the 1 liter of water?

   ____

   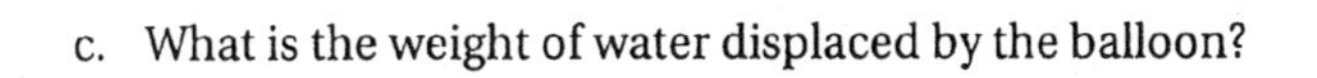

   c. What is the weight of water displaced by the balloon?

   ____

   d. What is the buoyant force on the balloon?

   ____

   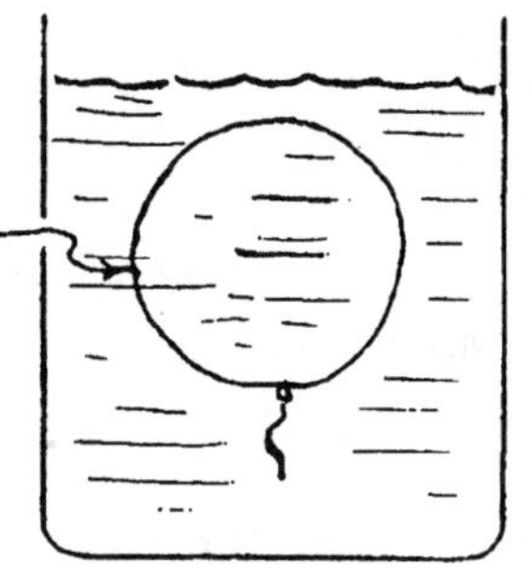

   Figure 1

   e. Sketch a pair of vectors in Figure 1: one for the weight of the balloon and the other for the buoyant force that acts on it. How do the size and directions of your vectors compare?

   ____

2. As a thought experiment, pretend we could remove the water from the balloon but still have it remain the same size of 1 liter. Then inside the balloon is a vacuum.

   a. What is the mass of the liter of nothing?

   ____

   b. What is the weight of the liter of nothing?

   ____

ANYTHING THAT DISPLACES 9.8 N OF WATER EXPERIENCES 9.8 N OF BUOYANT FORCE.

   c. What is the weight of water displaced by the massless balloon?

   ____

   d. What is the buoyant force on the massless balloon?

   ____

   CUZ IF YOU PUSH 9.8 N OF WATER ASIDE THE WATER PUSHES BACK ON YOU WITH 9.8 N!

   e. In which direction would the massless balloon be accelerated?

   ____

Hewitt Drew it!

3. Assume the balloon is replaced by a 0.5-kilogram piece of wood that has exactly the same volume (1000 cm^3), as shown in Figure 2. The wood is held in the same submerged position beneath the surface of the water.

   a. What volume of water is displaced by the wood? ______________________

   b. What is the mass of the water displaced by the wood? ______________________

   c. What is the weight of the water displaced by the wood? ______________________

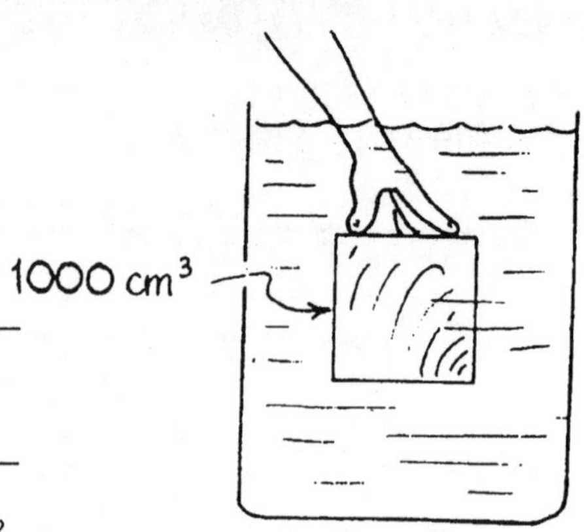

1000 cm³

Figure 2

   d. How much buoyant force does the surrounding water exert on the wood? ______________________

   e. When the hand is removed, what is the net force on the wood? ______________________

   f. In which direction does the wood accelerate when released? ______________________

THE BUOYANT FORCE ON A SUBMERGED OBJECT EQUALS THE WEIGHT OF WATER DISPLACED
... NOT THE WEIGHT OF THE OBJECT ITSELF!
... UNLESS IT IS FLOATING!

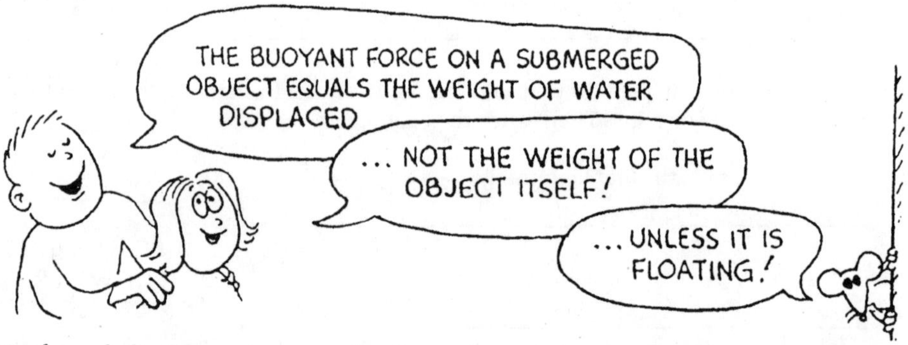

4. Repeat parts *a* through *f* in the previous question for a 5-kg rock that has the same volume (1000 cm^3), as shown in Figure 3. Assume the rock is suspended in the container of water by a string.

   a. ______________________

   b. ______________________

   c. ______________________

   d. ______________________

   e. ______________________

   f. ______________________

1000 cm³

Figure 3

WHEN THE WEIGHT OF AN OBJECT IS GREATER THAN THE BUOYANT FORCE EXERTED ON IT, IT SINKS!
Hewitt Drew it!

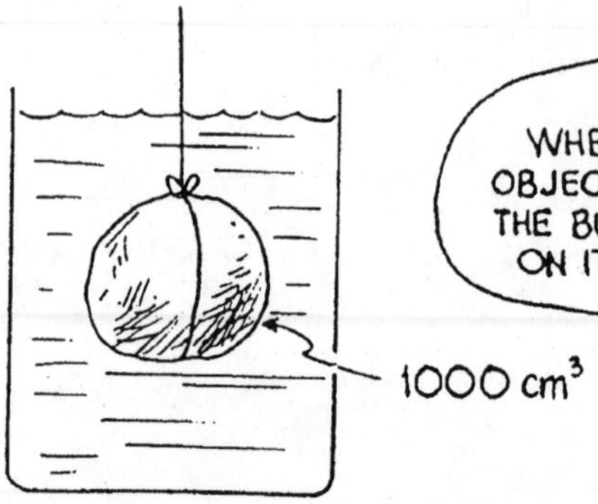

Name ______________________________ Date ____________

CONCEPTUAL *Physics* PRACTICE PAGE

Chapter 13 Liquids
Archimedes' Principle II

1. The water lines for the first three cases are shown. Sketch in the appropriate water lines for cases *d* and *e*, and make up your own for case *f*.

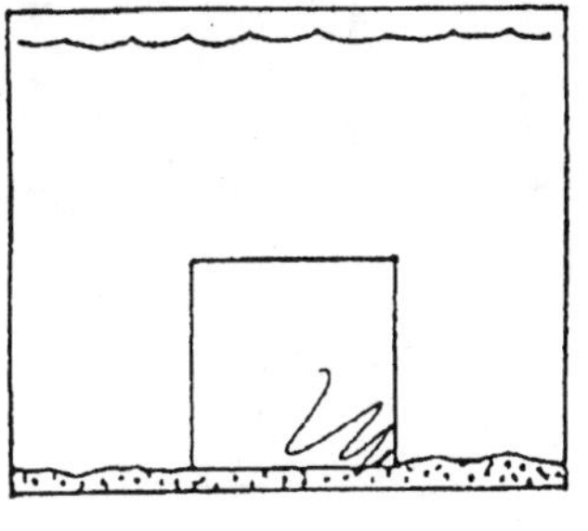

a. DENSER THAN WATER

b. SAME DENSITY AS WATER

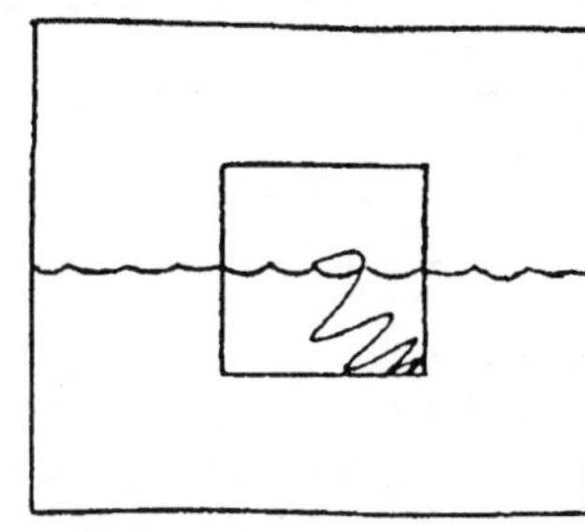

c. 1/2 AS DENSE AS WATER

d. 1/4 AS DENSE AS WATER

e. 3/4 AS DENSE AS WATER

f.______AS DENSE AS WATER

2. If the weight of a ship is 100 million N, then the water it displaces weighs______________________.
If cargo weighing 1000 N is put on board then the ship will sink down until an extra
__________________ of water is displaced.

3. The first two sketches below show the water line for an empty and a loaded ship. Draw in the appropriate water line for the third sketch.

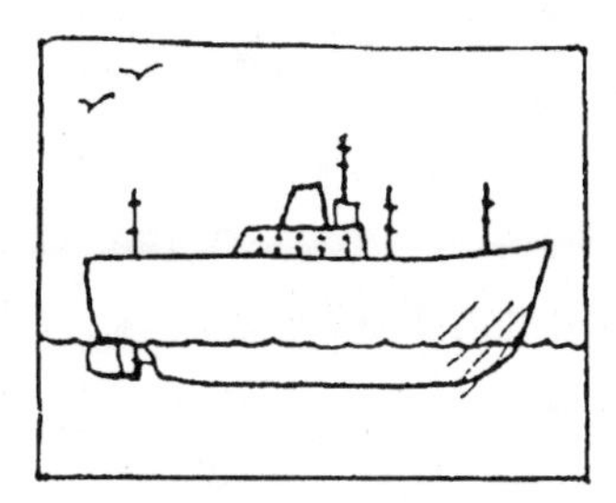

a. SHIP EMPTY

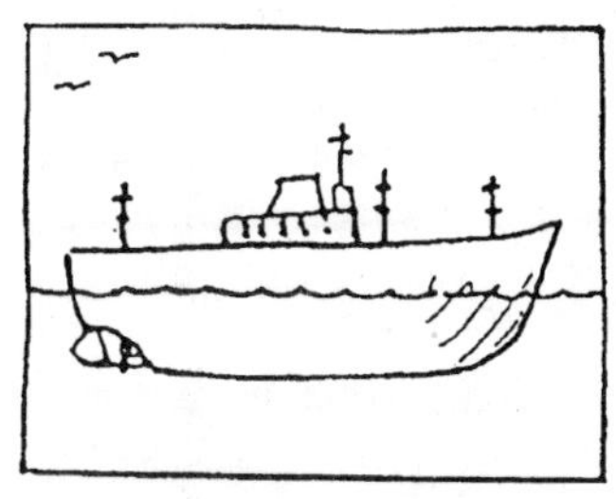

b. SHIP LOADED WITH 50 TONS OF IRON

c. SHIP LOADED WITH 50 TONS OF STYROFOAM

Hewitt Drew it!

4. Here is a glass of ice water with an ice cube floating in it. Draw the water line after the ice cube melts. (Will the water line rise, fall, or remain the same?)

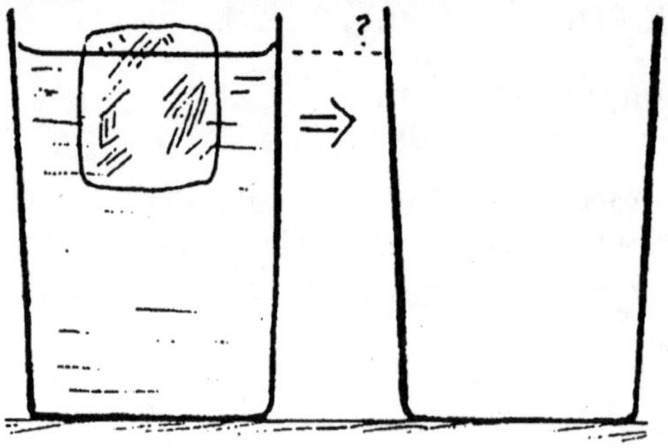

5. The air-filled balloon is weighted so it sinks in water. Near the surface, the balloon has a certain volume. Draw the balloon at the bottom (inside the dashed square) and show whether it is bigger, smaller, or the same size.

   a. Since the weighted balloon sinks, how does its overall density compare to the density of water?

   b. As the weighted balloon sinks, does its density increase, decrease, or remain the same?

   c. Since the weighted balloon sinks, how does the buoyant force on it compare to its weight?

   d. As the weighted balloon sinks deeper, does the buoyant force on it increase, decrease, or remain the same?

5. What would be your answers to Questions *a, b, c,* and *d* for a rock instead of an air-filled balloon?

   a.

   b.

   c.

   d.

Name ______________________ Date __________

CONCEPTUAL **Physics** PRACTICE PAGE

Chapter 14 Gases
Gas Pressure

1. A principle difference between a liquid and a gas is that when a liquid is under pressure, its volume

   (increases) (decreases) (doesn't change noticeably)

   and its density

   (increases) (decreases) (doesn't change noticeably)

   When a gas is under pressure, its volume

   (increases) (decreases) (doesn't change noticeably)

   and its density

   (increases) (decreases) (doesn't change noticeably)

2. The sketch shows the launching of a weather balloon at sea level. Make a sketch of the same weather balloon when it is high in the atmosphere. In words, what is different about its size and why?

   ____

   ____

   ____

   ____

HIGH-ALTITUDE SIZE

GROUND-LEVEL SIZE

3. A hydrogen-filled balloon that weighs 10 N must displace______N of air in order to float in air.

   If it displaces less than_______N it will be buoyed up with less than_______N and sink.

   If it displaces more than_______N of air it will move upward.

4. Why is the cartoon more humorous to physics types than to non-physics types? What physics has occurred?

   ____

   ____

   ____

   ____

RATS TO YOU TOO, DANIEL BERNOULLI!

Name ______________________________ Date __________

CONCEPTUAL **Physics** PRACTICE PAGE

Chapter 15 Temperature, Heat, and Expansion
Measuring Temperatures

1. Complete the table:

TEMPERATURE OF MELTING ICE	°C	32 °F	K
TEMPERATURE OF BOILING WATER	°C	212 °F	K

2. Suppose you apply a flame and heat one liter of water, raising its temperature 10°C. If you transfer the same heat energy to two liters, how much will the temperature rise? For three liters? *Record your answers on the blanks in the drawing at the right.*

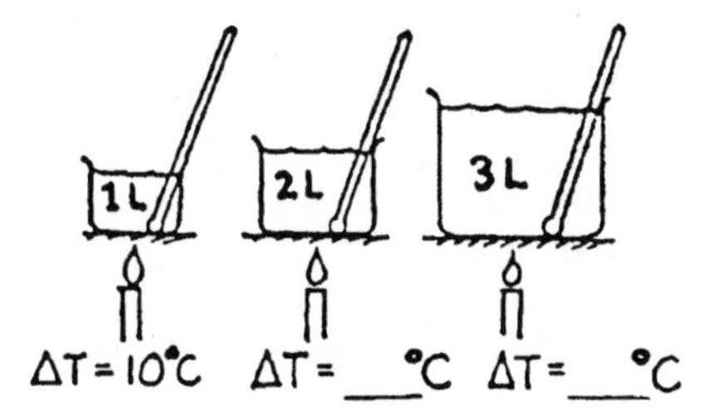

3. A thermometer is in a container half-filled with 20°C water.

   a. When an equal volume of 20°C water is added, the temperature of the mixture is

   (10°C) (20°C) (40°C)

   b. When instead an equal volume of 40°C water is added, the temperature of the mixture will be

   (20°C) (30°C) (40°C)

   c. When instead a small amount of 40°C water is added, the temperature of the mixture will be

   (20°C) (between 20°C and 30°C) (30°C) (more than 30°C)

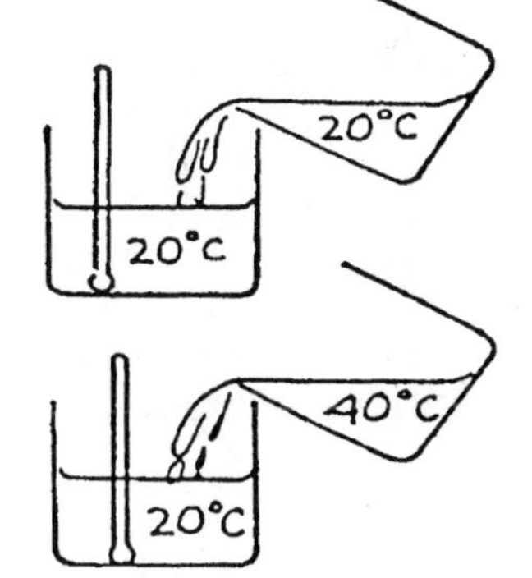

4. A red-hot piece of iron is put into a bucket of cool water. *Mark the following statements true (T) or false (F).* (Ignore heat transfer to the bucket.)

   a. The decrease in iron temperature equals the increase in the water temperature.__________

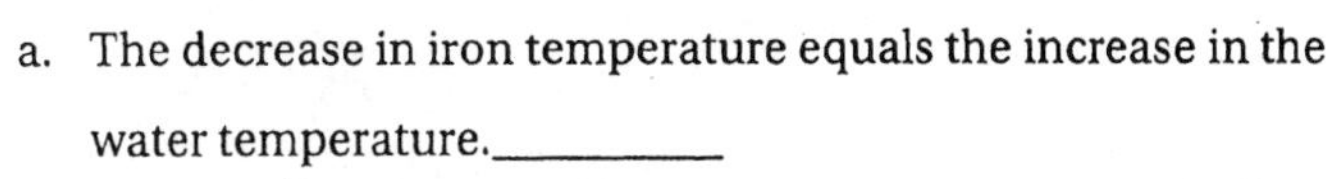

   b. The quantity of heat lost by the iron equals the quantity of heat gained by the water. __________

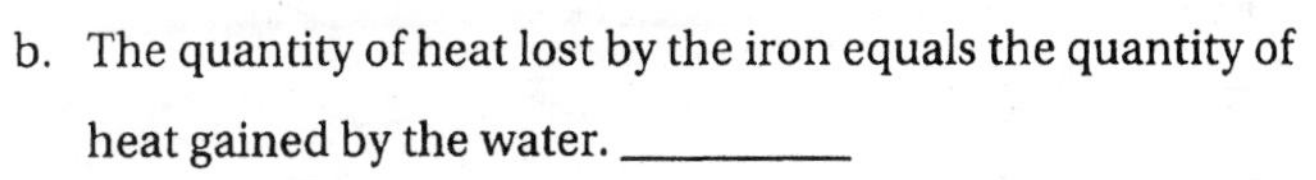

   c. The iron and water both will reach the same temperature. ________

   d. The final temperature of the iron and water is halfway between the initial temperatures of each. _________

CAN COMMON ICE BE COLDER THAN 0°C?

Hewitt Drew it!

CONCEPTUAL **Physics** PRACTICE PAGE

Thermal Expansion

1. The weight hangs above the floor from the copper wire. When a candle is moved along the wire and heats it, what happens to the height of the weight above the floor? Why?

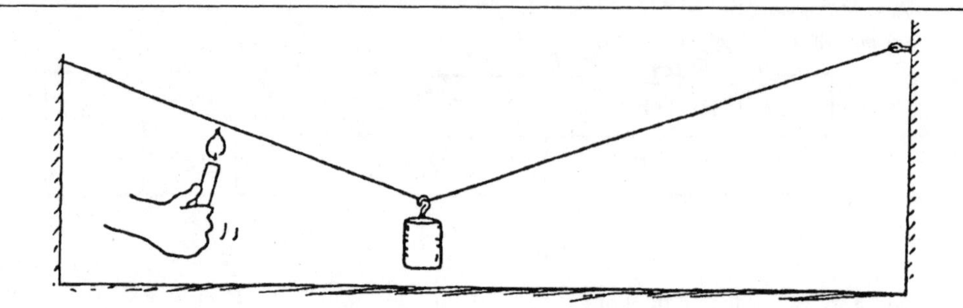

2. The levels of water at 0°C and 1°C are shown below in the first two flasks. At these temperatures there is microscopic slush in the water. There is slightly more slush at 0°C than at 1°C. As the water is heated, some of the slush collapses as it melts, and the level of the water falls in the tube. That's why the level of water is slightly lower in the 1°C-tube. Make rough estimates and sketch in the appropriate levels of water at the other temperatures shown. What is important about the level when the water reaches 4°C?

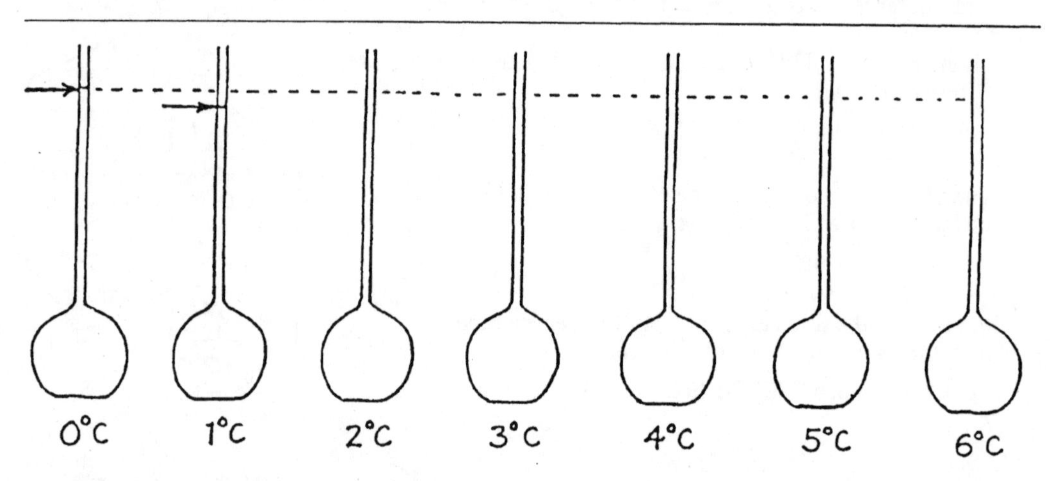

3. The diagram at right shows an ice-covered pond. Mark the probable temperatures of water at the top and bottom of the pond.

I CAN'T GET THIS METAL LID OFF THE JAR··· SHOULD I HEAT THE LID OR COOL IT? WHY?
CAP

ICE
°C
°C
WHICH WILL WEIGH MORE, 1 LITER OF ICE OR 1 LITER OF WATER?
HEWITT DREW IT!

Name ______________________________ Date __________

CONCEPTUAL Physics PRACTICE PAGE

Chapter 16 Heat Transfer
Transmission of Heat

1. The tips of both brass rods are held in the gas flame. *Mark the following true (T) or false (F).*

   a. Heat is conducted only along Rod A.______

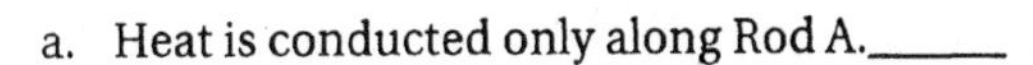

   b. Heat is conducted only along Rod B.______

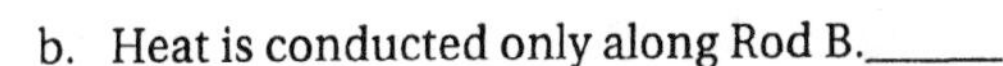

   c. Heat is conducted equally along both Rod A and Rod B.________

   d. The idea that "heat rises" applies to heat transfer by *convection*, not by *conduction*.

   ____

2. Why does a bird fluff its feathers to keep warm on a cold day?

   ____

3. Why does a down-filled sleeping bag keep you warm on a cold night? Why is it useless if the down is wet?

   ____

4. What does *convection* have to do with the holes in the shade of the desk lamp?

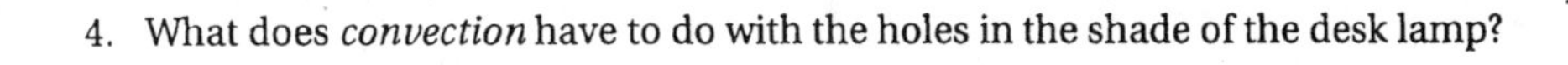

   ____

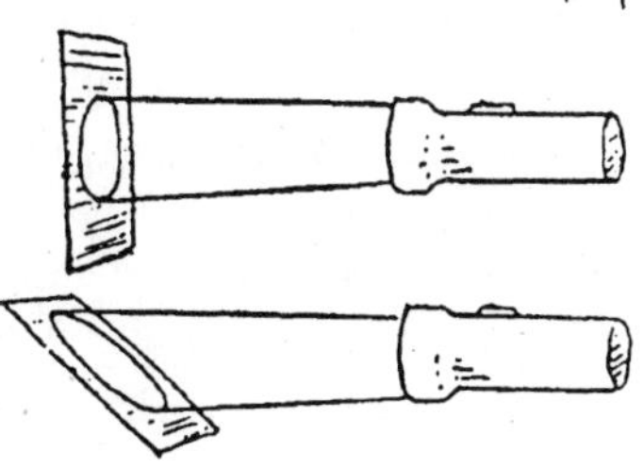

5. The warmth of equatorial regions and coldness of polar regions on the Earth can be understood by considering light from a flashlight striking a surface. If it strikes perpendicularly, light energy is more concentrated as it covers a smaller area; if it strikes at an angle, the energy spreads over a larger area. So the energy per unit area is less.

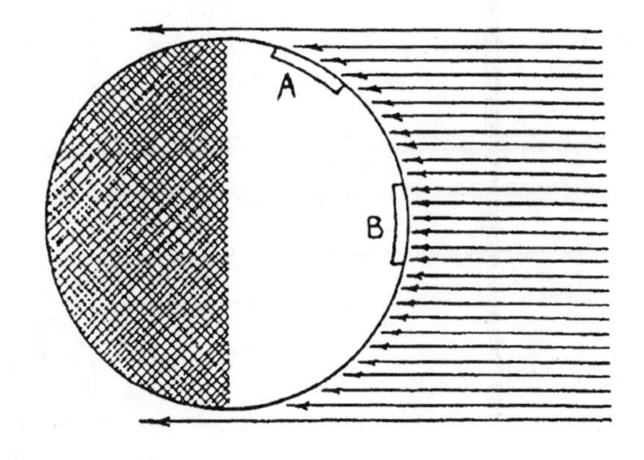

The arrows represent rays of light from the distant sun incident upon the Earth. Two areas of equal size are shown, Area A near the north pole and Area B near the equator. Count the rays that reach each area, and explain why B is warmer than A.

____

____

Hewitt Drew it!

6. The Earth's seasons result from the 23.5-degree tilt of the earth's daily spin axis as it orbits the sun. When the Earth is at the point shown on the right in the sketch below (not to scale), the Northern Hemisphere tilts toward the sun, and sunlight striking it is strong (more rays per area). Sunlight striking the Southern Hemisphere is weak (fewer rays per area). Days in the north are warmer, and daylight is longer. You can see this by imagining the Earth making its complete daily 24-hour spin.

   Do two things on the sketch: (1) Shade the part of the Earth in nighttime darkness for all positions, as is already done in the left position. (2) Label each position with the proper month — March, June, September, or December.

N
S
( )
BE SURE TO DO THE SHADING BEFORE YOU ANSWER THE QUESTIONS BELOW!

a. When the Earth is in any of the four positions shown, during one 24-hour spin a location at the equator receives sunlight half the time and is in darkness the other half the time.
   This means that regions at the equator always get about ________ hours of sunlight and
   _______ hours of darkness.

b. Can you see that in the June position regions farther north have longer daylight hours and shorter nights? Locations north of the Arctic Circle (dotted line in Northern Hemisphere) always face toward the sun as the Earth spins, so they get daylight ________ hours a day.

c. How many hours of light and darkness are there in June at regions south of the Antarctic Circle (dotted line in Southern Hemisphere)?

   ____

d. Six months later, when the Earth is at the December position, is the situation in the Antarctic the same or is it the reverse?

   ____

e. Why do South America and Australia enjoy warm weather in December instead of June?

   ____

   ____

Name ______________________ Date __________

CONCEPTUAL *Physics* PRACTICE PAGE

Chapter 17 Change of Phase
Ice, Water, and Steam

All matter can exist in the solid, liquid, or gaseous phases. The solid phase exists at relatively low temperatures, the liquid phase at higher temperatures, and the gaseous phase at still higher temperatures. Water is the most common example, not only because of its abundance but also because the temperatures for all three phases are common. Study "Energy and Changes of Phase" in your textbook and then answer the following:

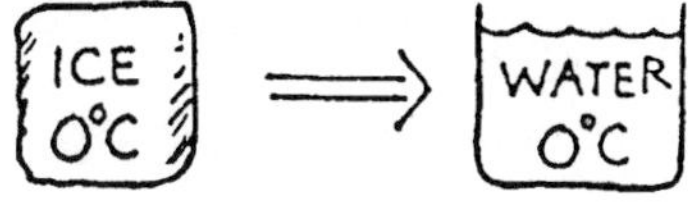

1. How many calories are needed to change 1 gram of 0˚C ice to water?

____

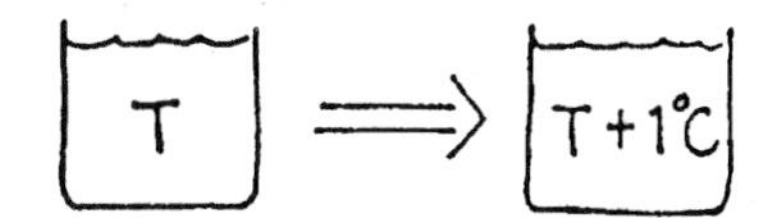

2. How many calories are needed to change the temperature of 1 gram of water by 1˚C?

____

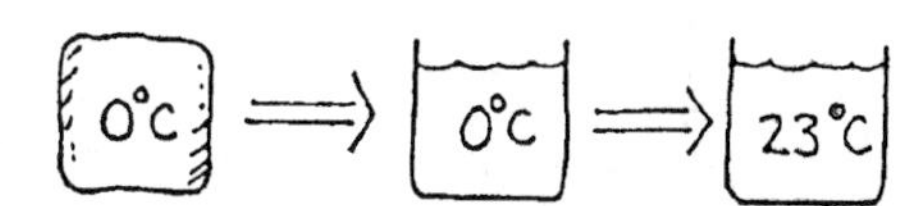

3. How many calories are needed to melt 1 gram of 0˚C ice and turn it to water at a room temperature of 23˚C?

____

4. A 50-gram sample of ice at 0°C is placed in a glass beaker that contains 200 g of water at 20˚C.

   a. How much heat is needed to melt the ice? ____________

   b. By how much would the temperature of the water change if it gave up this much heat to the ice? ____________

   c. What will be the final temperature of the mixture? (Disregard any heat absorbed by the glass or given off by the surrounding air.) _____

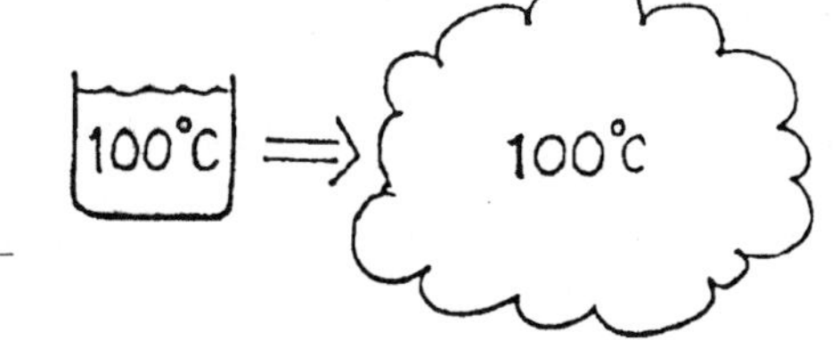

5. How many calories are needed to change 1 gram of 100˚C boiling water to 100˚C steam?

____

6. Fill in the number of calories at each step below for changing the state of 1 gram of 0˚C ice to 100˚C steam.

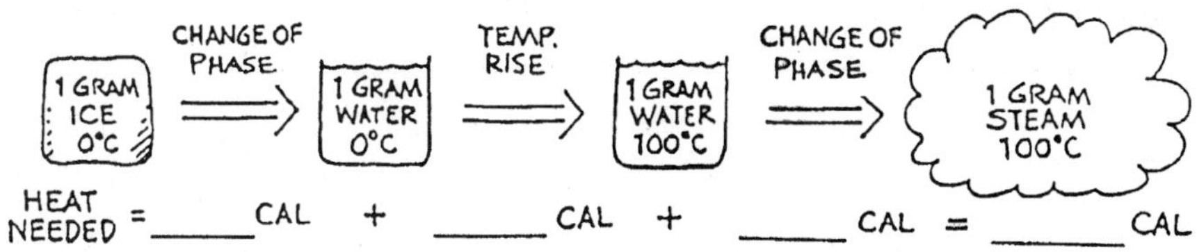

HEWITT DREW IT!

7. One gram of steam at 100°C condenses, and the water cools to 22°C.

   a. How much heat is released when the steam condenses?________________

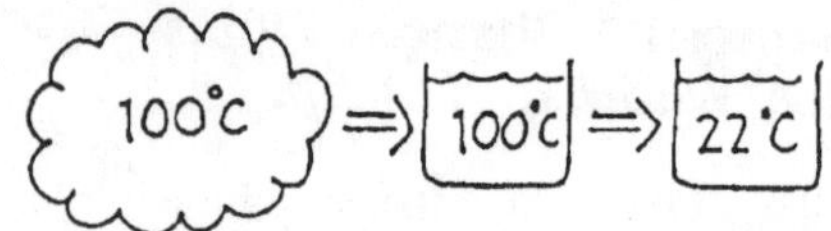

   b. How much heat is released when the water cools from 100°C to 22°C?

   ____

   c. How much heat is released altogether? ________________________

8. In a household radiator 1000 g of steam at 100°C condenses, and the water cools to 90°C.

   a. How much heat is released when the steam condenses?

   ____

   b. How much heat is released when the water cools from 100°C to 90°C?

   ____

   c. How much heat is released altogether?

   ____

9. Why is it difficult to make tea on the top of a high mountain?

   ____

10. How many calories are given up by 1 gram of 100°C steam that condenses to 100°C water?

   ____

11. How many calories are given up by 1 gram of 100°C steam that condenses and drops in temperature to 22°C water?

   ____

12. How many calories are given to a household radiator when 1000 grams of 100°C steam condenses, and drops in temperature to 90°C water?

   ____

13. To get water from the ground, even in the hot desert, dig a hole about a half meter wide and a half meter deep. Place a cup at the bottom. Spread a sheet of plastic wrap over the hole and place stones along the edge to hold it secure. Weight the center of the plastic with a stone so it forms a cone shape. Why will water collect in the cup? (Physics can save your life if you're ever stranded in a desert!)

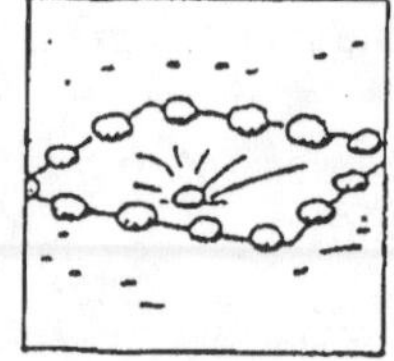

   ____

   ____

   ____

   ____

   ____

Hewitt Drew it!

Name ________________________________ Date ____________

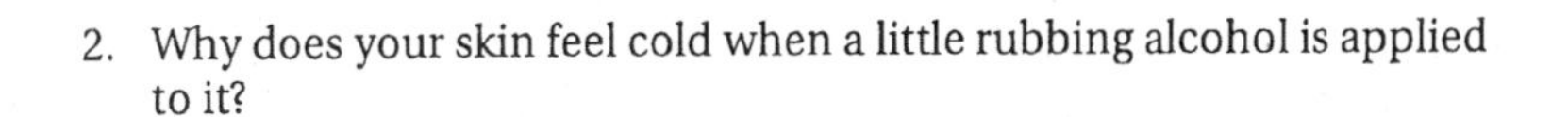

Chapter 17 Change of Phase
Evaporation

1. Why does it feel colder when you swim at a pool on a windy day?

____

2. Why does your skin feel cold when a little rubbing alcohol is applied to it?

____

3. Briefly explain from a molecular point of view why evaporation is a cooling process.

____

____

4. When hot water rapidly evaporates, the result can be dramatic. Consider 4 g of boiling water spread over a large surface so that 1 g rapidly evaporates. Suppose further that the surface and surroundings are very cold so that all 540 calories for evaporation come from the remaining 3 g of water.

a. How many calories are taken from each gram of water?

____

b. How many calories are released when 1 g of 100°C water cools to 0°C?

____

c. How many calories are released when 1 g of 0°C water changes to 0°C ice?

____

d. What happens in this case to the remaining 3 g of boiling water when 1 g rapidly evaporates?

____

____

Hewitt Drew it!

Name ______________________ Date __________

CONCEPTUAL *Physics* PRACTICE PAGE

Chapter 18 Thermodynamics
Absolute Zero

A mass of air is contained so that the volume can change but the pressure remains constant. Table I shows air volumes at various temperatures when the air is heated slowly.

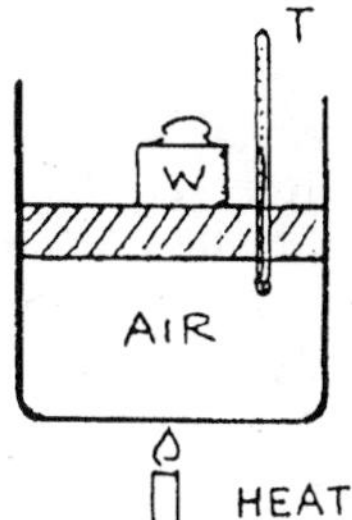

1. Plot the data in Table I on the graph, and connect the points.

Table I

TEMP. (°C)	VOLUME (mL)
0	50
25	55
50	60
75	65
100	70

VOLUME (mL): 70, 60, 50, 40, 30, 20, 10

TEMPERATURE (°C): -200, -100, 0, 50, 100

2. The graph shows how the volume of air varies with temperature at constant pressure. The straightness of the line means that the air expands uniformly with temperature. From your graph, you can predict what will happen to the volume of air when it is cooled.

   Extrapolate (extend) the straight line of your graph to find the temperature at which the volume of the air would become zero. Mark this point on your graph. Estimate this temperature: ____________

3. Although air would liquify before cooling to this temperature, the procedure suggests that there is a lower limit to how cold something can be. This is the absolute zero of temperature.
   Careful experiments show that absolute zero is ________ °C.

4. Scientists measure temperature in *kelvins* instead of degrees Celsius, where the absolute zero of temperature is 0 kelvins. If you relabeled the temperature axis on the graph in Question 1 so that it shows temperature in kelvins, would your graph look like the one below? ____________

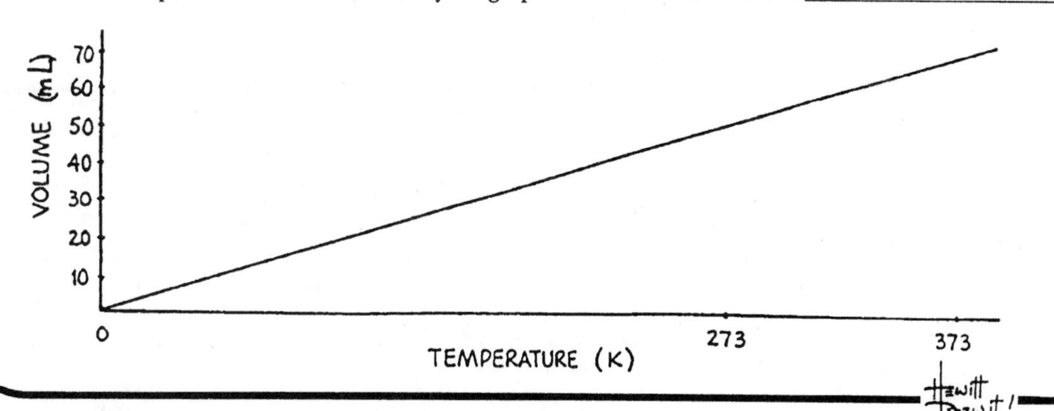

CONCEPTUAL **Physics** PRACTICE PAGE

Our Earth's Hot Interior

A major puzzle faced scientists in the 19th Century. Volcanoes showed that the Earth is molten beneath its crust. Penetration into the crust by bore holes and mines showed that the Earth's temperature increases with depth. Scientists knew that heat flows from the interior to the surface. They assumed that the source of the Earth's internal heat was primordial, the afterglow of its fiery birth. Measurements of cooling rates indicated a relatively young Earth—some 25 to 30 millions years in age. But geological evidence indicated an older Earth. This puzzle wasn't solved until the discovery of radioactivity. Then it was learned that the interior is kept hot by the energy of radioactive decay. We now know the age of the Earth is some 4.5 billions years—a much older Earth.

All rock contains trace amounts of radioactive minerals. Those in common granite release energy at the rate 0.03 J/kg y. Granite at the Earth's surface transfers this energy to the surroundings practically as fast as it is generated, so we don't find granite warm to the touch. But what if a sample of granite were thermally insulated? That is, suppose the increase of internal energy due to radioactivity were contained. Then it would get hotter. How much? Let's figure it out, using 790 joule/kilogram kelvin as the specific heat of granite.

Calculations to make:

1. How many joules are required to increase the temperature of 1 kg of granite by 1000 K?

____

2. How many years would it take radioactive decay in a kilogram of granite to produce this many joules?

____

Questions to answer:

1. How many years would it take a thermally insulated 1-kg chunk of granite to undergo a 1000 K increase in temperature?

____

2. How many years would it take a thermally insulated one-million-kilogram chunk of granite to undergo a 1000 K increase in temperature?

____

An electric toaster stays hot while electric energy is supplied, and doesn't cool until switched off. Similarly, do you think the energy source now keeping the earth hot will one day suddenly switch off like a disconnected toaster — or gradually decrease over a long time?

3. Why does the Earth's interior remain molten hot?

____

4. Rock has a higher melting temperature deep in the interior. Why?

____

5. Why doesn't the Earth just keep getting hotter until it all melts?

____

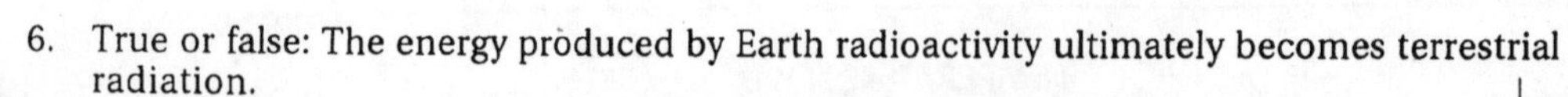

6. True or false: The energy produced by Earth radioactivity ultimately becomes terrestrial radiation.

Hewitt Drew it!

Name ______________________________ Date __________

CONCEPTUAL *Physics* PRACTICE PAGE

Chapter 19 Vibrations and Waves
Vibration and Wave Fundamentals

1. A sine curve that represents a transverse wave is drawn below. With a ruler, measure the wavelength and amplitude of the wave.

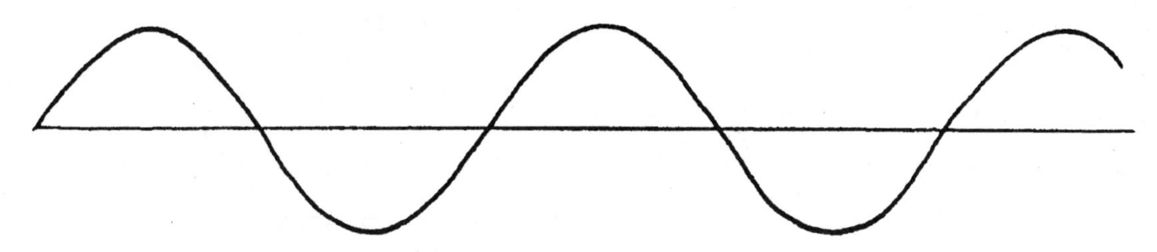

a. Wavelength = __________ b. Amplitude = __________

2. A kid on a playground swing makes a complete to-and-fro swing each 2 seconds. The frequency of swing is

(0.5 hertz) (1 hertz) (2 hertz)

and the period is

(0.5 second) (1 second) (2 seconds)

3. *Complete the statements.*

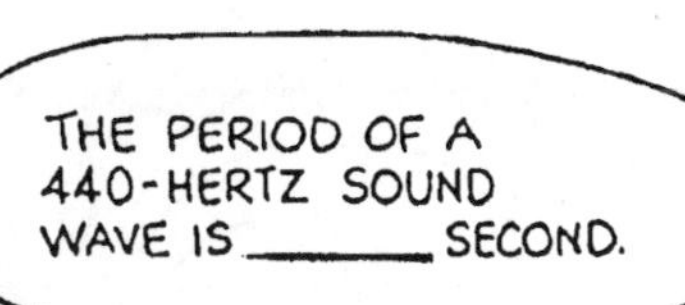

THE PERIOD OF A 440-HERTZ SOUND WAVE IS ______ SECOND.

A MARINE WEATHER STATION REPORTS WAVES ALONG THE SHORE THAT ARE 8 SECONDS APART. THE FREQUENCY OF THE WAVES IS THEREFORE ______________ HERTZ.

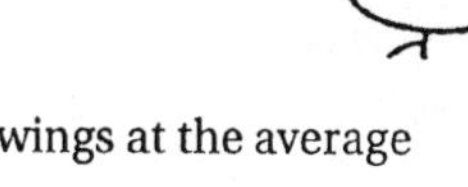

4. The annoying sound from a mosquito is produced when it beats its wings at the average rate of 600 wingbeats per second.

a. What is the frequency of the soundwaves?

____

b. What is the wavelength? (Assume the speed of sound is 340 m/s.)

____

Hewitt Drew it!

5. A machine gun fires 10 rounds per second. The speed of the bullets is 300 m/s.

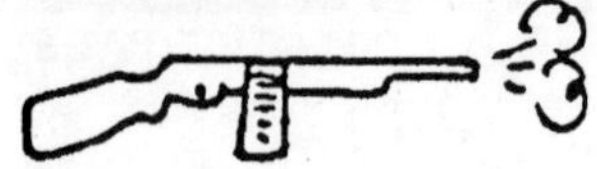

   a. What is the distance in the air between the flying bullets?______________

   b. What happens to the distance between the bullets if the rate of fire is increased?

   ____

6. Consider a wave generator that produces 10 pulses per second. The speed of the waves is 300 cm/s.

   a. What is the wavelength of the waves? __________

   b. What happens to the wavelength if the frequency of pulses is increased?

   ____

7. The bird at the right watches the waves. If the portion of a wave between 2 crests passes the pole each second, what is the speed of the wave?

   ____

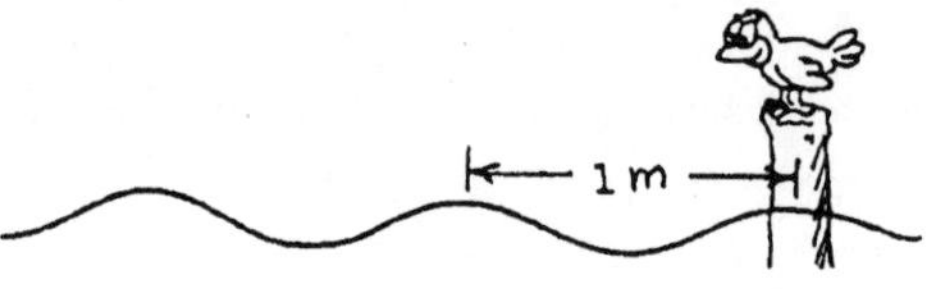

   What is its period?

   ____

8. If the distance between crests in the above question were 1.5 meters apart, and 2 crests pass the pole each second, what would be the speed of the wave?

   ____

   What would be its period?

   ____

9. When an automobile moves toward a listener, the sound of its horn seems relatively

   (low pitched) (normal)

   (high pitched)

   and when moving away from the listener, its horn seems

   (low pitched) (normal)

   (high pitched)

10. The changed pitch of the Doppler effect is due to changes in

   (wave speed) (wave frequency)

Name ______________________ Date __________

CONCEPTUAL **Physics** PRACTICE PAGE

Chapter 19 Vibrations and Waves

Shock Waves

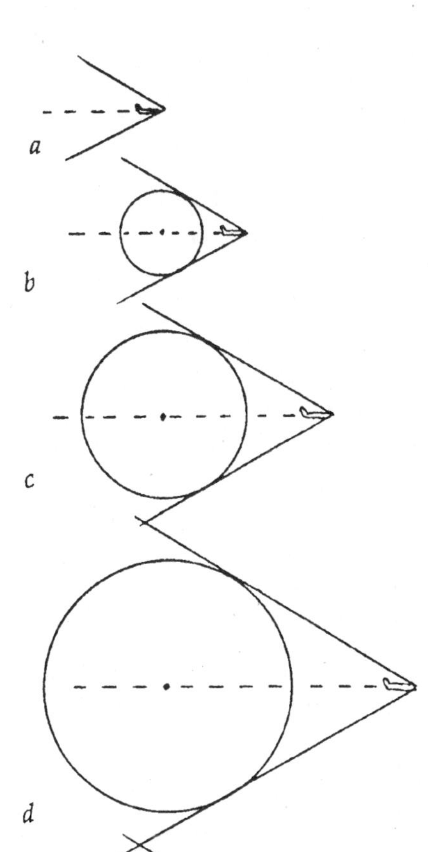

The cone-shaped shock wave produced by a supersonic aircraft is actually the result of overlapping spherical waves of sound, as indicated by the overlapping circles in Figure 18.19 in your textbook. Sketches *a, b, c, d,* and *e,* at the left show the "animated" growth of only one of the many spherical sound waves (shown as an expanding circle in the two-dimensional sketch). The circle originates when the aircraft is in the position shown in *a.* Sketch *b* shows both the growth of the circle and position of the aircraft at a later time. Still later times are shown in *c, d,* and *e.* Note that the circle grows and the aircraft moves farther to the right. Note also that the aircraft is moving farther than the sound wave. This is because the aircraft is moving faster than sound.

Careful examination will reveal how fast the aircraft is moving compared to the speed of sound. Sketch *e* shows that in the same time the sound travels from O to A, the aircraft has traveled from O to B — twice as far. You can check this with a ruler.

Circle the answer.

1. Inspect sketches *b* and *d.* Has the aircraft traveled twice as far as sound in the same time in these postions also?

   (yes) (no)

2. For greater speeds, the angle of the shock wave would be

   (wider) (the same) (narrower)

O A B

e

DURING THE TIME THAT SOUND TRAVELS FROM O TO A, THE PLANE TRAVELS TWICE AS FAR --- FROM O TO B.

SO IT'S FLYING AT TWICE THE SPEED OF SOUND!

HEWITT DREW IT!

3. Use a ruler to estimate the speeds of the aircraft that produce the shock waves in the two sketches below.

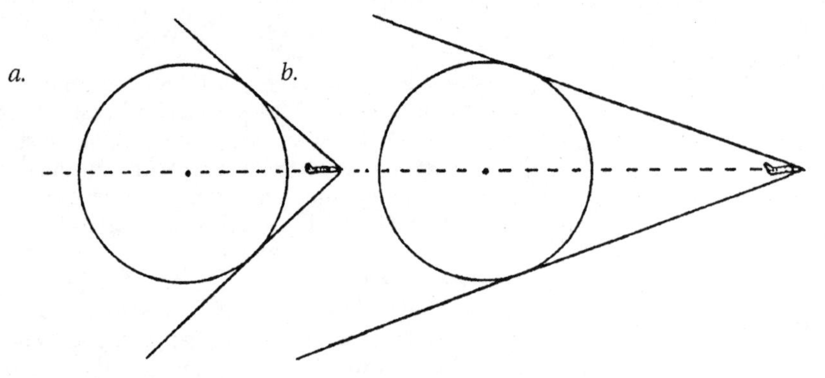

Aircraft *a* is traveling about _______ times the speed of sound.

Aircraft *b* is traveling about _______ times the speed of sound.

4. Draw your own circle (anywhere) and estimate the speed of the aircraft to produce the shock wave shown below.

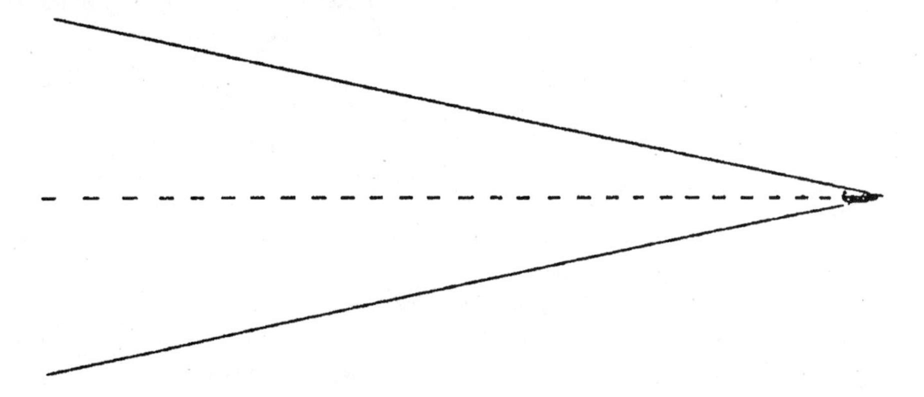

The speed is about _______ times the speed of sound.

5. In the space below, draw the shock wave made by a supersonic missile that travels at four times the speed of sound.

Hewitt Drew it!

Name ______________________ Date ________

CONCEPTUAL **Physics** PRACTICE PAGE

Chapter 20 Sound
Wave Superposition

A pair of pulses travel toward each at equal speeds. The composite waveforms,as they pass through each other and interfere, are shown at 1-second intervals. In the left column note how the pulses interfere to produce the composite waveform (solid line). Make a similar construction for the two wave pulses in the right column. Like the pulses in the first column, they each travel at 1 space per second.

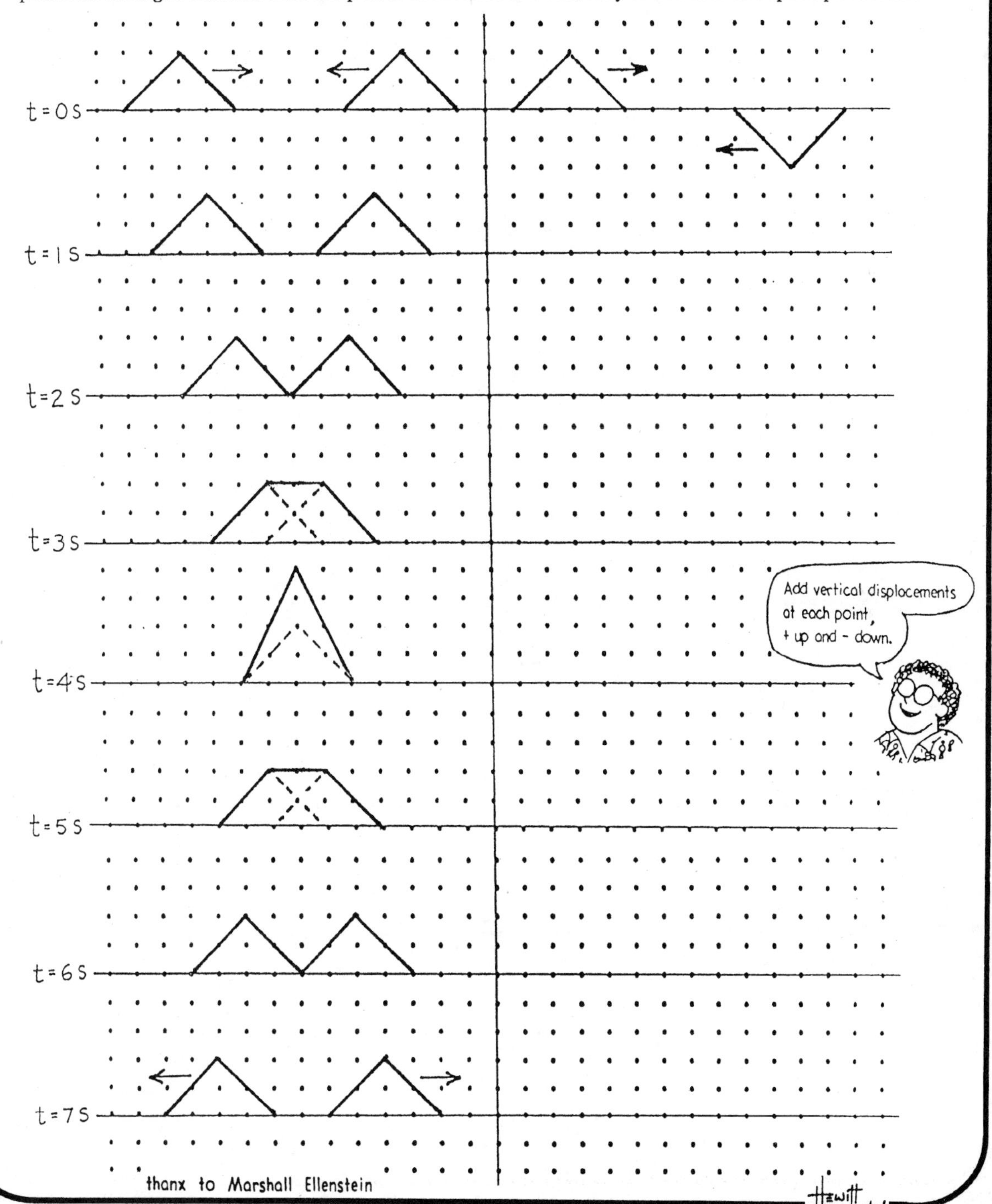

Construct the composite waveforms at 1-second intervals for the two waves traveling toward each other at equal speed.

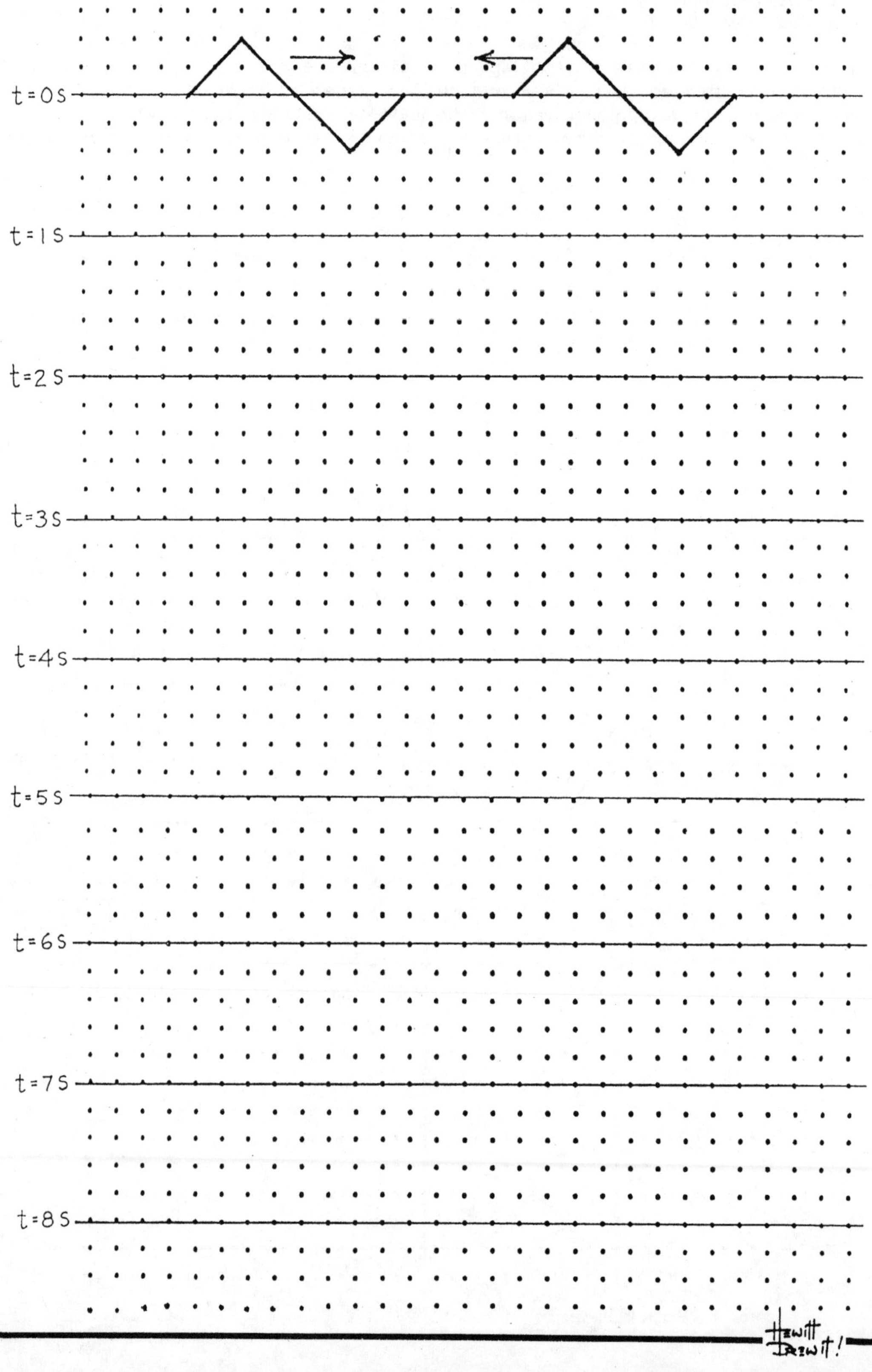

t=0s

t=1s

t=2s

t=3s

t=4s

t=5s

t=6s

t=7s

t=8s

Hewitt Drew it!

Name ______________________ Date ________

CONCEPTUAL **Physics** PRACTICE PAGE

Chapter 22 Electrostatics
Static Charge

1. Consider the diagrams below. (a) A pair of insulated metal spheres, A and B, touch each other, so in effect they form a single uncharged conductor. (b) A positively charged rod is brought near A, but not touching, and electrons in the metal sphere are attracted toward the rod. Charges in the spheres have redistributed, and the negative charge is labeled. Draw the appropriate + signs that are repelled to the far side of B. (c) Draw the signs of charge in (c), when the spheres are separated while the rod is still present, and in (d) after the rod has been removed. Your completed work should be similar to Figure 21.7 in the textbook. The spheres have been charged by *induction.*

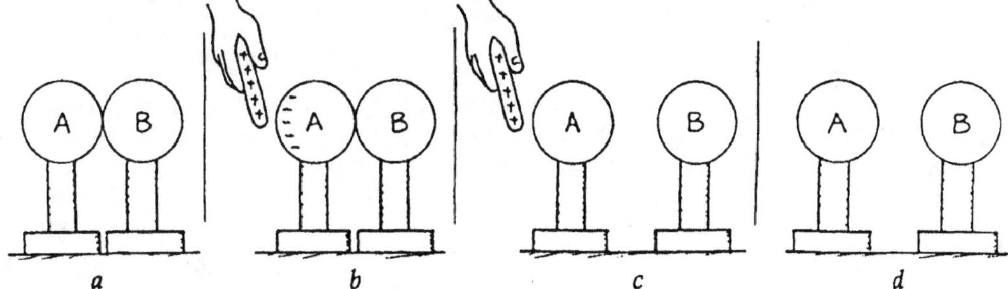

2. Consider below a single metal insulated sphere, (a) initially uncharged. When a negatively charged rod is nearby, (b), charges in the metal are separated. Electrons are repelled to the far side. When the sphere is touched with your finger, (c), electrons flow out to the sphere to the earth through the hand. The sphere is "grounded." Note the positive charge left (d) while the rod is still present and your finger removed, and (e) when the rod is removed. This is an example of *charge induction by grounding.* In this procedure the negative rod "gives" a positive charge to the sphere.

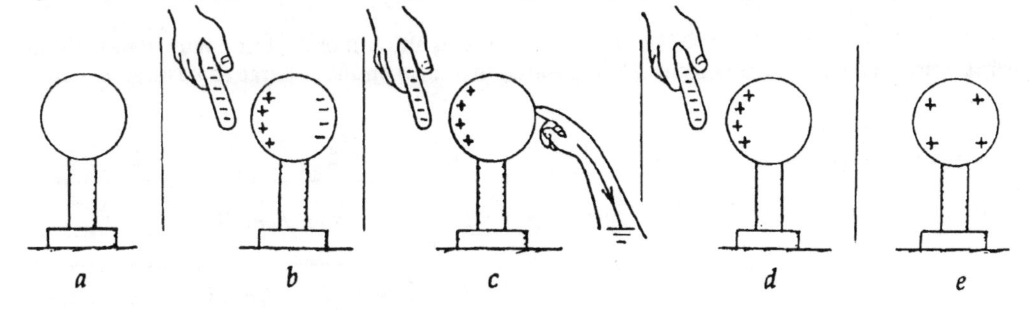

The diagrams below show a similar procedure with a positive rod. Draw the correct charges in the diagrams.

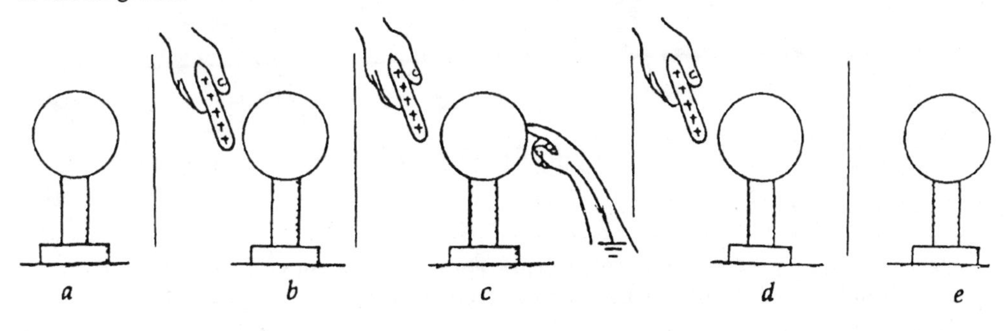

Hewitt Drew it!

Electric Potential

1\.

Just as PE (potential energy) transforms to KE (kinetic energy) for a mass lifted against the gravitational field (left), the electric PE of an electric charge transforms to other forms of energy when it changes location in an electric field (right). When released, how does the KE acquired by each compare to the decrease in PE?

____

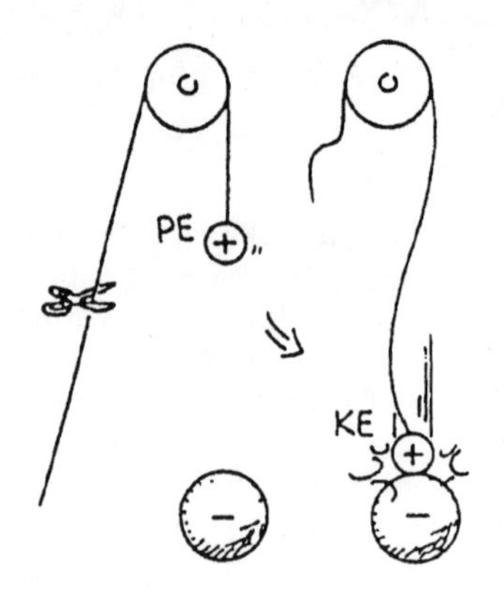

2\. *Complete the statements.*

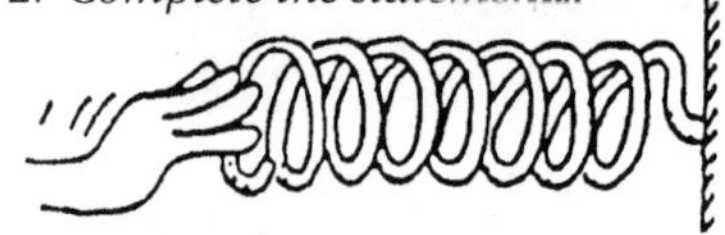

A force compresses the spring. The work done in compression is the product of the average force and the distance moved. W = Fd. This work increases the PE of the spring.

Similarly, a force pushes the charge (call it a test charge) closer to the charged sphere. The work done in moving the test charge is the product of the average ________ and the _________ moved. W = ______. This work ____________ the PE of the test charge.

If the test charge is released, it will be repelled and fly past the starting point. Its gain in KE at this point is ____________________ to its decrease in PE.

At any point, a greater quantity of test charge means a greater amount of PE, but not a greater amount of PE *per quantity* of charge. The quantities PE (measured in joules) and PE/charge (measured in volts) are different concepts.

By definition: **Electric Potential = PE/charge.** 1 volt = 1 joule/1coulomb.

3\. *Complete the statements.*

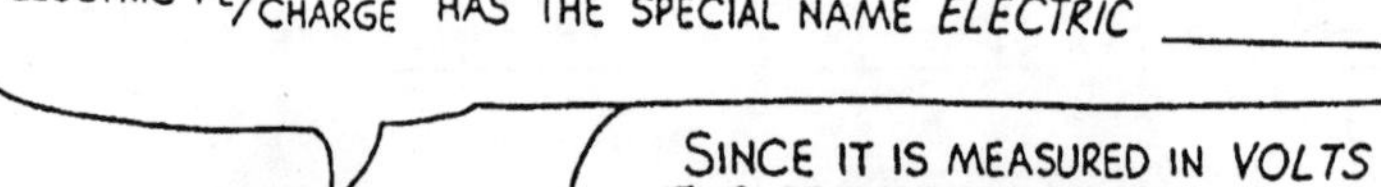

ELECTRIC PE/CHARGE HAS THE SPECIAL NAME *ELECTRIC* __________

SINCE IT IS MEASURED IN *VOLTS* IT IS COMMONLY CALLED __________

4\. If a conductor connected to the terminal of a battery has a potential of 12 volts, then each coulomb of charge on the conductor has a PE of ___________ J.

5\. Some people get mixed up between force and pressure. Recall that pressure is force *per area.* Similarly, some people get mixed up between electric PE and voltage. According to this chapter, voltage is electric PE *per* _________________.

Hewitt Drew it!

Name ____________________ Date ____________

CONCEPTUAL **Physics** PRACTICE PAGE

Chapter 23 Electric Current
Flow of Charge

1. Water doesn't flow in the pipe when (a) both ends are at the same level. Another way of saying this is that water will not flow in the pipe when both ends have the same potential energy (PE). Similarly, charge will not flow in a conductor if both ends of the conductor are at the same electric potential. But tip the water pipe and increase the PE of one side so there is a difference in PE across the ends of the pipe, as in (b), and water will flow. Similarly, increase the electric potential of one end of an electric conductor so there is a potential difference across the ends, and charge will flow.

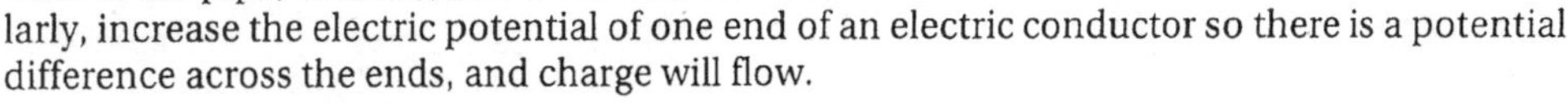

a

b

   a. The units of electric potential difference are

      (volts) (amperes) (ohms) (watts)

   b. It is common to call electric potential difference

      (voltage) (amperage) (wattage)

   c. The flow of electric charge is called electric

      (voltage) (current) (power),

      and is measured in

      (volts) (amperes) (ohms) (watts)

A VOLT IS A UNIT OF ________
AND AN AMPERE IS A UNIT OF ________

DOES VOLTAGE CAUSE CURRENT, OR DOES CURRENT CAUSE VOLTAGE? WHICH IS THE CAUSE AND WHICH IS THE EFFECT?

2. Complete the statements:

   a. A current of 1 ampere is a flow of charge at the rate of__________coulomb per second.
   b. When a charge of 15 C flows through any area in a circuit each second, the current is

      ________________A.

   c. One volt is the potential difference between two points if 1 joule of energy is needed to move

      __________coulomb of charge between the two points.

   d. When a lamp is plugged into a 120-V socket, each coulomb of charge that flows in the circuit

      is raised to a potential energy of__________joules.

   e. Which offers more resistance to water flow, a wide pipe or a narrow pipe? ____________
      Similarly, which offers more resistance to the flow of charge, a thick wire or a thin wire?

      ____

Hewitt Drew it!

Chapter 23 Electric Current

Ohm's Law

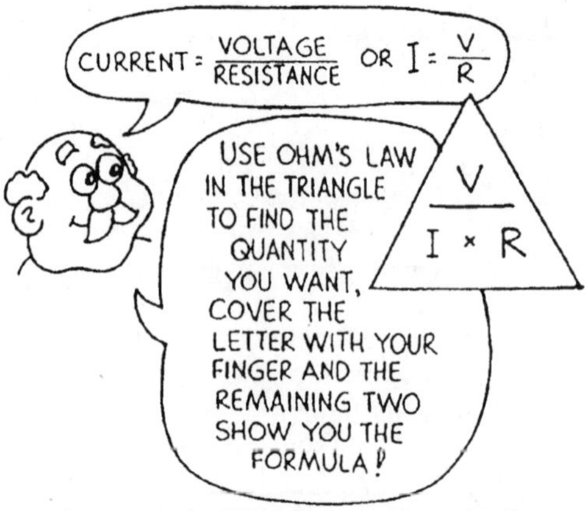

CURRENT = $\frac{\text{VOLTAGE}}{\text{RESISTANCE}}$ OR $I = \frac{V}{R}$

USE OHM'S LAW IN THE TRIANGLE TO FIND THE QUANTITY YOU WANT, COVER THE LETTER WITH YOUR FINGER AND THE REMAINING TWO SHOW YOU THE FORMULA!

$\frac{V}{I \times R}$

CONDUCTORS AND RESISTORS HAVE RESISTANCE TO THE CURRENT IN THEM.

1. How much current flows in a 1000-ohm resistor when 1.5 volts are impressed across it?

   ____

2. If the filament resistance in an automobile headlamp is 3 ohms, how many amps does it draw when connected to a 12-volt battery?

   ____

3. The resistance of the side lights on an automobile are 10 ohms. How much current flows in them when connected to 12 volts?

   ____

4. What is the current in the 30-ohm heating coil of a coffee maker that operates on a 120-volt circuit?

   ____

5. During a lie detector test, a voltage of 6 V is impressed across two fingers. When a certain question is asked, the resistance between the fingers drops from 400 000 ohms to 200 000 ohms. What is the current (a) initially through the fingers, and (b) when the resistance between them drops?

   (a) ____________________ (b) ____________________

6. How much resistance allows an impressed voltage of 6 V to produce a current of 0.006 A?

   ____

7. What is the resistance of a clothes iron that draws a current of 12 A at 120 V?

   ____

OHM MY GOODNESS !

8. What is the voltage across a 100-ohm circuit element that draws a current of 1 A?

   ____

9. What voltage will produce 3 A through a 15-ohm resistor?

   ____

10. The current in an incandescent lamp is 0.5 A when connected to a 120-V circuit, and 0.2 A when connected to a 10-V source. Does the resistance of the lamp change in these cases? Explain your answer and defend it with numerical values.

    ____

    ____

Hewitt Drew it!

Name ______________________ Date __________

CONCEPTUAL **Physics** PRACTICE PAGE

Chapter 23 Electric Current

Electric Power

Recall that the rate energy is converted from one form to another is *power.*

$$\text{power} = \frac{\text{energy converted}}{\text{time}} = \frac{\text{voltage x charge}}{\text{time}} = \text{voltage x } \frac{\text{charge}}{\text{time}} = \text{voltage x current}$$

The unit of power is the *watt* (or *kilowatt*). So in units form,

Electric power (*watts*) = current (*amperes*) x voltage (*volts*),

where 1 *watt = 1 ampere x 1 volt.*

AND CHARGE/TIME = CURRENT ; NEAT ;

1. What is the power when a voltage of 120 V drives a 2-A current through a device?

____

2. What is the current when a 60-W lamp is connected to 120 V?

____

3. How much current does a 100-W lamp draw when connected to 120 V?

____

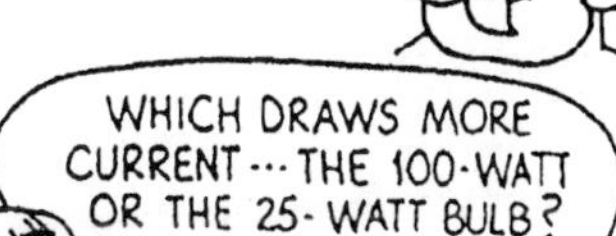

WHICH DRAWS MORE CURRENT··· THE 100-WATT OR THE 25-WATT BULB?

4. If part of an electric circuit dissipates energy at 6 W when it draws a current of 3 A, what voltage is impressed across it?

____

5. The equation

$$\text{power} = \frac{\text{energy converted}}{\text{time}}$$

rearranged gives

energy converted =

WATT'S HAPPENING ?

6. Explain the difference between a kilowatt and a kilowatt-hour.

____

7. One deterrent to burglary is to leave your front porch light on all the time. If your fixture contains a 60-W bulb at 120 V, and your local power utility sells energy at 10 cents per kilowatt-hour, how much will it cost to leave the bulb on for the whole month? Show your work on the other side of this page.

Hewitt Drew it!

Name ______________________ Date __________

CONCEPTUAL Physics PRACTICE PAGE

Chapter 23 Electric Current
Series Circuits

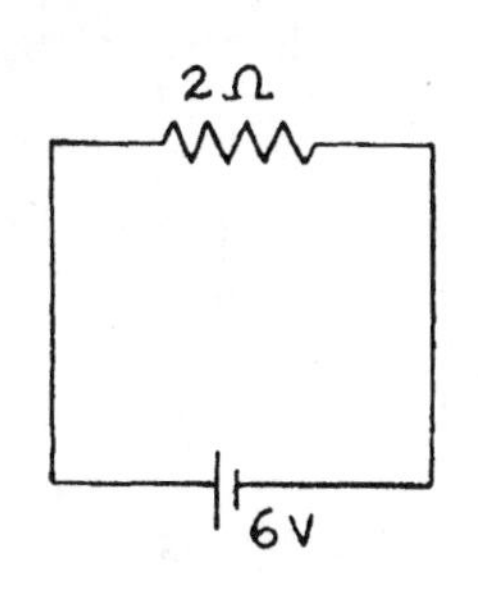

THE EQUIVALENT RESISTANCE OF RESISTORS IN SERIES IS SIMPLY THEIR SUM!

1. In the circuit shown at the right, a voltage of 6 V pushes charge through a single resistor of 2 Ω. According to Ohm's law, the current in the resistor (and therefore in the whole circuit) is

   __________ A.

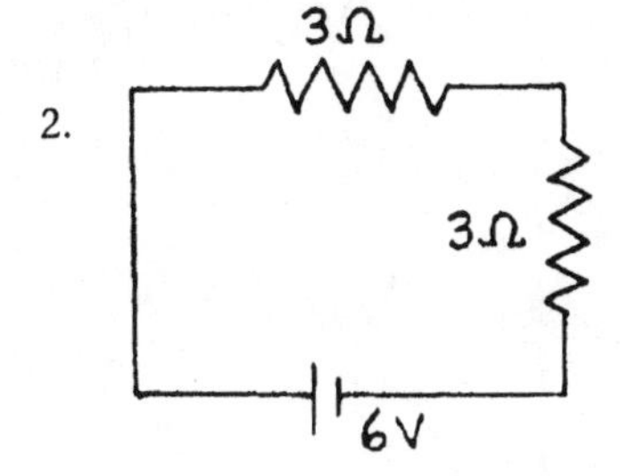

2. If a second identical lamp is added, as on the left, the 6-V battery must push charge through a total resistance of ______ Ω. The current in the circuit is then ____ A.

3. The equivalent resistance of three 4-Ω resistors in series is __________ Ω.

4. Does current flow *through* a resistor, or *across* a resistor? __________

   Is voltage established *through* a resistor, or *across* a resistor? __________

5. Does current in the lamps occur simultaneously, or does charge flow first through one lamp, then the other, and finally the last in turn?

   ____

6. Circuits *a* and *b* below are identical with all bulbs rated at equal wattage (therefore equal resistance). The only difference between the circuits is that Bulb 5 has a short circuit, as shown.

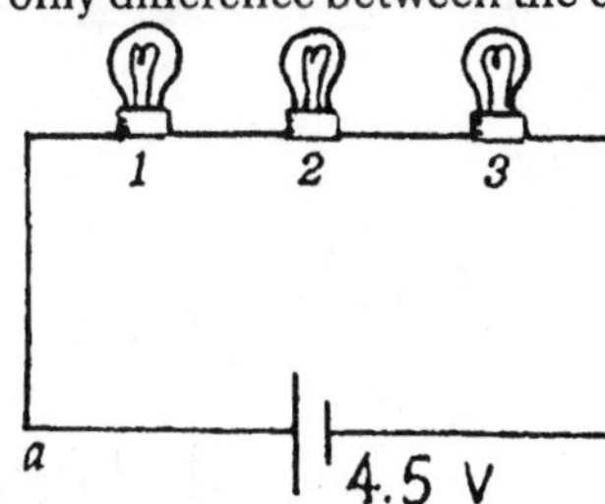

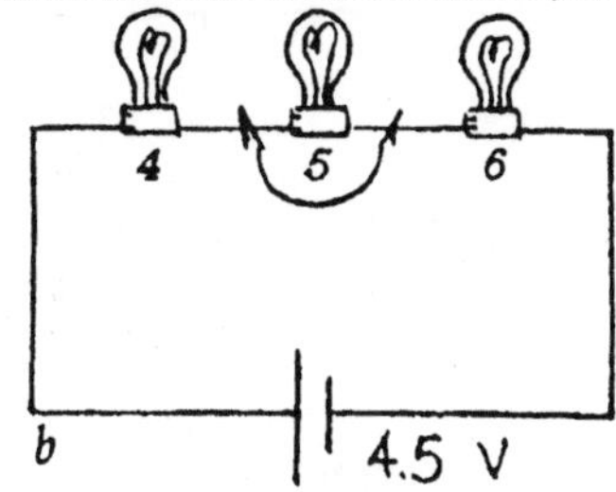

   a. In which circuit is the current greater? __________

   b. In which circuit are all three bulbs equally bright? __________

   c. What bulbs are the brightest? __________

   d. What bulb is the dimmest? __________

   e. What bulbs have the largest voltage drops across them? __________

   f. Which circuit dissipates more power? __________

   g. What circuit produces more light? __________

Hewitt Drew it!

CONCEPTUAL **Physics** PRACTICE PAGE

Parallel Circuits

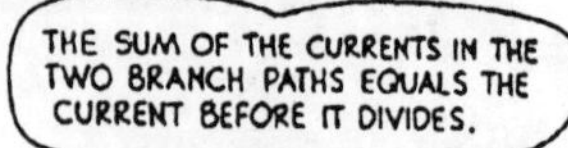

THE SUM OF THE CURRENTS IN THE TWO BRANCH PATHS EQUALS THE CURRENT BEFORE IT DIVIDES.
WATER FLOW

1. In the circuit shown below, there is a voltage drop of 6 V across *each* 2-Ω resistor.

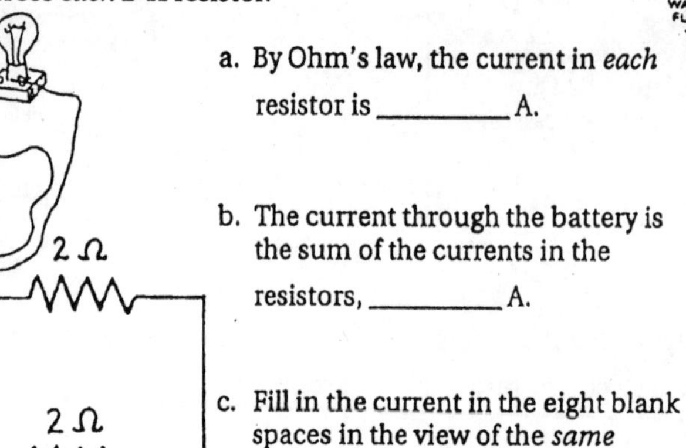

   a. By Ohm's law, the current in *each* resistor is ________ A.

   b. The current through the battery is the sum of the currents in the resistors, ________ A.

   c. Fill in the current in the eight blank spaces in the view of the *same circuit* shown again at the right.

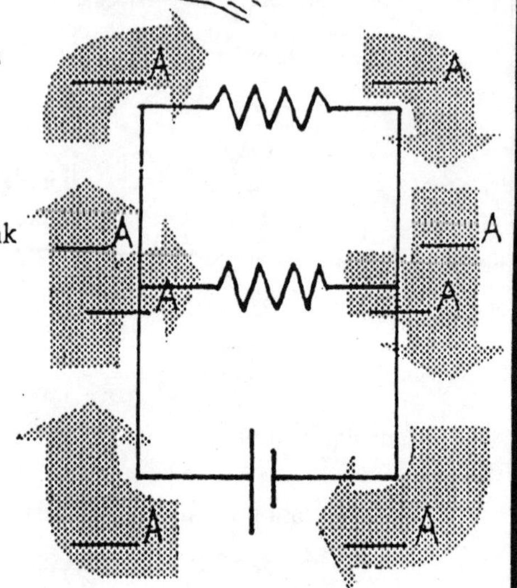

2. Cross out the circuit below that is *not* equivalent to the circuit above.

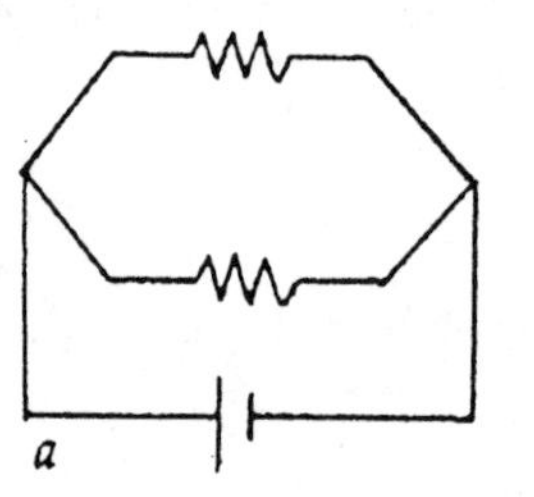
a

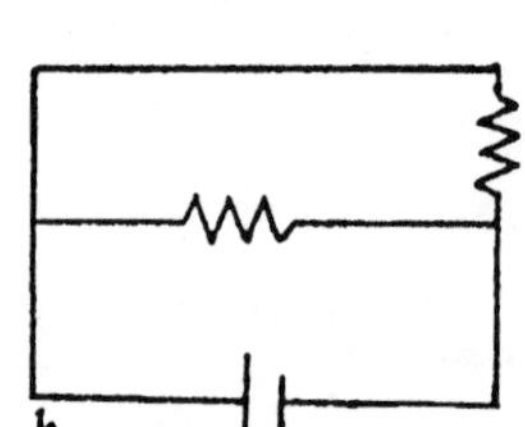
b

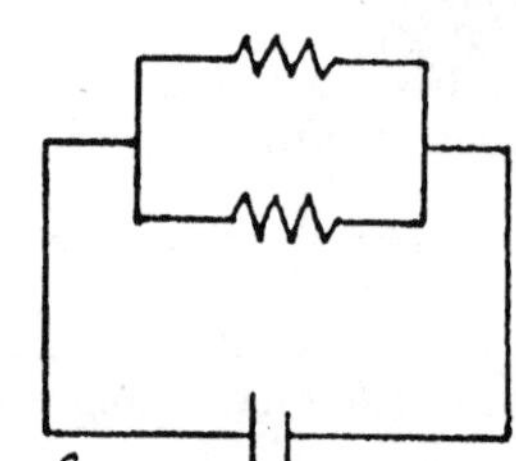
c

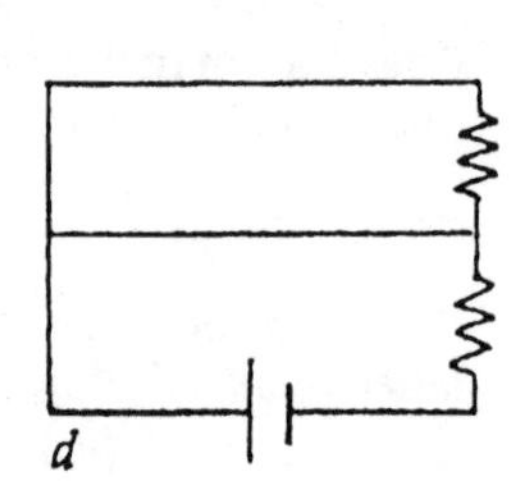
d

3. Consider the parallel circuit at the right.

   a. The voltage drop across each resistor is ________ V.

   b. The current in each branch is:

   2-Ω resistor ______ A

   2-Ω resistor ______ A

   1-Ω resistor ______ A

   b. The current through the battery equals the sum of the currents which equals ______ A.

   c. The equivalent resistance of the circuit equals ______ Ω.

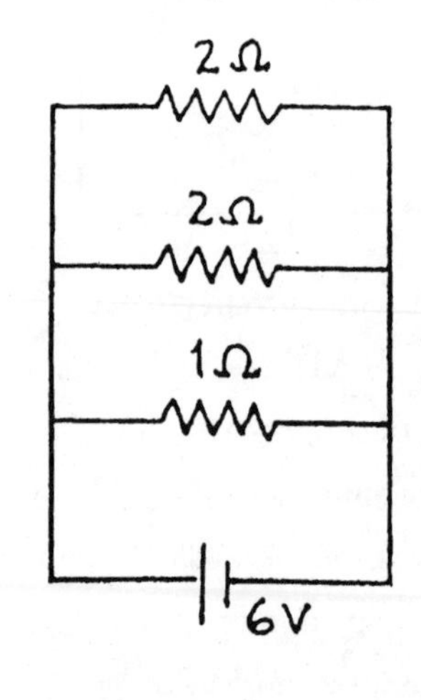

THE EQUIVALENT RESISTANCE OF A PAIR OF RESISTORS IN PARALLEL IS THEIR PRODUCT DIVIDED BY THEIR SUM!

Hewitt Drew it!

Name ______________________________ Date __________

CONCEPTUAL **Physics** PRACTICE PAGE

Chapter 23 Electric Current
Circuit Resistance

Figure what the resistances are, then show their values in the blanks to the left of each lamp.

All circuits below have the same lamp A, with resistance of 6 Ω, and the same 12-volt battery with negligible resistance. The unknown resistances of lamps B through L are such that the current in lamp A remains 1 ampere.

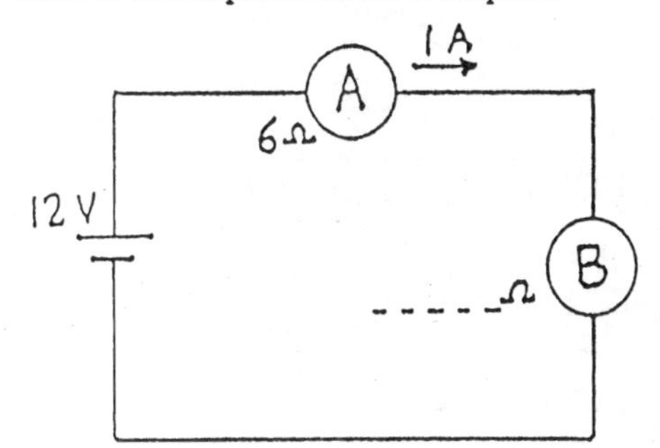

Circuit 1: How much current flows through the battery?

____ A

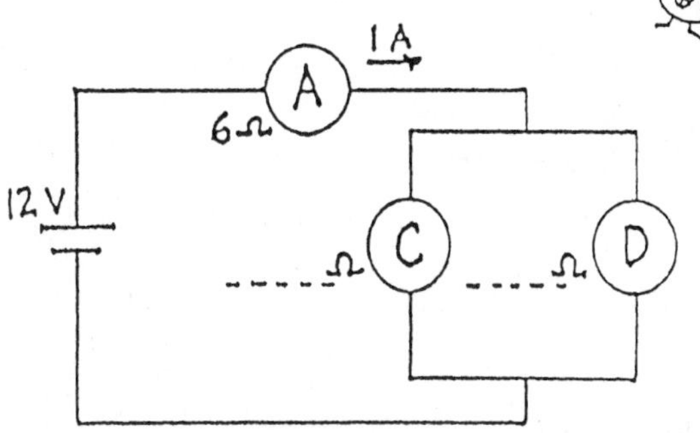

Circuit 2: Assume lamps C and D are identical. Current through lamp D is

____ A

Handy rule: For a pair of resistors in parallel:

$$\text{Equivalent resistance} = \frac{\text{product of resistances}}{\text{sum of resistances}}$$

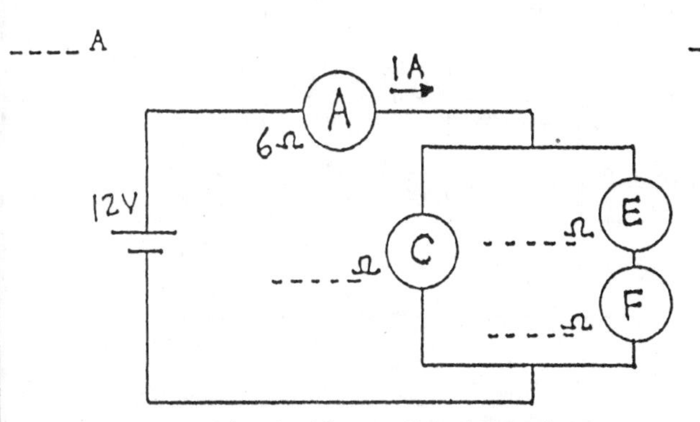

Circuit 3: Here identical lamps E and F replace Lamp D. Current through lamp C is

____ A

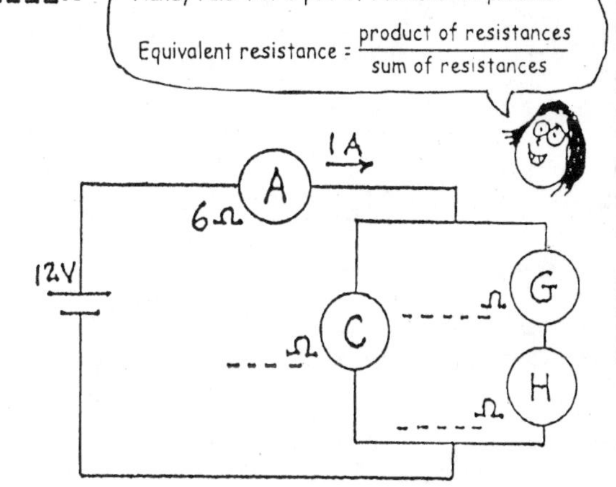

Circuit 4: Here lamps G and H replace lamps E and F, and the resistance of lamp G is twice that of lamp H. Current through lamp H is

____ A

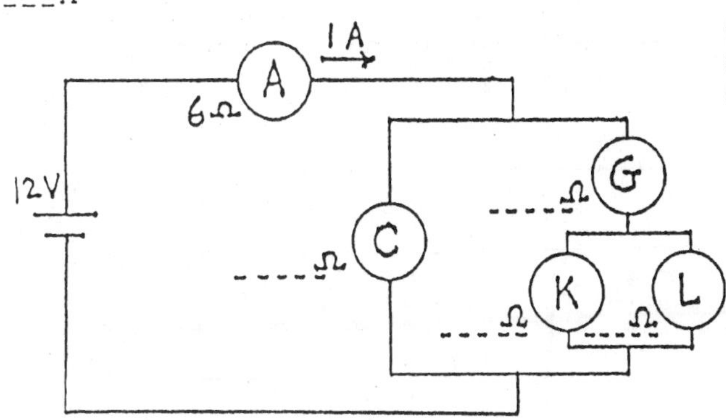

Circuit 5: Identical lamps K and L replace lamp H. Current through lamp L is

____ A

The equivalent resistance of a circuit is the value of a single resistor that will replace all the resistors of the circuit to produce the same load on the battery. How do the equivalent resistances of the circuits 1-5 compare?

Hewitt Drew it!

Chapter 23 Electric Current
Electric Power

The table beside circuit *a* below shows the current through each resistor, the voltage across each resistor, and the power dissipated as heat in each resistor. Find the similar correct values for circuits *b*, *c*, and *d*, and put your answers in the tables shown.

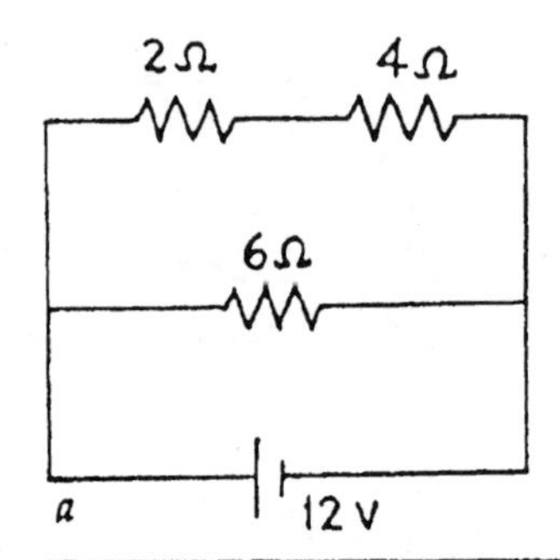

RESISTANCE	CURRENT ×	VOLTAGE =	POWER
2 Ω	2 A	4 V	8 W
4 Ω	2 A	8 V	16 W
6 Ω	2 A	12 V	24 W

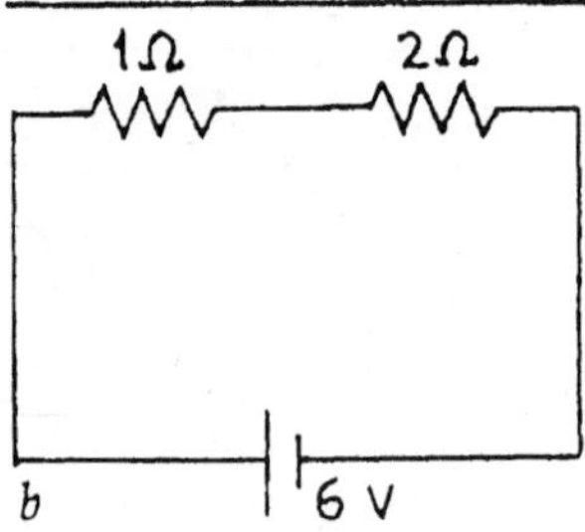

RESISTANCE	CURRENT ×	VOLTAGE =	POWER
1 Ω			
2 Ω			

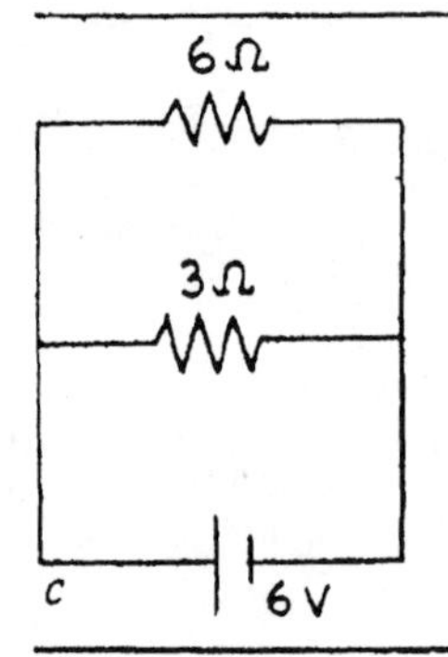

RESISTANCE	CURRENT ×	VOLTAGE =	POWER
6 Ω			
3 Ω			

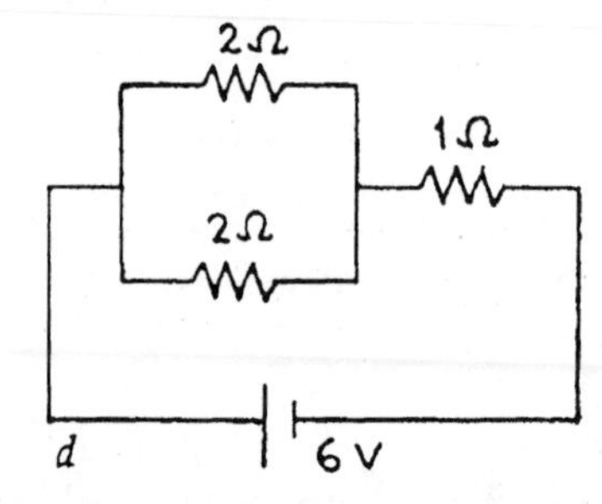

RESISTANCE	CURRENT ×	VOLTAGE =	POWER
2 Ω			
2 Ω			
1 Ω			

Hewitt Drew it!

Name ______________________________ Date __________

CONCEPTUAL Physics PRACTICE PAGE

Chapter 24 Magnetism
Magnetic Fundamentals

Fill in each blank with the appropriate word.

1. Attraction or repulsion of charges depends on their *signs*, positives or negatives. Attraction or repulsion of magnets depends on their magnetic ______________: ______________ or ______________.

2. Opposite poles attract; like poles ____________.

3. A magnetic field is produced by the ____________ of electric charge.

4. Clusters of magnetically aligned atoms are magnetic____________________.

5. A magnetic ____________ surrounds a current-carrying wire.

6. When a current-carrying wire is made to form a coil around a piece of iron, the result is an ____

7. A charged particle moving in a magnetic field experiences a deflecting ________________ that is maximum when the charge moves ________________________ to the field.

8. A current-carrying wire experiences a deflecting ________________________that is maximum when the wire and magnetic field are ________________________ to one another.

9. A simple instrument designed to detect electric current is the________________________; when calibrated to measure current, it is an ________________________; when calibrated to measure voltage, it is a ________________________.

10. The largest size magnet in the world is the ________________ itself.

11. The illustration below is similar to Figure 24.2 in your textbook. Iron filings trace out patterns of magnetic field lines about a bar magnet. In the field are some magnetic compasses. The compass needle in only one compass is shown. Draw in the needles with proper orientation in the other compasses.

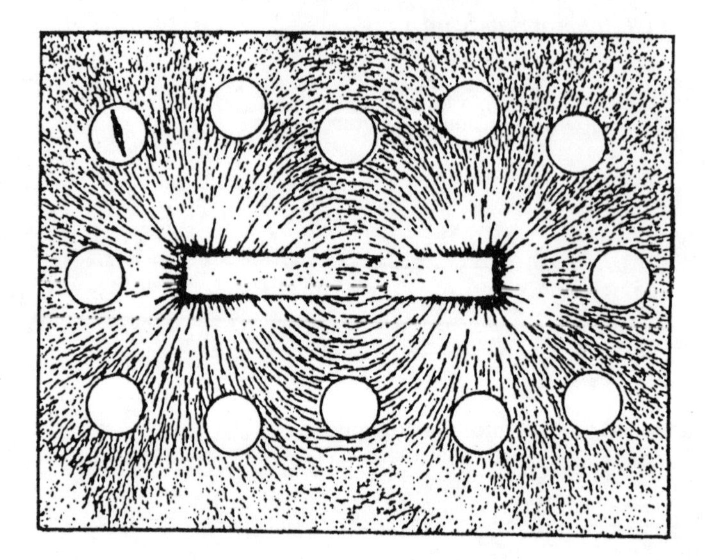

12. The illustration below is similar to Figure 24.10 (center) in your textbook. Iron filings trace out the magnetic field pattern about the loop of current-carrying wire. Draw in the compass needle orientations for all the compasses.

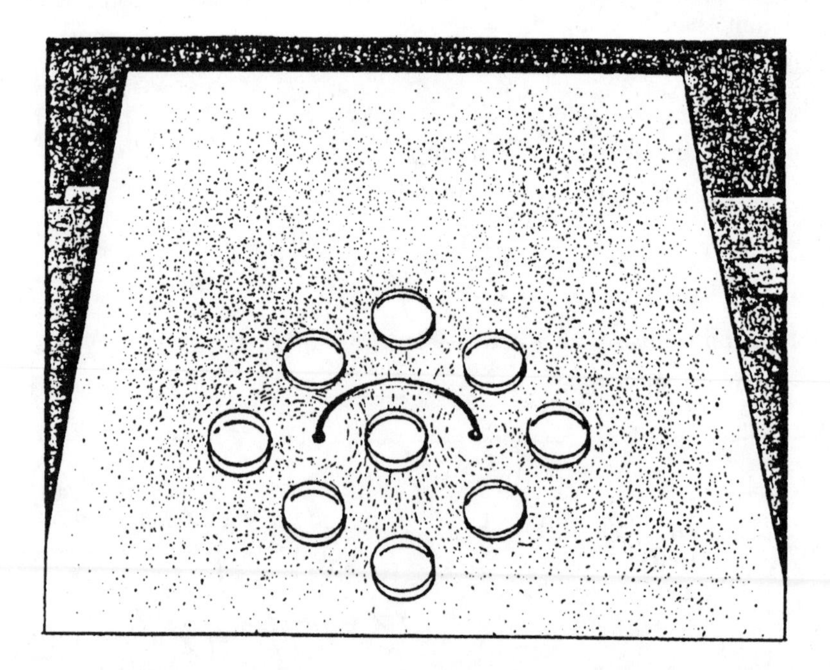

Hewitt Drew it!

Name .... Date ....

CONCEPTUAL *Physics* PRACTICE PAGE

Chapter 25 Electromagnetic Induction

Faraday's Law

1. Hans Christian Oersted discovered that magnetism and electricity are

   (related) (independent of each other).

   Magnetism is produced by

   (batteries) (the motion of electric charges).

Faraday and Henry discovered that electric current can be produced by

(batteries) (motion of a magnet).

More specifically, voltage is induced in a loop of wire if there is a change in the

(batteries) (magnetic field in the loop).

This phenomenon is called

(electromagnetism) (electromagnetic induction).

2. When a magnet is plunged in and out of a coil of wire, voltage is induced in the coil. If the rate of the in-and-out motion of the magnet is doubled, the induced voltage

   (doubles) (halves) (remains the same).

   If instead the number of loops in the coil is doubled, the induced voltage

   (doubles) (halves) (remains the same).

3. A rapidly changing magnetic field in any region of space induces a rapidly changing

   (electric field) (magnetic field) (gravitational field)

   which in turn induces a rapidly changing

   (magnetic field) (electric field) (baseball field).

   This generation and regeneration of electric and magnetic fields makes up

   (electromagnetic waves) (sound waves) (both of these).

PHYSICS SIGH

Hewitt Drew it!

CONCEPTUAL **Physics** PRACTICE PAGE

Transformers

Consider a simple transformer that has a 100-turn primary coil and a 1000-turn secondary coil. The primary is connected to a 120-V AC source and the secondary is connected to an electrical device with a resistance of 1000 ohms.

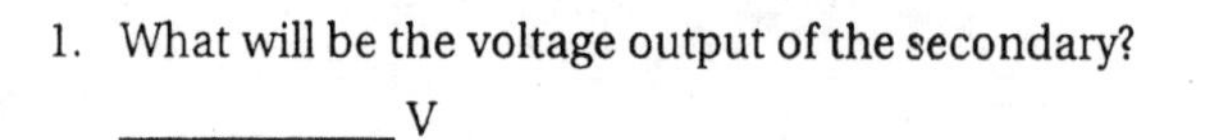

1. What will be the voltage output of the secondary?

   ____________ V

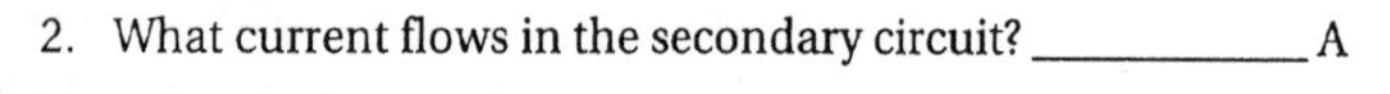

2. What current flows in the secondary circuit? ____________ A

3. Now that you know the voltage and the current, what is the power in the secondary coil? ____________ W

4. Neglecting small heating losses, and knowing that energy is conserved, what is the power in the primary coil? ____________ W

5. Now that you know the power and the voltage across the primary coil, what is the current drawn by the primary coil? ____________ A

Circle the correct answers:

6. The results show voltage is stepped (up) (down) from primary to secondary, and that current is correspondingly stepped (up) (down).

7. For a step-up transformer, there are (more) (fewer) turns in the secondary coil than the primary: For such a transformer, there is (more) (less) current in the secondary than in the primary.

8. A transformer can step up (voltage) (energy and power), but in no way can it step up (voltage) (energy and power).

9. If 120 V is used to power a toy electric train that operates on 6 V, then a (step up) (step down) transformer should be used that has a primary to secondary turns ratio of (1/20) (20/1).

10. A transformer operates on (dc) (ac) because the magnetic field within the iron core must (continually change) (remain steady).

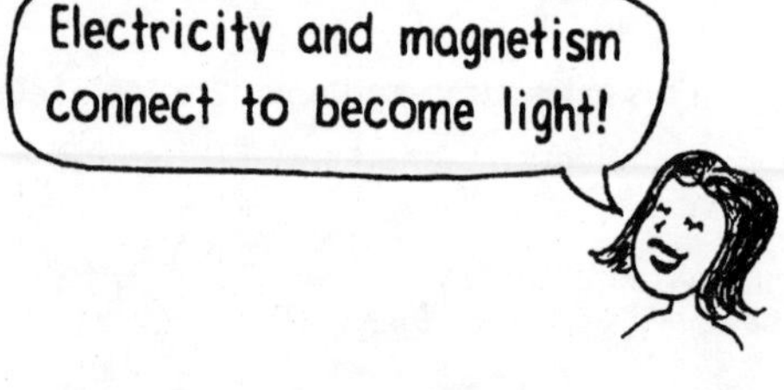

Electricity and magnetism connect to become light!

Name ______________________ Date ______________

CONCEPTUAL Physics PRACTICE PAGE

Chapter 26 Properties of Light
Speed, Wavelength, and Frequency

1. The first investigation that led to a determination of the speed of light was performed in about 1675 by the Danish astronomer Olaus Roemer. He made careful measurements of the period of Io, a moon about the planet Jupiter, and was surprised to find an irregularity in Io's observed period. While the Earth was moving away from Jupiter, the measured periods were slightly longer than average. While the Earth approached Jupiter, they were shorter than average. Roemer estimated that the cumulative discrepancy amounted to about 16.5 minutes. Later interpretations showed that what occurs is that light takes about 16.5 minutes to travel the extra 300,000,000-km distance across the Earth's orbit. Aha! We have enough information to calculate the speed of light!

a. Write a formula for speed in terms of the distance traveled and the time taken to travel that distance.

____

b. Using Roemer's data, and changing 16.5 min to seconds, calculate the speed of light. (Which wasn't a bad estimate at the time!)

____

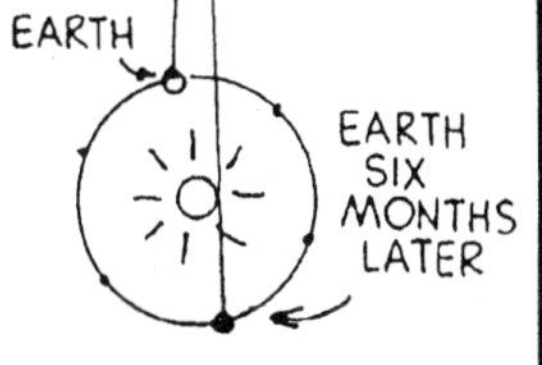

2. Study Figure 26.3 (Conceptual Physics, 9th Ed.) and answer the following:

a. Which has the longer *wavelengths,* radio waves or light waves?

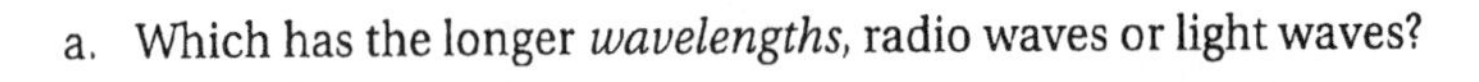

b. Which has the longer *wavelengths,* light waves or gamma rays?

____

c. Which has the higher *frequencies,* ultraviolet or infrared waves?

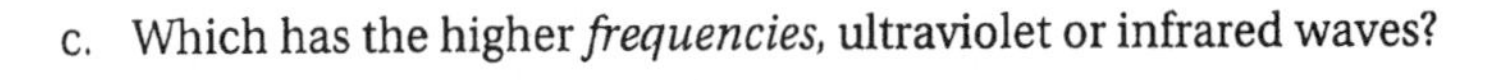

d. Which has the higher *frequencies,* ultraviolet waves or gamma rays?

____

3. Carefully study the section "Transparent Materials" in your textbook and answer the following:

a. Exactly what do vibrating electrons emit?

____

b. When ultraviolet light shines on glass, what does it do to electrons in the glass structure?

____

c. When energetic electrons in the glass structure vibrate against neighboring atoms, what happens to the energy of vibration?

____

d. What happens to the energy of a vibrating electron that does not collide with neighboring atoms?

____

e. Light in which range of frequencies, visible or ultraviolet, is absorbed in glass?

____

f. Light in which range of frequencies, visible or ultraviolet, is transmitted through glass?

____

g. How is the speed of light in glass affected by the succession of time delays that accompany the absorption and re-emission of light from atom to atom in the glass?

____

h. How does the speed of light compare in water, glass, and diamond?

____

4. The sun normally shines on both the Earth and on the moon. Both cast shadows. Sometimes the moon's shadow falls on the Earth, and at other times the Earth's shadow falls on the moon.

a. The sketch shows the sun and the Earth. Draw the moon at a position for a solar eclipse.

b. This sketch also shows the sun and the Earth. Draw the moon at a position for a lunar eclipse.

5. The diagram shows the limits of light rays when a large lamp makes a shadow of a small object on a screen. Make a sketch of the shadow on the screen, shading the umbra darker than the penumbra. In what part of the shadow could an ant see part of the lamp?

____

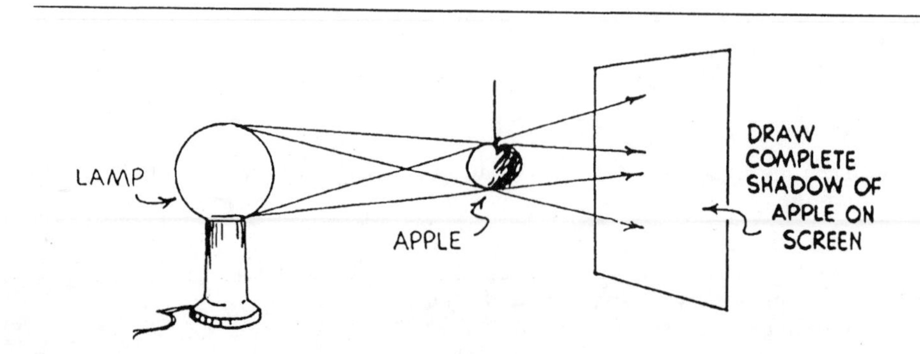

Hewitt Drew it!

Name ________________ Date ________

CONCEPTUAL **Physics** PRACTICE PAGE

Chapter 27 Color
Color Addition

The sketch to the right shows the shadow of an instructor in front of a white screen in a dark room. The light source is red, so the screen looks red and the shadow looks black. Color the sketch, or label the colors with pen or pencil.

A green lamp is added and makes a second shadow. The shadow cast by the red lamp is no longer black, but is illuminated by green light. So it is green. Color or mark it green. The shadow cast by the green lamp is not black because it is illuminated by the red lamp. Indicate its color. Do the same for the background, which receives a mixture of red and green light.

A blue lamp is added and three shadows appear. Indicate the appropriate colors of the shadows and the background.

The lamps are placed closer together so the shadows overlap. Indicate the colors of all screen areas.

If you have colored markers, have a go at these.

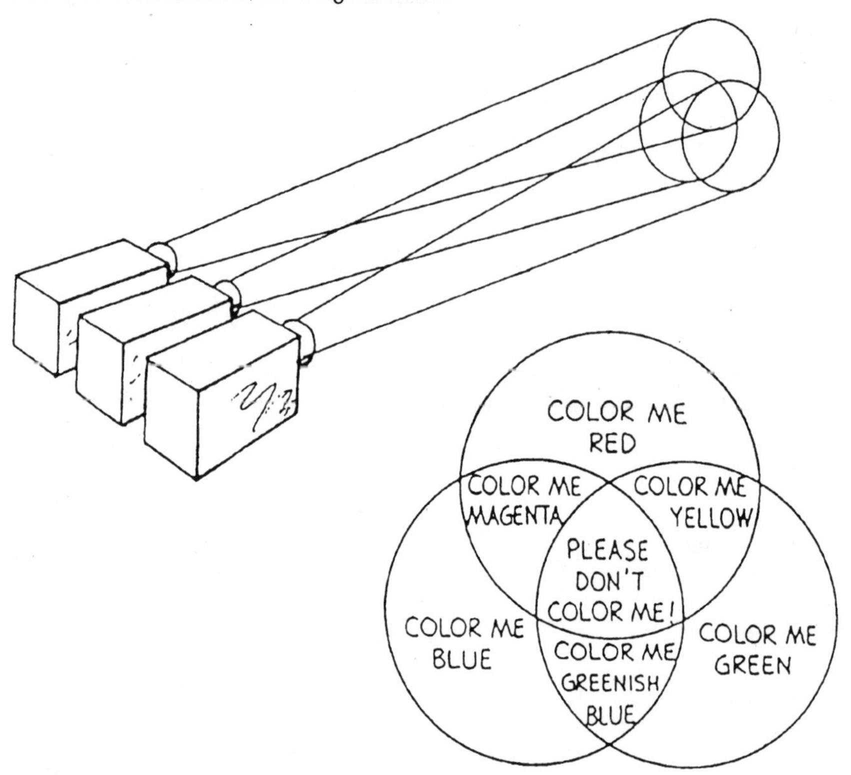

Name ______________________ Date __________

CONCEPTUAL **Physics** PRACTICE PAGE

Chapter 28 Reflection and Refraction

Pool Room Optics

The law of reflection for optics is useful in playing pool. A ball bouncing off the bank of a pool table behaves like a photon reflecting off a mirror. As the sketch shows, angles become straight lines with the help of mirrors. The diagram to the right shows a top view of this, with a flattened "mirrored" region. Note that the angled path on the table appears as a straight line (dashed) in the mirrored region.

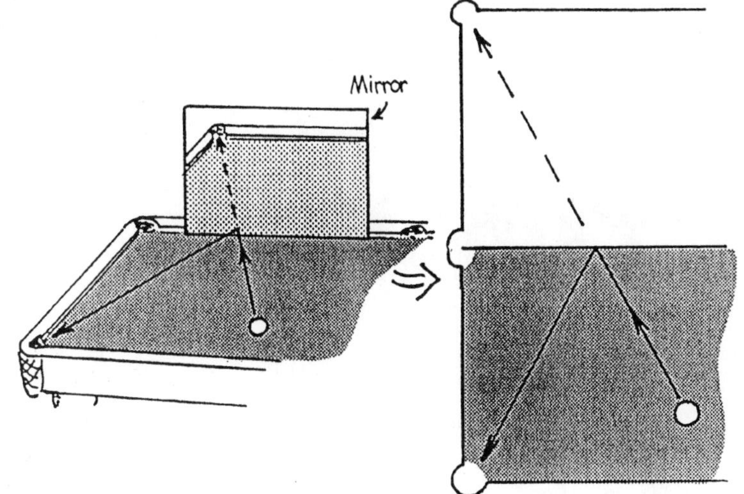

1. Consider a one-bank shot (one reflection) from the ball to the north bank and then into side pocket E.

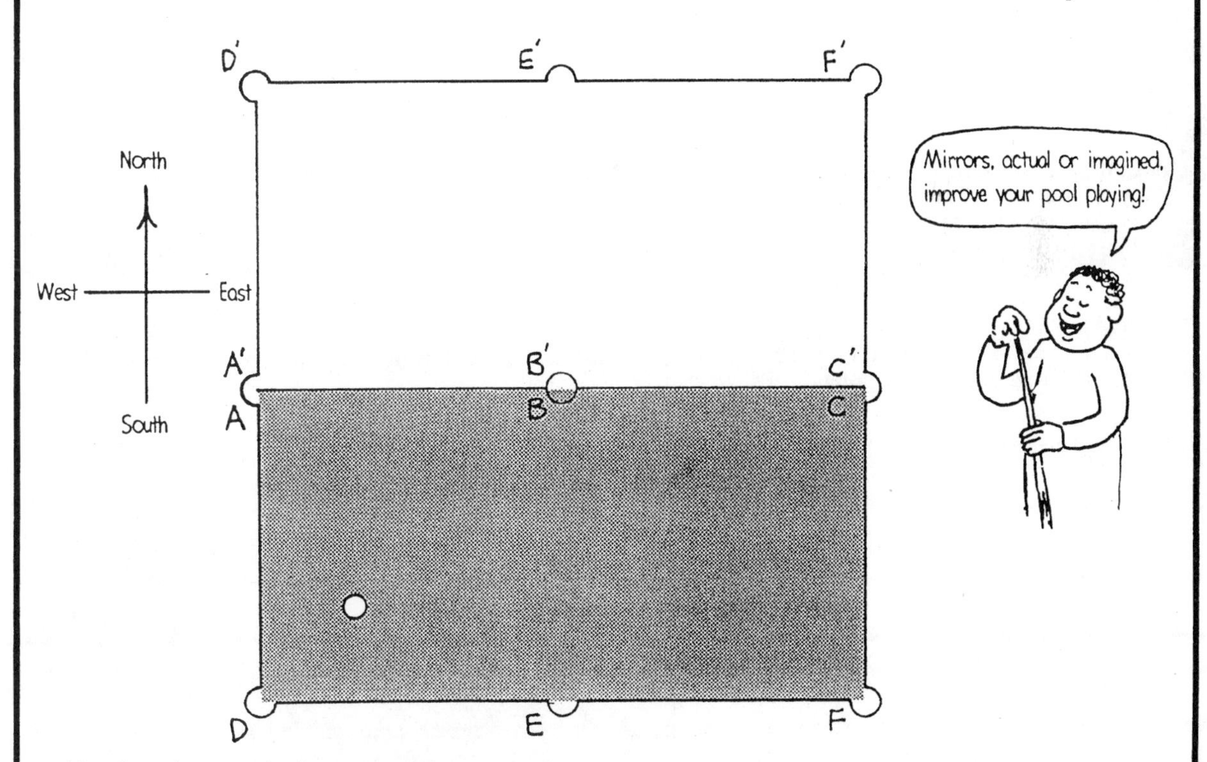

a. Use the mirror method to construct a straight line path to mirrored E'. Then construct the actual path to E.

b. Without using off-center strokes or other tricks, can a one-bank shot off the north bank put the ball in corner pocket F? ______________________ Show why or why not using the diagram.

2. Consider below a two-bank shot (two reflections) into corner pocket F. Here we use two mirrored regions. Note the straight line of sight to F", and how the north-bank impact point matches the intersection between B' and C'.
   a. Construct the similar path for a similar two-bank shot to get the ball in the side pocket E.

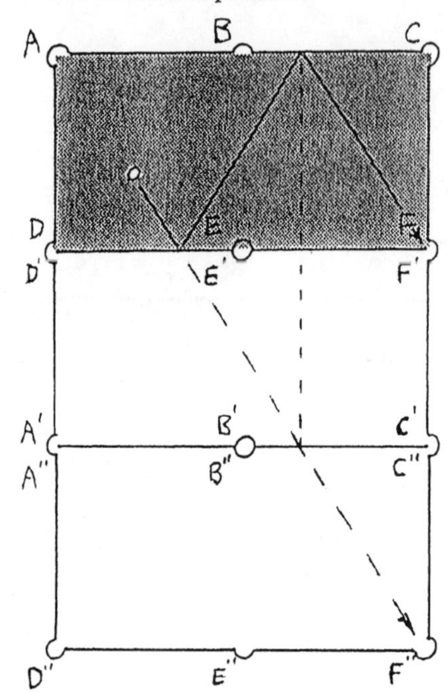

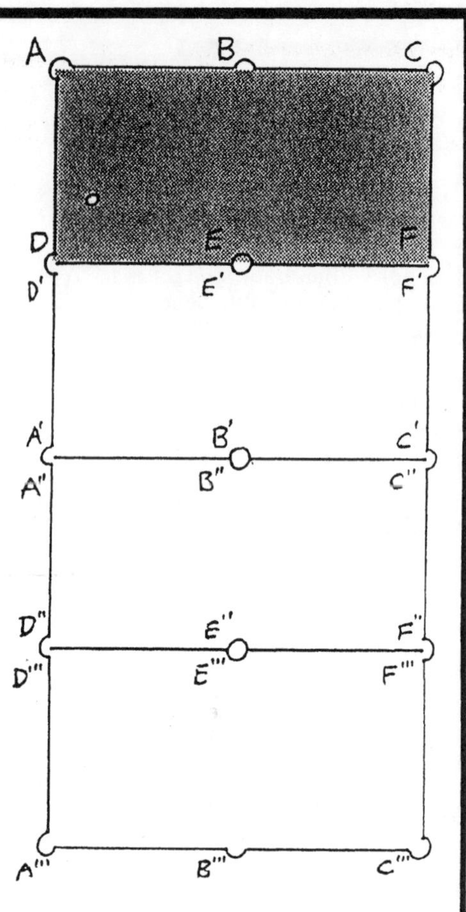

3. Consider above right a three-bank shot into corner pocket C, first bouncing against the south bank, then to the north, again to the south, and into pocket C.
   a. Construct the path. (First construct the single dashed line to C''').
   b. Construct the path to make a three-bank shot into side pocket B.

4. Let's try banking from adjacent banks of the table. Consider a two-bank shot to corner pocket F (first off the west bank, then to and off the north bank, then into F). Note how our two mirrored regions permit a straight-line path from the ball to F".

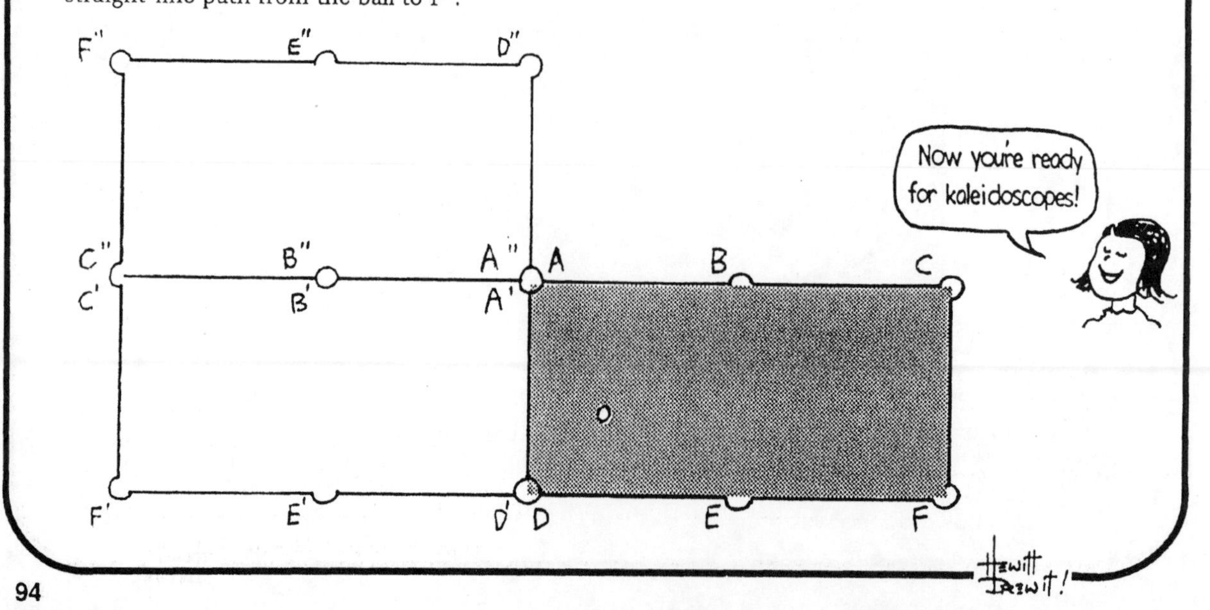

Name ______________________ Date __________

CONCEPTUAL **Physics** PRACTICE PAGE

Chapter 28 Reflection and Refraction

Reflection

←MIRROR

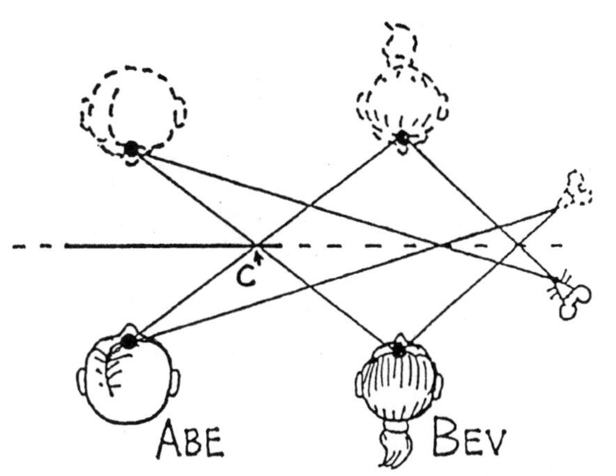

Abe and Bev both look in a plane mirror directly in front of Abe (left, top view). Abe can see himself while Bev cannot see herself—but can Abe see Bev, and can Bev see Abe? To find the answer we construct their artificial locations "through" the mirror, the same distance behind as Abe and Bev are in front (right, top view). If straight-line connections intersect the mirror, as at point C, then each sees the other. The mouse, for example, cannot see or be seen by Abe and Bev.

Here we have eight students in front of a small plane mirror. Their positions are shown in the diagram below. Make appropriate straight-line constructions to answer the following:

←MIRROR

ABE BEV CIS DON EVA FLO GUY HAN

Who can Abe see? ______________ Who can Abe not see? ______________
Who can Bev see? ______________ Who can Bev not see? ______________
Who can Cis see? ______________ Who can Cis not see? ______________
Who can Don see? ______________ Who can Don not see? ______________
Who can Eva see? ______________ Who can Eva not see? ______________
Who can Flo see? ______________ Who can Flo not see? ______________
Who can Guy see? ______________ Who can Guy not see? ______________
Who can Han see? ______________ Who can Han not see? ______________

thanx to Marshall Ellenstein Hewitt Drew it!

CONCEPTUAL Physics PRACTICE PAGE

Six of our group are now arranged differently in front of the same mirror. Their positions are shown below. Make appropriate constructions for this more interesting arrangement, and answer the questions below.

←MIRROR

ABE

EVA

BEV

FLO

CIS

DON

Who can Abe see? ______________ Who can Abe not see? ______________

Who can Bev see? ______________ Who can Bev not see? ______________

Who can Cis see? ______________ Who can Cis not see? ______________

Who can Don see? ______________ Who can Don not see? ______________

Who can Eva see? ______________ Who can Eva not see? ______________

Who can Flo see? ______________ Who can Flo not see? ______________

Harry Hotshot views himself in a full-length mirror (right). Construct straight lines from Harry's eyes to the image of his feet, and to the top of his head. Mark the mirror to indicate the minimum area Harry uses to see a full view of himself.

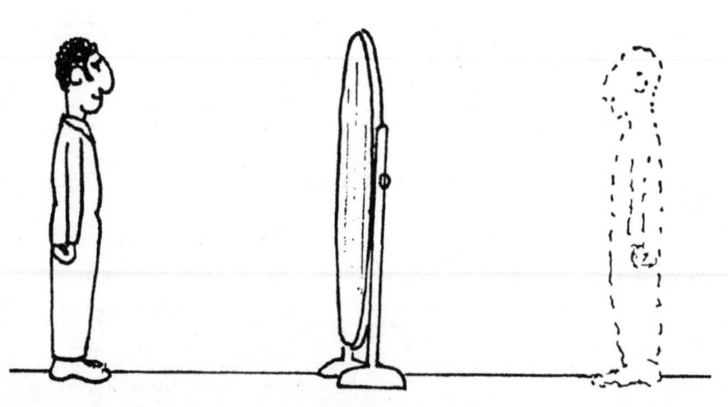

Does this region of the mirror depend on Harry's distance from the mirror?

Hewitt Drew it!

Chapter 28 Reflection and Refraction

Reflected Views

1. The ray diagram below shows the extension of one of the reflected rays from the plane mirror. Complete the diagram by (1) carefully drawing the three other reflected rays, and (2) extending them behind the mirror to locate the image of the flame. (Assume the candle and image are viewed by an observer on the left.)

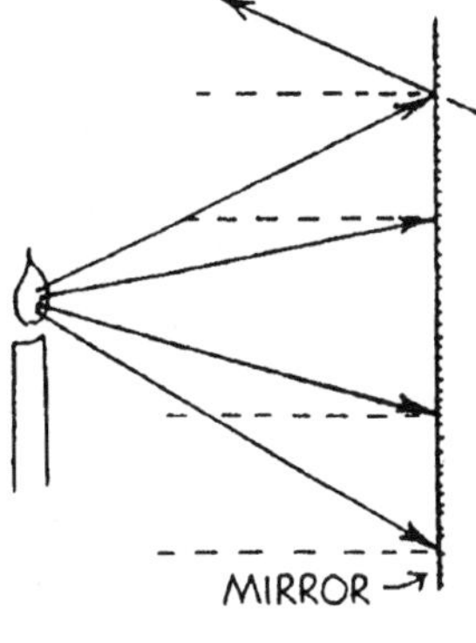

2. A girl takes a photograph of the bridge as shown. Which of the two sketches correctly shows the reflected view of the bridge? Defend your answer.

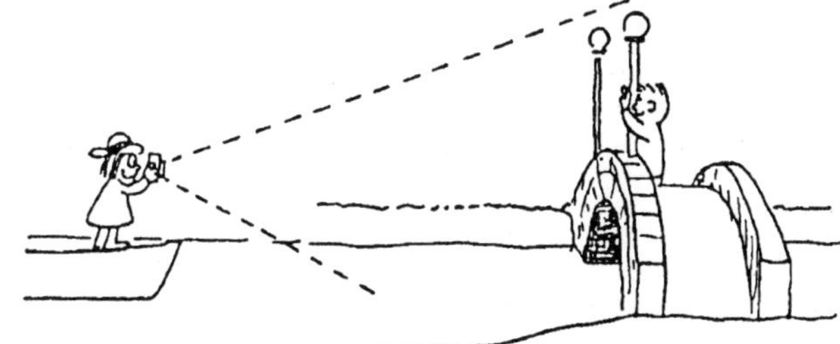

Name ______________________________ Date __________

CONCEPTUAL **Physics** PRACTICE PAGE

Chapter 28 Reflection and Refraction

Refraction

1. A pair of toy cart wheels are rolled obliquely from a smooth surface onto two plots of grass — a rectangular plot as shown at the left, and a triangular plot as shown at the right. The ground is on a slight incline so that after slowing down in the grass, the wheels speed up again when emerging on the smooth surface. Finish each sketch and show some positions of the wheels inside the plots and on the other side. Clearly indicate their paths and directions of travel.

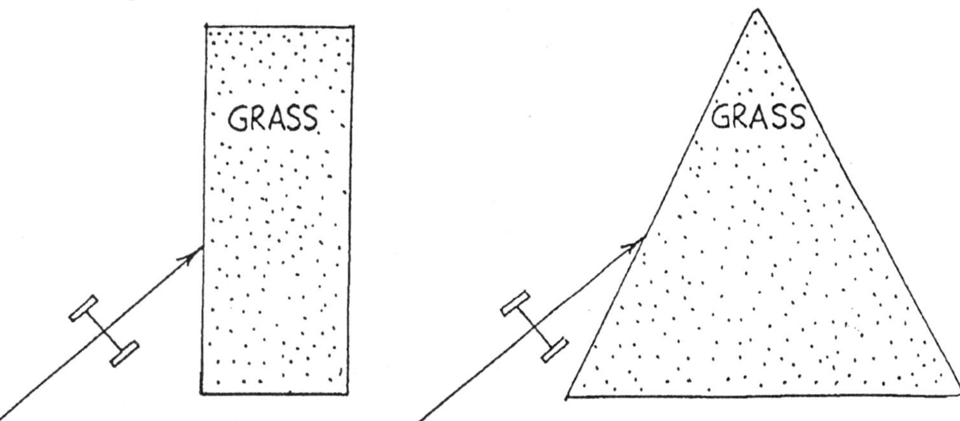

2. Red, green, and blue rays of light are incident upon a glass prism as shown. The average speed of red light in the glass is less than in air, so the red ray is refracted. When it emerges into the air it regains its original speed and travels in the direction shown. Green light takes longer to get through the glass. Because of its slower speed it is refracted as shown. Blue light travels even slower in glass. Complete the diagram by estimating the path of the blue ray.

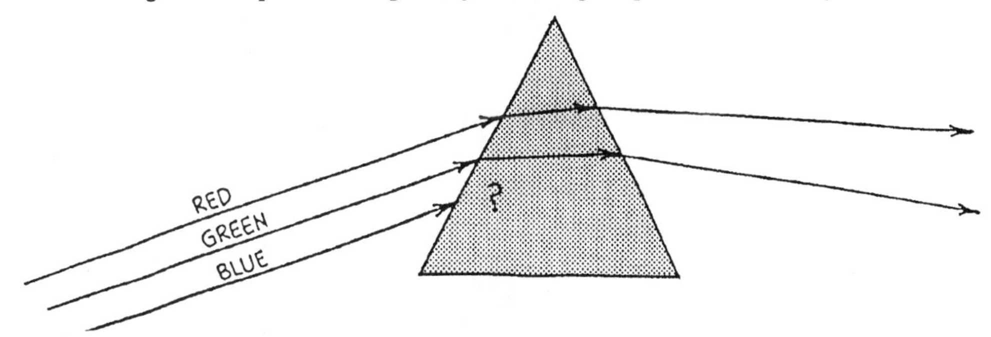

3. Below we consider a prism-shaped hole in a piece of glass — that is, an "air prism." Complete the diagram, showing likely paths of the beams of red, green, and blue light as they pass through this "prism" and back to glass.

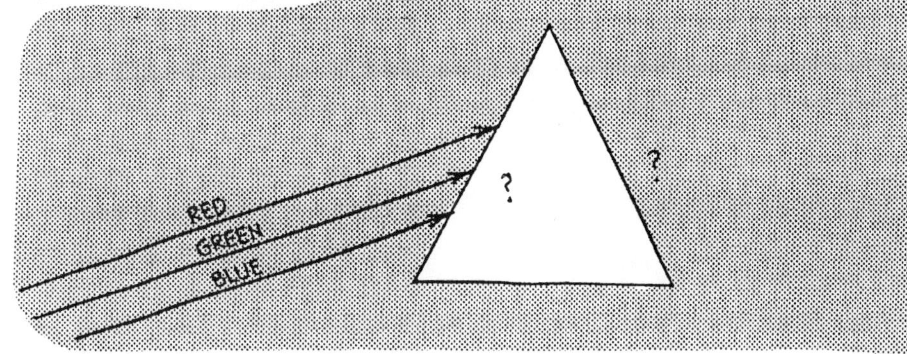

Hewitt Drew it!

4. Light of different colors diverges when emerging from a prism. Newton showed that with a second prism he could make the diverging beams become parallel again. Which placement of the second prism will do this?

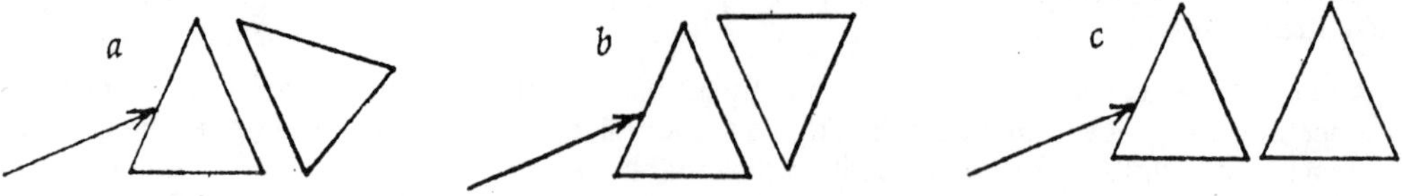

5. The sketch shows that due to refraction, the man sees the fish closer to the water surface than it actually is.

   a. Draw a ray beginning at the fish's eye to show the line of sight of the fish when it looks upward at 50° to the normal at the water surface. Draw the direction of the ray after it meets the surface of the water.

   b. At the 50° angle, does the fish see the man, or does it see the reflected view of the starfish at the bottom of the pond? Explain.

   ____

   ____

   c. To see the man, should the fish look higher or lower than the 50° path?

   ____

   d. If the fish's eye were barely above the water surface, it would see the world above in a 180° view, horizon to horizon. The fisheye view of the world above as seen beneath the water, however, is very different. Due to the 48° critical angle of water, the fish sees a normally 180° horizon-to-horizon view compressed within an angle of ________.

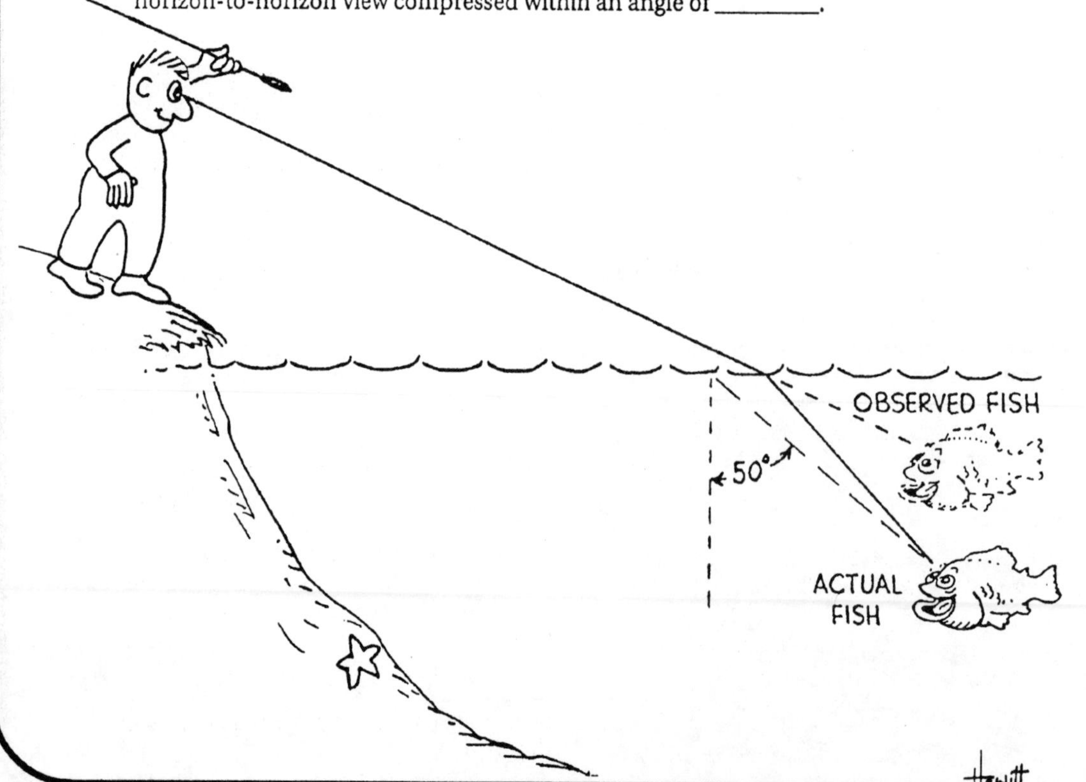

Name ______________________________ Date __________

CONCEPTUAL **Physics** PRACTICE PAGE

Chapter 28 Reflection and Refraction

More Refraction

1. The sketch to the right shows a light ray moving from air into water, at 45° to the normal. Which of the three rays indicated with capital letters is most likely the light ray that continues inside the water?

____

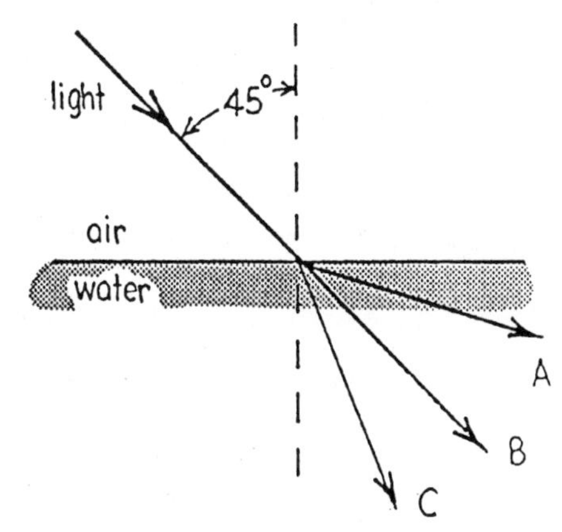

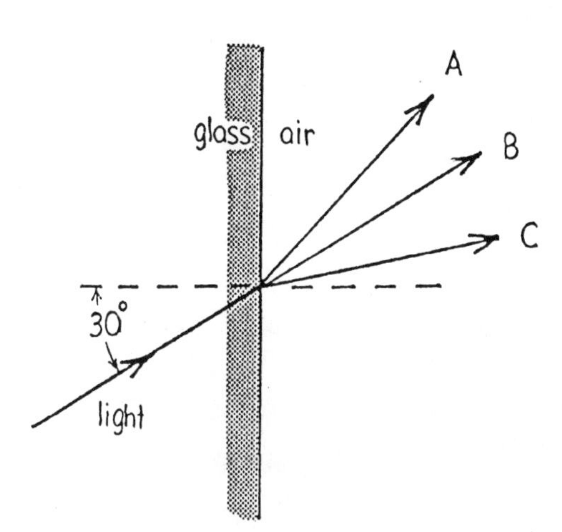

2. The sketch on the left shows a light ray moving from glass into air, at 30° to the normal. Which of the three is most likely the light ray that continues in the air?

____

3. To the right, a light ray is shown moving from air into a glass block, at 40° to the normal. Which of the three rays is most likely the light ray that travels in the air after emerging from the opposite side of the block?

____

Sketch the path the light would take inside the glass.

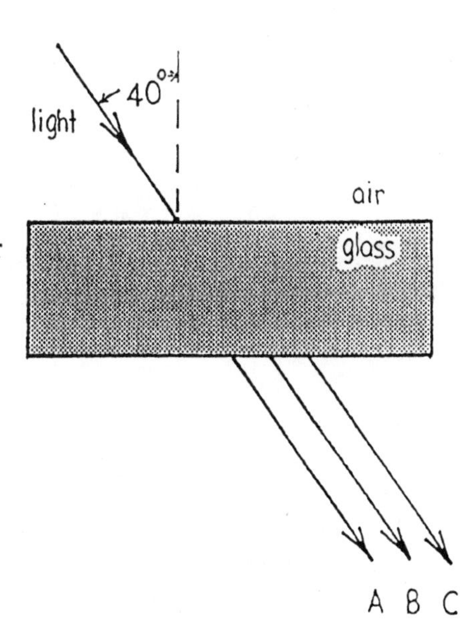

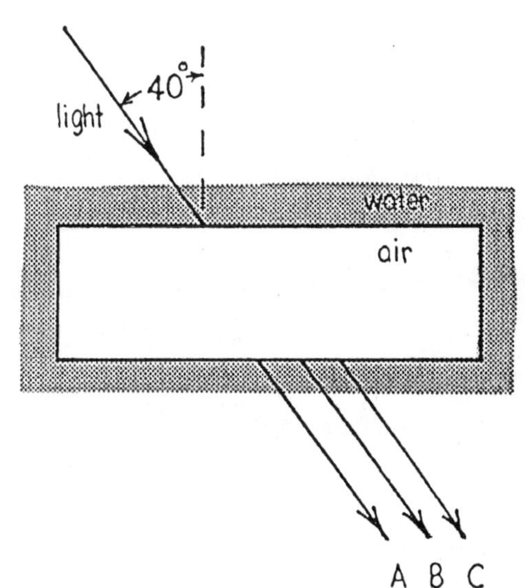

4. To the left, a light ray is shown moving from water into a rectangular block of air (inside a thin-walled plastic box), at 40° to the normal. Which of the three rays is most likely the light ray that continues into the water on the opposite side of the block?

____

Sketch the path the light would take inside the air.

thanx to Clarence Bakken

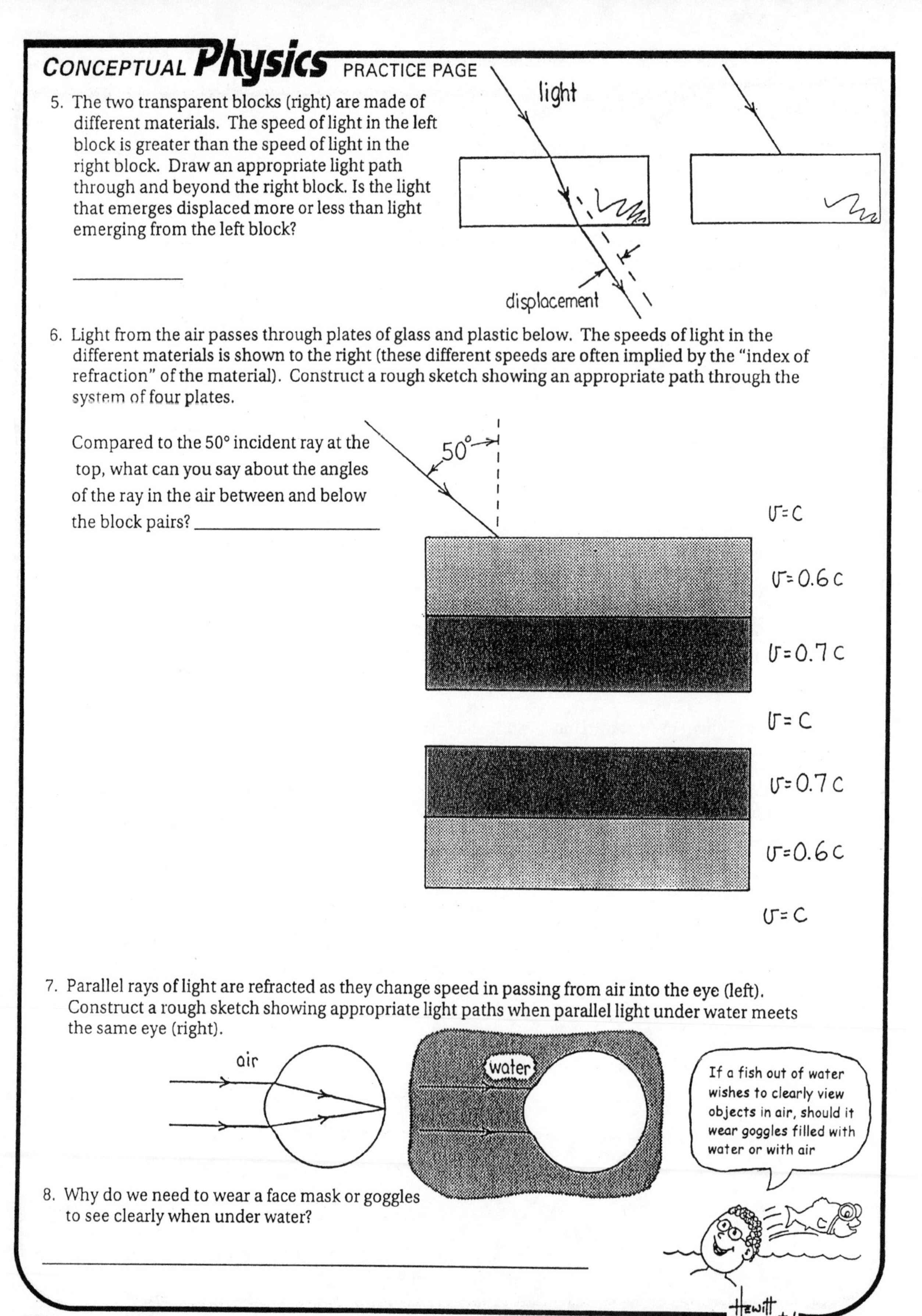

CONCEPTUAL **Physics** PRACTICE PAGE

5. The two transparent blocks (right) are made of different materials. The speed of light in the left block is greater than the speed of light in the right block. Draw an appropriate light path through and beyond the right block. Is the light that emerges displaced more or less than light emerging from the left block?

   ____

6. Light from the air passes through plates of glass and plastic below. The speeds of light in the different materials is shown to the right (these different speeds are often implied by the "index of refraction" of the material). Construct a rough sketch showing an appropriate path through the system of four plates.

   Compared to the 50° incident ray at the top, what can you say about the angles of the ray in the air between and below the block pairs? ______________

7. Parallel rays of light are refracted as they change speed in passing from air into the eye (left). Construct a rough sketch showing appropriate light paths when parallel light under water meets the same eye (right).

8. Why do we need to wear a face mask or goggles to see clearly when under water?

   ____

Name ______________________ Date __________

CONCEPTUAL Physics PRACTICE PAGE

Chapter 28 Reflection and Refraction

Lenses

Rays of light bend as shown when passing through the glass blocks.

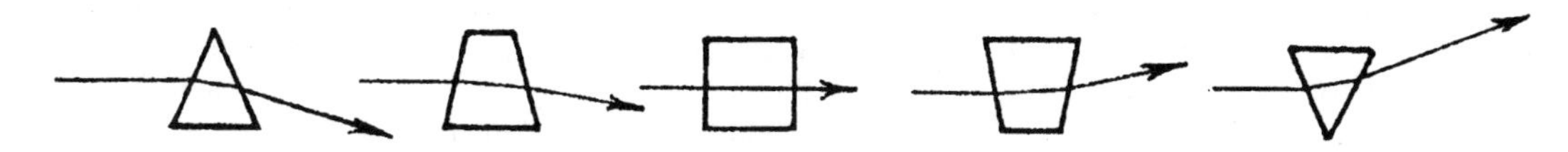

1. Show how light rays bend when they pass through the arrangement of glass blocks shown below.

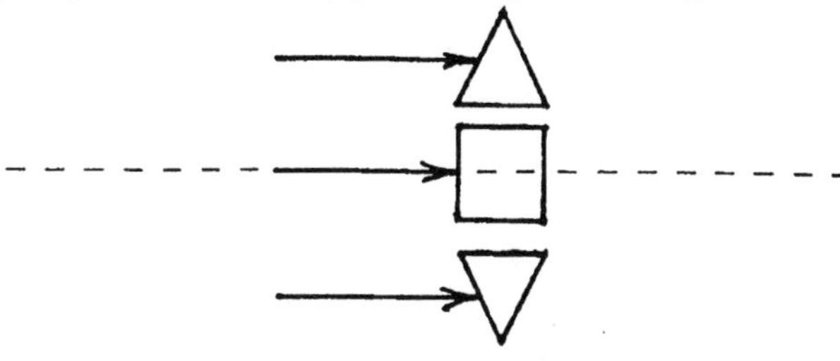

2. Show how light rays bend when they pass through the lens shown below. Is the lens a converging or a diverging lens? What is your evidence?

____

____

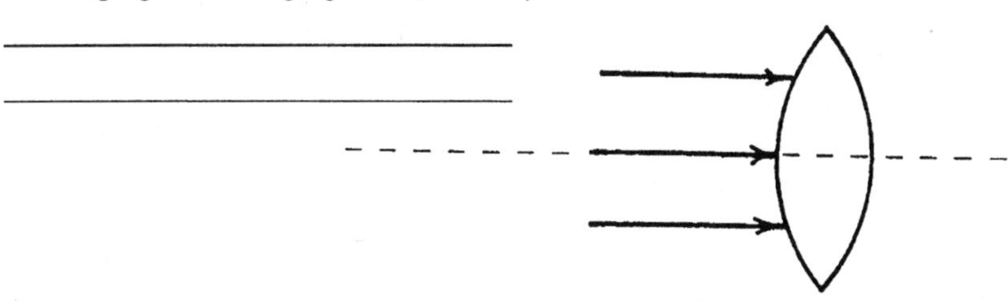

3. Show how light rays bend when they pass through the arrangement of glass blocks shown below.

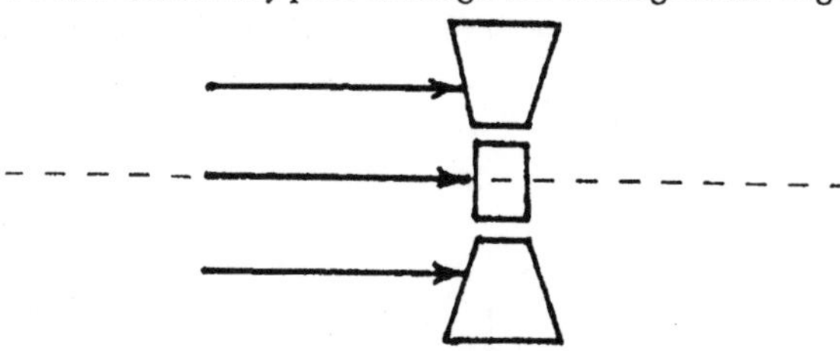

4. Show how light rays bend when they pass through the lens shown below. Is the lens a converging or a diverging lens? What is your evidence?

____

____

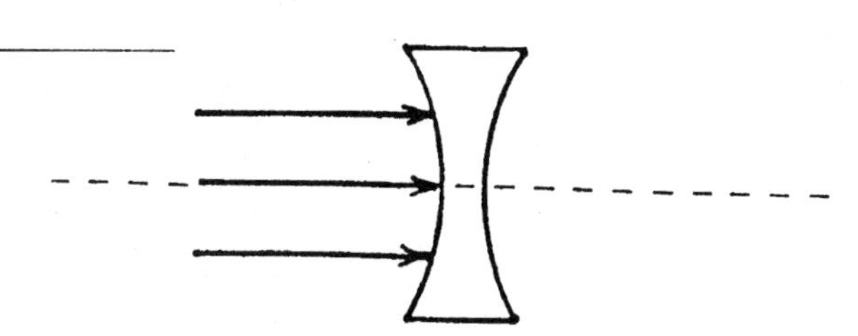

Hewitt Drew it!

5. Which type of lens is used to correct farsightedness? ______________________

   Nearsightedness? ______________________

6. Construct rays to find the location and relative size of the arrow's image for each of the lenses below. Rays that pass through the middle of a lens continue undeviated. In a converging lens, rays from the tip of the arrow that are parallel to the optic axis extend through the far focal point after going through the lens. Rays that go through the near focal point go parallel to the axis after going through the lens. In a diverging lens, rays parallel to the axis diverge and appear to originate from the near focal point after passing though the lens. Have fun!

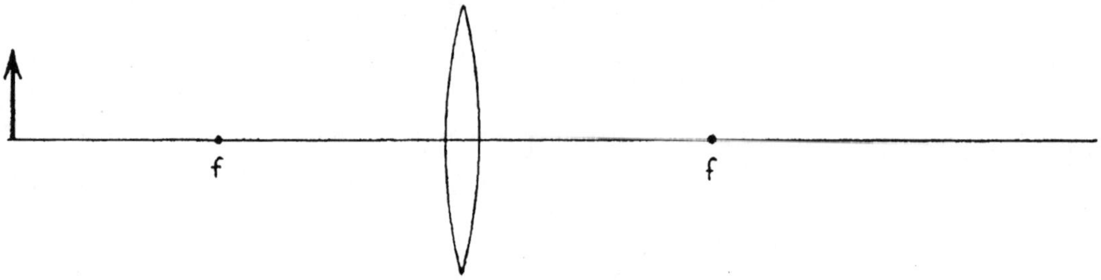

f f

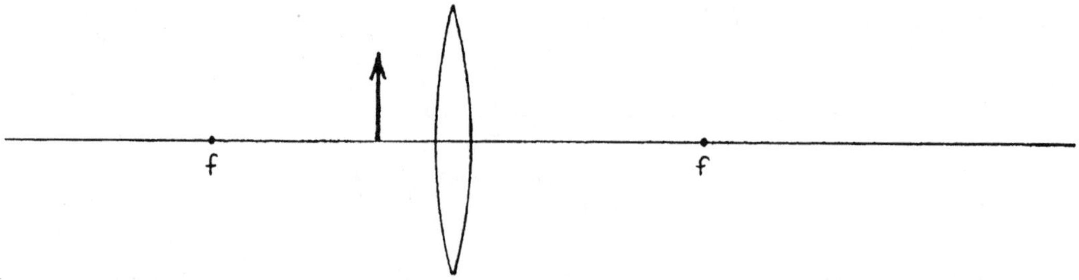

f f

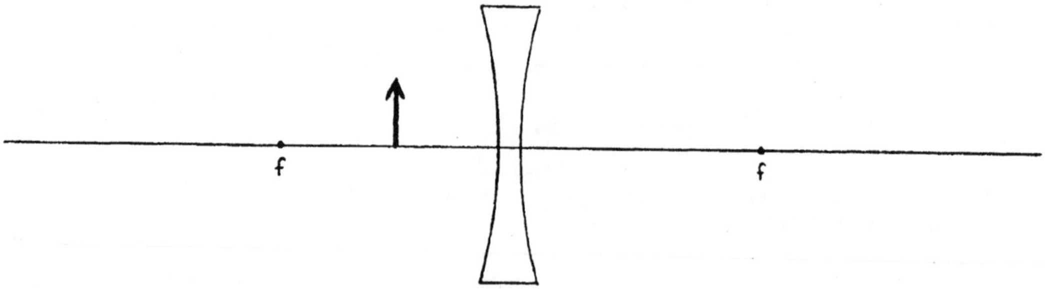

f f

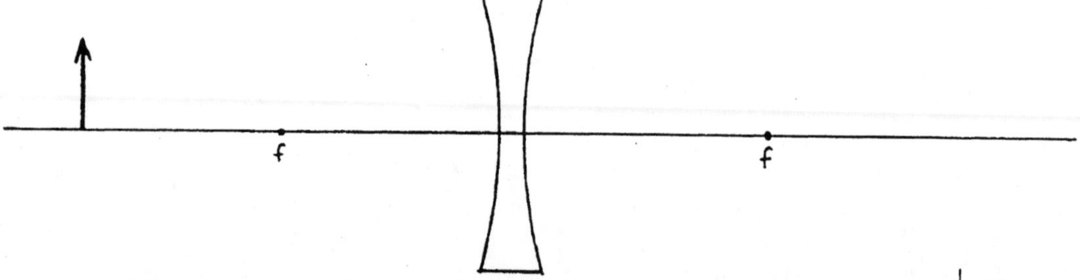

f f

Hewitt Drew it!

Name ______________________ Date __________

CONCEPTUAL *Physics* PRACTICE PAGE

Chapter 29 Light waves
Diffraction and Interference

1. Shown below are concentric solid and dashed circles, each different in radius by 1 cm. Consider the circular pattern a top view of water waves, where the solid circles are crests and the dashed circles are troughs.

   a. Draw another set of the same concentric circles with a compass. Choose any part of the paper for your center (except the present central point). Let the circles run off the edge of the paper.

   b. Find where a dashed line crosses a solid line and draw a large dot at the intersection. Do this for ALL places where a solid and dashed line intersect.

   c. With a wide felt marker, connect the dots with smooth lines. These *nodal lines* lie in regions where the waves have cancelled — where the crest of one wave overlaps the trough of another (see Figures 29.15 and 29.16 in your textbook).

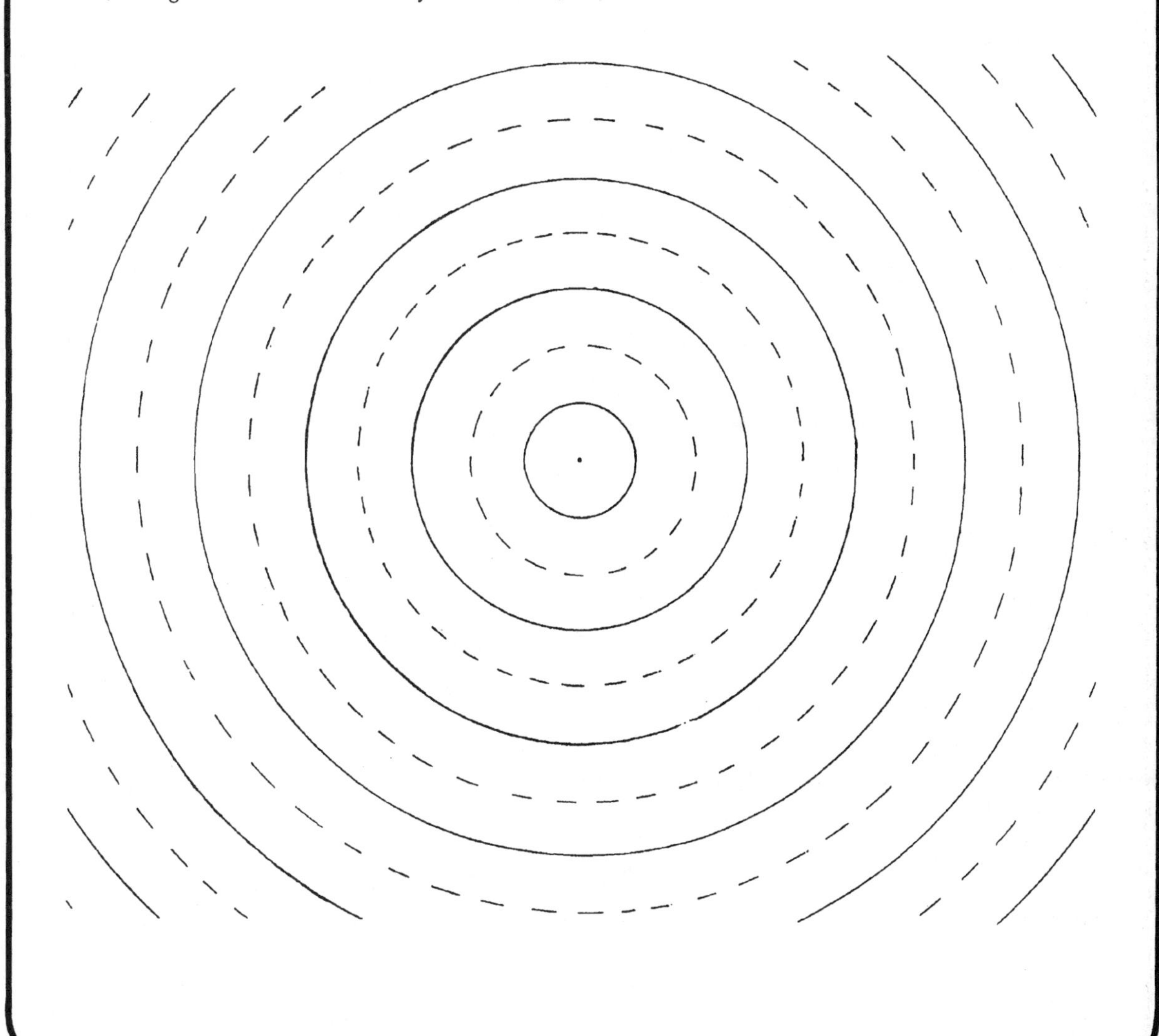

Hewitt Drew it!

2. Look at the construction of overlapping circles on your classmates' papers. Some will have more nodal lines than others, due to different starting points. How does the number of nodal lines in a pattern relate to the distance between centers of circles, (or sources of waves)?

____

____

3. Figure 28.19 from your text is repeated below. Carefully count the number of wavelengths (same as the number of wave crests) along the following paths between the slits and the screen.

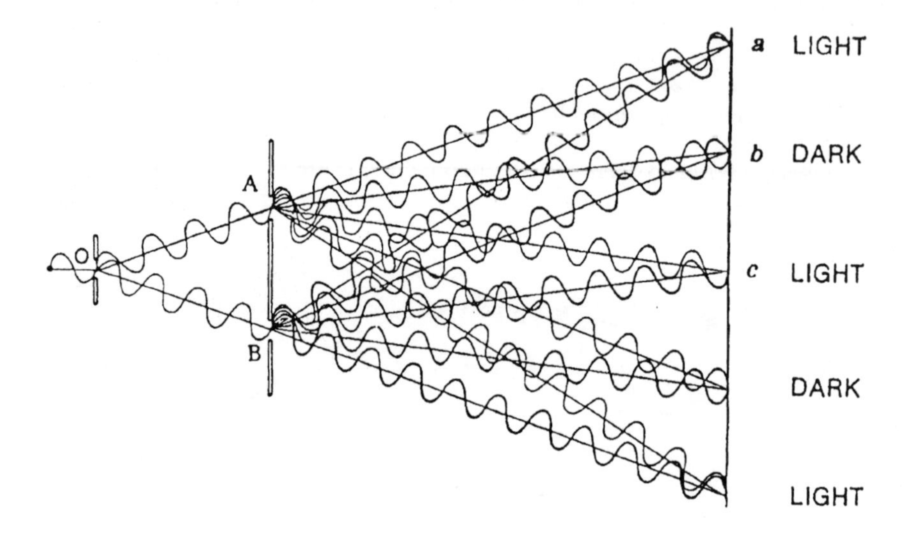

a. Number of wavelengths between slit A and point a = ____________

b. Number of wavelengths between slit B and point a = ____________

c. Number of wavelengths between slit A and point b = ____________

d. Number of wavelengths between slit B and point b = ____________

e. Number of wavelengths between slit A and point c = ____________

f. Number of wave crests between slit B and point c = ____________

When the number of wavelengths along each path is the same or differs by one or more whole wavelengths, interference is

(constructive) (destructive).

and when the number of wavelengths differ by a half wavelength (or odd multiples of a half wavelength), interference is

(constructive) (destructive).

It's nice how knowing some physics really changes the way we see things!

Hewitt Drew it!

Name ______________________ Date ______________

CONCEPTUAL **Physics** PRACTICE PAGE

Chapter 29 Light waves
Polarization

The amplitude of a light wave has magnitude and direction, and can be represented by a vector. Polarized light vibrates in a single direction and is represented by a single vector. To the left the single vector represents vertically polarized light. The vibrations of non-polarized light are equal in all directions. There are as many vertical components as horizontal components. The pair of perpendicular vectors to the right represents non-polarized light.

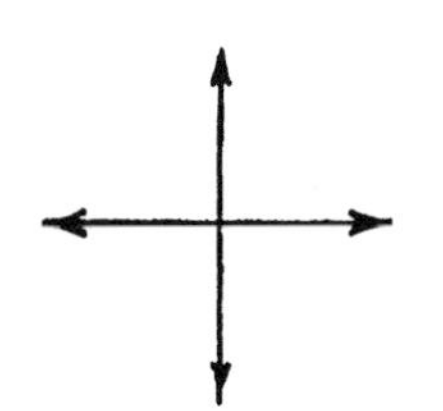

1. In the sketch below non-polarized light from a flashlight strikes a pair of Polaroid filters.

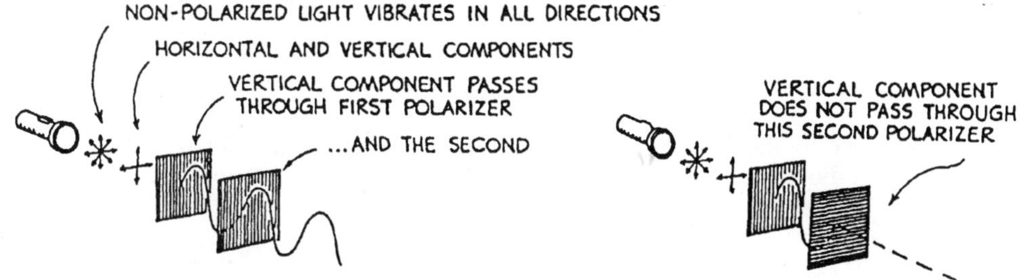

NON-POLARIZED LIGHT VIBRATES IN ALL DIRECTIONS
HORIZONTAL AND VERTICAL COMPONENTS
VERTICAL COMPONENT PASSES THROUGH FIRST POLARIZER
...AND THE SECOND
VERTICAL COMPONENT DOES NOT PASS THROUGH THIS SECOND POLARIZER

a. Light is transmitted by a pair of Polaroids when their axes are

(aligned) (crossed at right angles)

and light is blocked when their axes are

(aligned) (crossed at right angles)

b. Transmitted light is polarized in a direction

(the same as) (different than) the polarization axis of the filter.

2. Consider the transmission of light through a pair of Polaroids with polarization axes at 45° to each other. Although in practice the Polaroids are one atop the other, we show them spread out side by side below. From left to right: (*a*) Nonpolarized light is represented by its horizontal and vertical components. (*b*) These components strike filter A. (*c*) The vertical component is transmitted, and (*d*) falls upon filter B. This vertical component is not aligned with the polarization axis of filter B, but it has a component that is — component *t*; (*e*) which is transmitted.

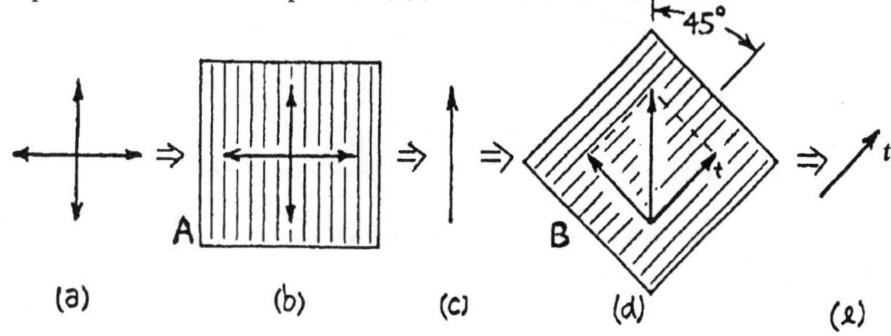

45°
A
B
t
(a)
(b)
(c)
(d)
(e)

a. The amount of light that gets through Filter B, compared to the amount that gets through Filter A is

(more) (less) (the same)

b. The component perpendicular to *t* that falls on Filter B is

(also transmitted) (absorbed)

Hewitt Drew it!

3. Below are a pair of Polaroids with polarization axes at 30° to each other. Carefully draw vectors and appropriate components (as in Question 2) to show the vector that emerges at *e.*

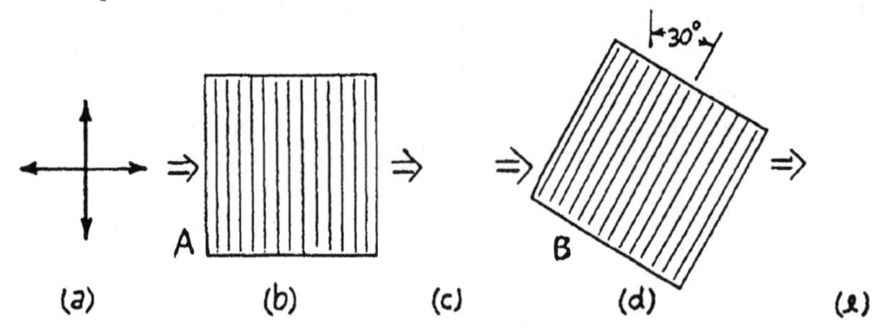

a. The amount of light that gets through the Polaroids at 30°, compared to the amount that gets though the 45° Polaroids is

(less) (more) (the same)

4. Figure 29.35 in your textbook shows the smile of Ludmila Hewitt emerging through three Polaroids. Use vector diagrams to complete steps *b* through *g* below to show how light gets through the three-Polaroid system.

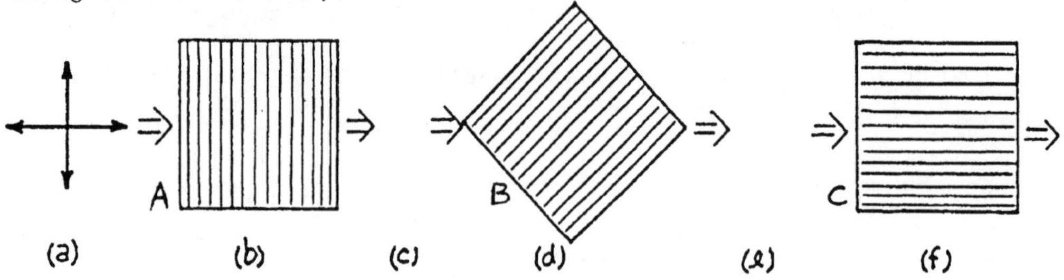

5. A novel use of polarization is shown below. How do the polarized side windows in these next-to-each-other houses provide privacy for the occupants? (Who can see what?)

Hewitt Drew it!

Name ______________________________ Date ____________

CONCEPTUAL Physics PRACTICE PAGE

Chapters 31 and 32 Light Quanta, and The Atom and the Quantum
Light Quanta

1. To say that light is quantized means that light is made up of

   (elemental units) (waves)

2. Compared to photons of low-frequency light, photons of higher-frequency light have more

   (energy) (speed) (quanta)

3. The photoelectric effect supports the

   (wave model of light) (particle model of light)

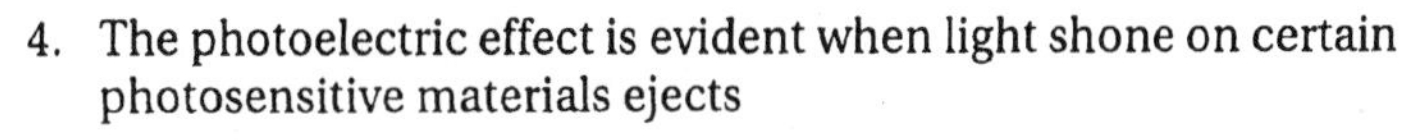

4. The photoelectric effect is evident when light shone on certain photosensitive materials ejects

   (photons) (electrons)

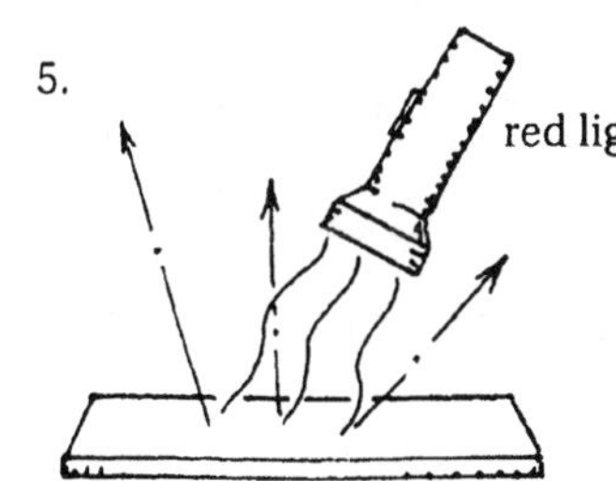

5. The photoelectric effect is more effective with violet light than with red light because the photons of violet light

   (resonate with the atoms in the material)

   (deliver more energy to the material)

   (are more numerous)

6. According to De Broglie's wave model of matter, a beam of light and a beam of electrons

   (are fundamentally different) (are similar)

7. According to De Broglie, the greater the speed of an electron beam, the

   (greater is its wavelength) (shorter is its wavelength)

8. The discreteness of the energy levels of electrons about the atomic nucleus is best understood by considering the electron to be a

   (wave) (particle)

9. Heavier atoms are not appreciably larger in size than lighter atoms. The main reason for this is that the greater nuclear charge

   (pulls surrounding electrons into tighter orbits)

   (holds more electrons about the atomic nucleus) (produces a denser atomic structure)

10. Whereas in the everyday macroworld the study of motion is called *mechanics* in the microworld the study of quanta is called

    (Newtonian mechanics) (quantum mechanics)

A QUANTUM MECHANIC!

Hewitt Drewit!

Chapter 33 Atomic Nucleus and Radioactivity
Radioactivity

1. *Complete the following statements.*

   a. A lone neutron spontaneously decays into a proton plus an ________________.

   b. Alpha and beta rays are made of streams of particles, whereas gamma rays are streams of ________________.

   c. An electrically charged atom is called an ________________.

   d. Different ________________ of an element are chemically identical but differ in the number of neutrons in the nucleus.

   e. Transuranic elements are those beyond atomic number ____________.

   f. If the amount of a certain radioactive sample decreases by half in four weeks, in four more weeks the amount remaining should be ________________ the original amount.

   g. Water from a natural hot spring is warmed by ________________ inside the earth.

2. The gas in the little girl's balloon is made up of former alpha and beta particles produced by radioactive decay.

   a. If the mixture is electrically neutral, how many more beta particles than alpha particles are in the balloon?

   ____

   ____

   b. Why is your answer not "same"?

   ____

   ____

   c. Why are the alpha and beta particles no longer harmful to the child?

   ____

   ____

   d. What element does this mixture make?

   ____

Name ______________________ Date ______

CONCEPTUAL **Physics** PRACTICE PAGE

Chapter 33 Atomic Nucleus and Radioactivity

Natural Transmutation

Fill in the decay-scheme diagram below, similar to that shown in Figure 33.13 in your textbook, but beginning with U-235 and ending up with an isotope of lead. Use the table at the left, and identify each element in the series with its chemical symbol.

Step	Particle emitted
1	Alpha
2	Beta
3	Alpha
4	Alpha
5	Beta
6	Alpha
7	Alpha
8	Alpha
9	Beta
10	Alpha
11	Beta
12	Stable

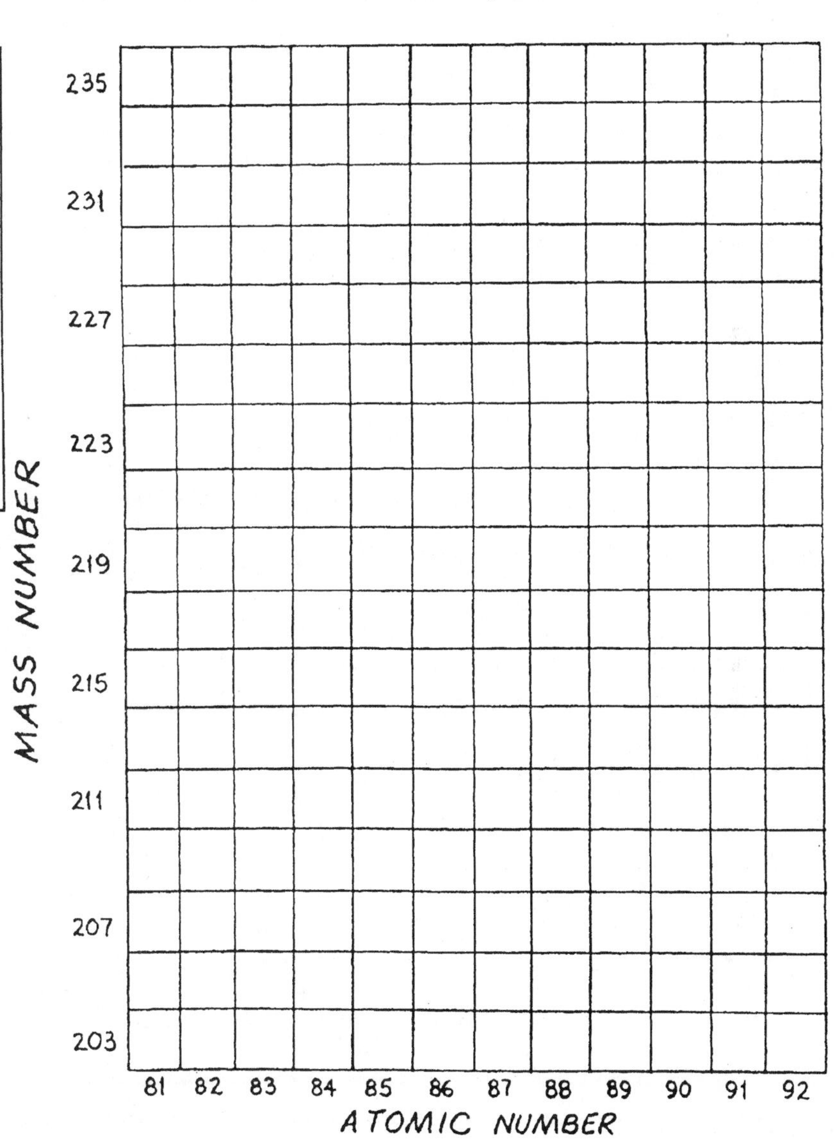

What isotope is the final product? ______________________

Hewitt Drew it!

CONCEPTUAL **Physics** PRACTICE PAGE

Chapter 33 Atomic Nucleus and Radioactivity

Nuclear Reactions

Complete these nuclear reactions.

THORIUM LATE, I OVERTHLEPT!

1. ${}^{238}_{92}U \rightarrow {}^{234}_{90}Th + {}^{4}_{2}$ ___

2. ${}^{234}_{90}Th \rightarrow {}^{234}_{91}Pa + {}^{0}_{-1}$ ___

3. ${}^{234}_{91}Pa \rightarrow$ ___ $+ {}^{4}_{2}He$

4. ${}^{220}_{86}Rn \rightarrow$ ___ $+ {}^{4}_{2}He$

5. ${}^{216}_{84}Po \rightarrow$ ___ $+ {}^{0}_{-1}e$

6. ${}^{216}_{84}Po \rightarrow$ ___ $+ {}^{4}_{2}He$

7. ${}^{210}_{83}Bi \rightarrow$ ___ $+ {}^{0}_{-1}e$

8. ${}^{1}_{0}n + {}^{10}_{5}B \rightarrow$ ___ $+ {}^{4}_{2}He$

NUCLEAR PHYSICS--- IT'S THE SAME TO ME WITH THE FIRST TWO LETTERS INTERCHANGED!

Hewitt Drew it!

Name ______________________ Date __________

CONCEPTUAL Physics PRACTICE PAGE

Chapter 34 Nuclear Fission and Fusion
Nuclear Reactions

1. Complete the table for a chain reaction in which two neutrons from each step individually cause a new reaction.

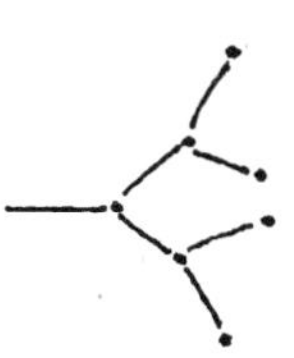

EVENT	1	2	3	4	5	6	7
NO. OF REACTIONS	1	2	4				

2. Complete the table for a chain reaction where three neutrons from each reaction cause a new reaction.

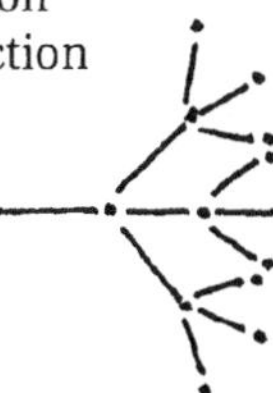

EVENT	1	2	3	4	5	6	7
NO. OF REACTIONS	1	3	9				

3. Complete these beta reactions, which occur in a breeder reactor.

$$^{239}_{92}U \rightarrow \_\_\_\_ + ^{0}_{-1}e$$

$$^{239}_{93}Np \rightarrow \_\_\_\_ + ^{0}_{-1}e$$

4. Complete the following fission reactions.

$$^{1}_{0}n + ^{235}_{92}U \rightarrow ^{143}_{54}Xe + ^{90}_{38}Sr + \_\_\_(^{1}_{0}n)$$

$$^{1}_{0}n + ^{235}_{92}U \rightarrow ^{152}_{60}Nd + \_\_\_\_ + 4(^{1}_{0}n)$$

$$^{1}_{0}n + ^{239}_{94}Pu \rightarrow \_\_\_\_ + ^{97}_{40}Zr + 2(^{1}_{0}n)$$

5. Complete the following fusion reactions.

$$^{2}_{1}H + ^{2}_{1}H \rightarrow ^{3}_{2}He + \_\_\_\_$$

$$^{2}_{1}H + ^{3}_{1}H \rightarrow ^{4}_{2}He + \_\_\_\_$$

Hewitt Drew it!

Name ______________________________ Date ____________

CONCEPTUAL Physics PRACTICE PAGE

Chapter 35 Special Theory of Relativity
Time Dilation

Chapter 35 in your textbook discusses *The Twin Trip,* in which a traveling twin takes a 2-hour journey while a stay-at-home brother records the passage of 2 1/2 hours. Quite remarkable! Times in both frames of reference are marked by flashes of light, sent each 6 minutes from the spaceship, and received on earth at 12-minute intervals for the ship going away, and 3-minute intervals for the ship returning. Read this section in the book carefully, and fill in the clock readings aboard the spaceship when each flash is emitted, and the clock reading on earth when each flash is received.

SHIP LEAVING EARTH		
FLASH	TIME ON SHIP WHEN FLASH SENT	TIME ON EARTH WHEN FLASH SEEN
0	12:00	12:00
1	12:06	
2		
3		
4		
5		
6		
7		
8		
9		
10		

SHIP APPROACHING EARTH		
FLASH	TIME ON SHIP WHEN FLASH SEEN	TIME ON EARTH WHEN FLASH SEEN
11		
12		
13		
14		
15		
16		
17		
18		
19		
20		

THIS CHECKS: FOR $v = 0.6c$

$$t = \frac{t_0}{\sqrt{1-(\frac{v}{c})^2}} = \frac{2\text{ HR}}{\sqrt{1-(\frac{0.6c}{c})^2}} = 2.5\text{ HR}$$

Hewitt Drew it!

Answers to the Practice Pages

Compare your responses to the previous pages with my responses in the reduced ones that follow. You have the choice of taking a shortcut and looking at my responses first — or you can be nice to yourself and work out your own without first looking. In working through these pages on your own or with friends, without looking at my responses until after you've given them a good college try, you may experience the exhilaration that comes with doing a good thing well.

Learning can be enjoyable!
PGH

Name ______________________ Date __________

CONCEPTUAL **Physics** PRACTICE PAGE

Chapter 1 About Science
Making Hypotheses

WHICH IS AN EDUCATED GUESS... A HYPOTHESIS OR A THEORY?

WHICH RESULTS FROM A LARGE BODY OF KNOWLEDGE?

The word science comes for Latin, meaning "to know."
The word *hypothesis* comes from Greek, "under an idea."
A hypothesis (an educated guess) often leads to new knowledge and may help to establish a theory.

Examples:

I CUT A DISK FROM THIS IRON PLATE. WHEN I HEAT THE PLATE, WILL THE HOLE GET BIGGER, OR SMALLER?

1. It is well known that objects generally expand when heated. An iron plate gets slightly bigger, for example, when put in a hot oven. But what of a hole in the middle of the plate? Will the hole get bigger or smaller when expansion occurs? One friend may say the size of the hole will increase, and another says it will decrease.

   a. What is your hypothesis about hole size, and if you are wrong, is there a test for finding out?

   HYP 1: HOLE GETS BIGGER. HYP 2: SMALLER. HYP 3 NO CHANGE

   TEST: HEAT IT IN AN OVEN, THEN MEASURE! (HYP 1 IS CORRECT)

   WHAT HAPPENS IF HE PLUGS THE DISK BACK INTO THE HOLE BEFORE HEATING EVERYTHING?

   b. There are often several ways to test a hypothesis. For example, you can perform a physical experiment and witness the results yourself, or you can use the library to find the reported results of other investigators. Which of these two methods do you favor, and why?

   (IT DEPENDS ON THE SITUATION — MOST RESEARCH INVOLVES BOTH.)

2. Before the time of the printing press, books were hand-copied by scribes, many of whom were monks in monasteries. There is the story of the scribe who was frustrated to find a smudge on an important page he was copying. The smudge blotted out part of the sentence that reported the number of teeth in the head of a donkey. The scribe was very upset and didn't know what to do. He consulted with other scribes to see if any of their books stated the number of teeth in the head of a donkey. After many hours of fruitless searching through the library, it was agreed that the best thing to do was to send a messenger by donkey to the next monastery and continue the search there. What would be your advice?

ACTUALLY LOOK IN THE DONKEY'S MOUTH AND COUNT! (WATCH FOR MISSING TEETH)

Making Distinctions

Many people don't seem to see the difference between a thing and the *abuse* of the thing. For example, a city council that bans skateboarding may not distinguish between skateboarding and reckless skateboarding. A person who advocates that a particular technology be banned may not distinguish between that technology and the abuses of that technology. There's a difference between a thing and the abuse of the thing.

On a separate sheet of paper, list other examples where use and abuse are often not distinguished. Compare your list with others in your class.

Hewitt Drew it!

CONCEPTUAL **Physics** PRACTICE PAGE

Pinhole Image Formation

Look carefully at the round spots of light on the shady ground beneath trees. These are *sunballs*, which are images of the sun. They are cast by openings between leaves in the trees that act as pinholes. (Did you make a pinhole "camera" back in middle school?) Large sunballs, several centimeters in diameter or so, are cast by openings that are relatively high above the ground, while small ones are produced by closer "pinholes." The interesting point is that the ratio of the diameter of the sunball to its distance from the pinhole is the same as the ratio of the sun's diameter to its distance from the pinhole. We know the sun is approximately 150,000,000 km from the pinhole, so careful measurements of the ratio of diameter/distance for a sunball leads you to the diameter of the sun. That's what this page is about. Instead of measuring sunballs under the shade of trees on a sunny day, make your own easier-to-measure sunball.

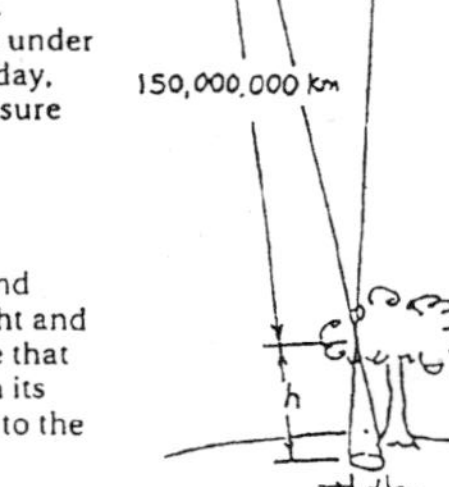

1. Poke a small hole in a piece of card. Perhaps an index card will do, and poke the hole with a sharp pencil or pen. Hold the card in the sunlight and note the circular image that is cast. This is an image of the sun. Note that its size doesn't depend on the size of the hole in the card, but only on its distance. The image is a circle when cast on a surface perpendicular to the rays — otherwise it's "stretched out" as an ellipse.

2. Try holes of various shapes; say a square hole, or a triangular hole. What is the shape of the image when its distance from the card is large compared with the size of the hole? Does the shape of the pinhole make a difference?

   IMAGE IS ALWAYS A CIRCLE. SHAPE OF PINHOLE IS NOT THE SHAPE OF THE IMAGE CAST THROUGH IT.

3. Measure the diameter of a small coin. Then place the coin on a viewing area that is perpendicular to the sun's rays. Position the card so the image of the sunball exactly covers the coin. Carefully measure the distance between the coin and the small hole in the card. Complete the following:

   $$\frac{\text{Diameter of sunball}}{\text{Distance to pinhole}} = \frac{d}{h} \approx \frac{1}{110} \quad \left(\text{SO SUN'S DIAM} = \frac{1}{110} \times 150{,}000{,}000 \text{ km}\right)$$

   With this ratio, estimate the diameter of the sun. Show your work on a separate piece of paper.

4. If you did this on a day when the sun is partially eclipsed, what shape of image would you expect to see?

   UPSIDE-DOWN CRESCENT, IMAGE OF THE PARTIALLY-ECLIPSED SUN

WHAT SHAPE DO SUNBALLS HAVE DURING A PARTIAL ECLIPSE OF THE SUN?

Hewitt Drew it!

Name ______________________ Date __________

CONCEPTUAL **Physics** PRACTICE PAGE

Chapter 2 Newton's First Law of Motion—Inertia
Static Equilibrium

1. Little Nellie Newton wishes to be a gymnast and hangs from a variety of positions as shown. Since she is not accelerating, the net force on her is zero. This means the upward pull of the rope(s) equals the downward pull of gravity. She weighs 300 N. Show the scale reading for each case.

2. When Burl the painter stands in the exact middle of his staging, the left scale reads 600 N. Fill in the reading on the right scale. The total weight of Burl and staging must be 1200 N.

3. Burl stands farther from the left. Fill in the reading on the right scale.

4. In a silly mood, Burl dangles from the right end. Fill in the reading on the right scale.

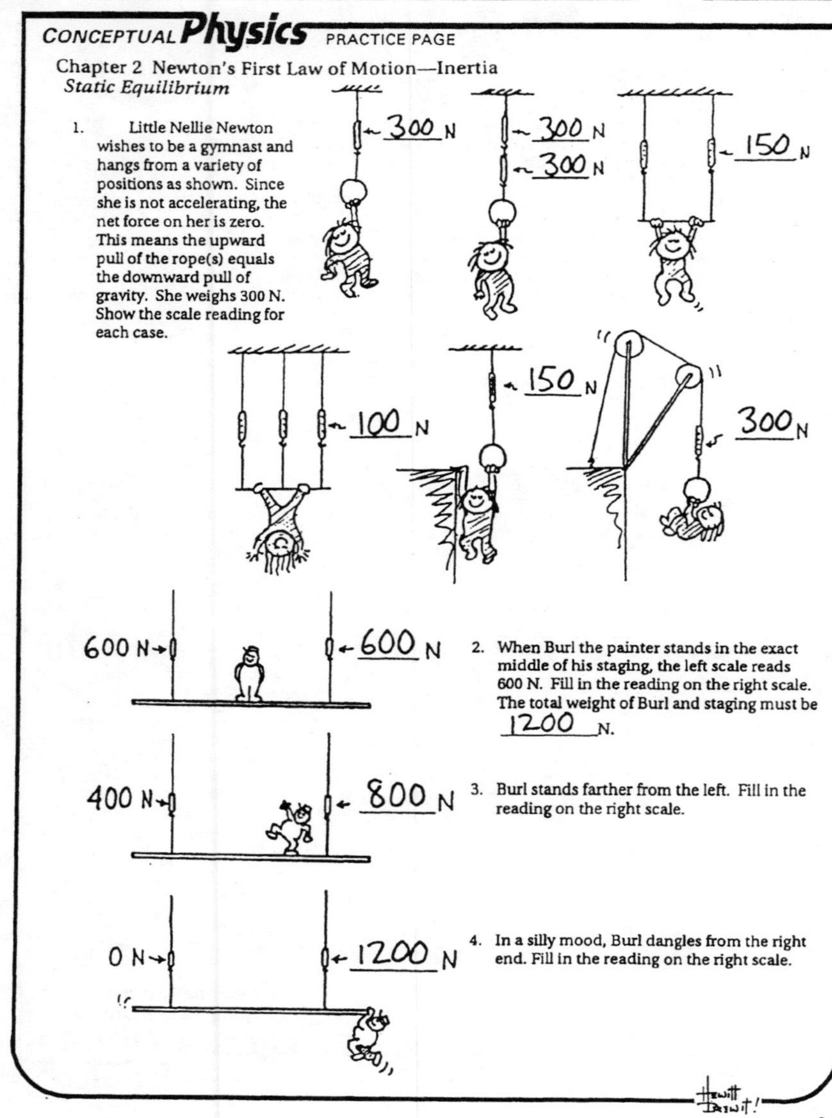

CONCEPTUAL **Physics** PRACTICE PAGE

Chapter 2 Newton's First Law of Motion—Inertia
The Equilibrium Rule: $\Sigma F = 0$

1. Manuel weighs 1000 N, and stands in the middle of a board that weighs 200 N. The ends of the board rest on bathroom scales. (We can assume the weight of the board acts at its center). Fill in the correct weight reading on each scale.

2. When Manuel moves to the left as shown, the scale closest to him reads 850 N. Fill in the weight reading for the far scale.

3. A 12-ton truck is one-quarter the way across a bridge that weighs 20 tons. A 13-ton force supports the right side of the bridge as shown. How much support force is on the left side?

4. A 1000-N crate resting on a horizontal surface is connected to a 500-N iron block through a frictionless pulley as shown. Friction between the crate and surface is enough to keep the system at rest. The arrows show the forces that act on the crate and the block. Fill in the magnitude of each force.

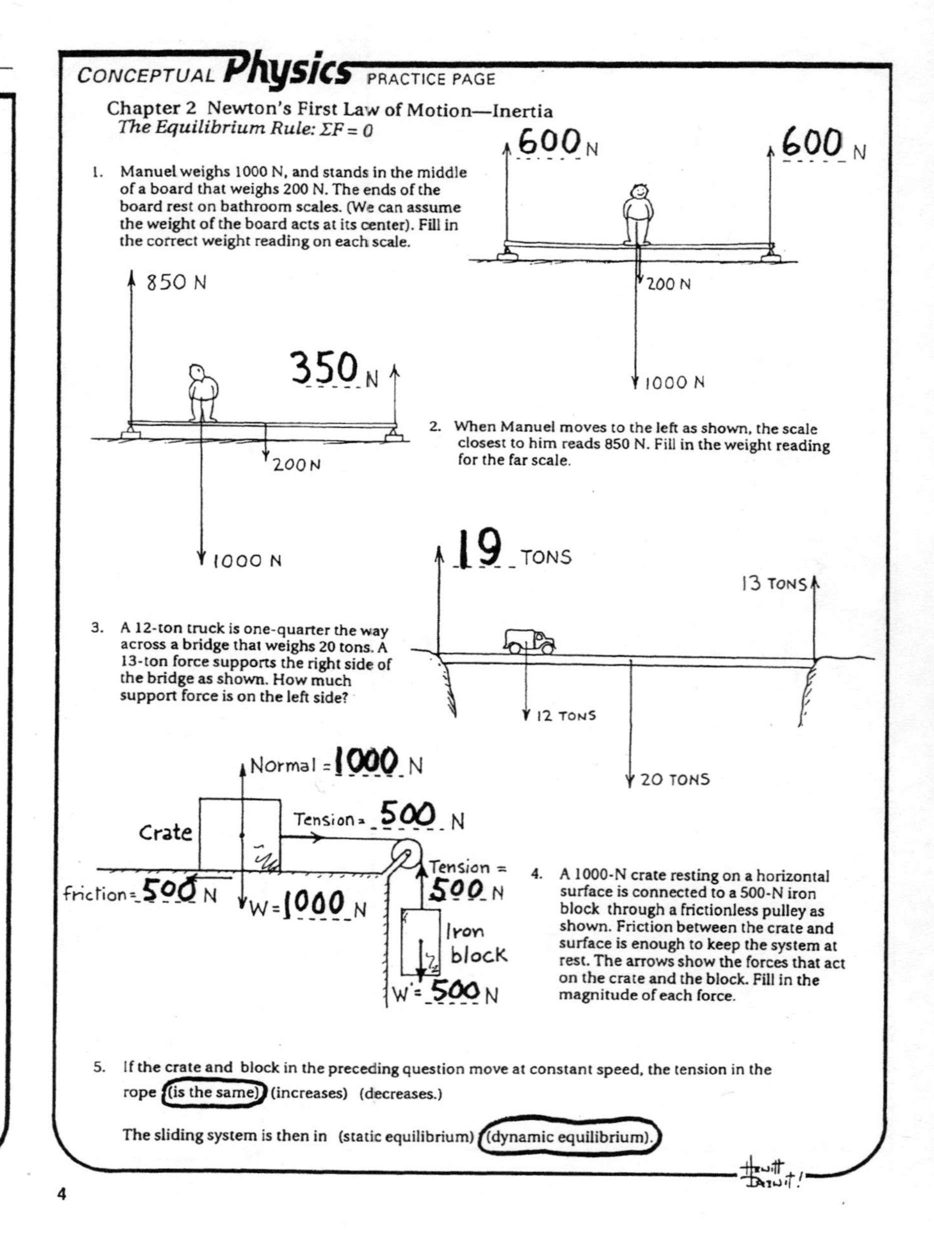

5. If the crate and block in the preceding question move at constant speed, the tension in the rope (is the same) (increases) (decreases.)

The sliding system is then in (static equilibrium) (dynamic equilibrium).

Name ______________________ Date __________

CONCEPTUAL **Physics** PRACTICE PAGE

Chapter 3 Linear Motion
Free Fall Speed

1. Aunt Minnie gives you \$10 per second for 4 seconds. How much money do you have after 4 seconds? **\$40**

2. A ball dropped from rest picks up speed at 10 m/s per second. After it falls for 4 seconds, how fast is it going? **40 m/s**

3. You have \$20, and Uncle Harry gives you \$10 each second for 3 seconds. How much money do you have after 3 seconds? **\$50**

4. A ball is thrown straight down with an initial speed of 20 m/s. After 3 seconds, how fast is it going? **50 m/s**

5. You have \$50 and you pay Aunt Minnie \$10/second. When will your money run out? **5 s**

6. You shoot an arrow straight up at 50 m/s. When will it run out of speed? **5 s**

7. So what will be the arrow's speed 5 seconds after you shoot it? **0 m/s**

8. What will its speed be 6 seconds after you shoot it? 7 seconds? **10 m/s** **20 m/s**

Free Fall Distance

1. Speed is one thing; distance another. *Where* is the arrow you shoot up at 50 m/s when it runs out of speed? **125 m**

2. How high will the arrow be 7 seconds after being shot up at 50 m/s? **105 m**

3 a. Aunt Minnie drops a penny into a wishing well and and it falls for 3 seconds before hitting the water. How fast is it going when it hits? **30 m/s**

b. What is the penny's average speed during its 3-second drop? **15 m/s**

c. How far down is the water surface? **45 m**

FROM REST, $v = 10t$, $d = 5t^2$

4. Aunt Minnie didn't get her wish, so she goes to a deeper wishing well and throws a penny straight down into it at 10 m/s. How far does this penny go in 3 seconds? **75 m**

$\bar{v} = \frac{v_0 + v}{2} = \frac{v_0 + (v_0 + 10t)}{2}$ THEN $d = \bar{v}t$

Distinguish between "how fast," "how far," and "how long"!

CONCEPTUAL **Physics** PRACTICE PAGE

Chapter 3 Linear Motion
Acceleration of Free Fall

A rock dropped from the top of a cliff picks up speed as it falls. Pretend that a speedometer and odometer are attached to the rock to show readings of speed and distance at 1-second intervals. Both speed and distance are zero at time = zero (see sketch). Note that after falling 1 second the speed reading is 10 m/s and the distance fallen is 5 m. The readings for succeeding seconds of fall are not shown and are left for you to complete. So draw the position of the speedometer pointer and write in the correct odometer reading for each time. Use $g = 10\ m/s^2$ and neglect air resistance.

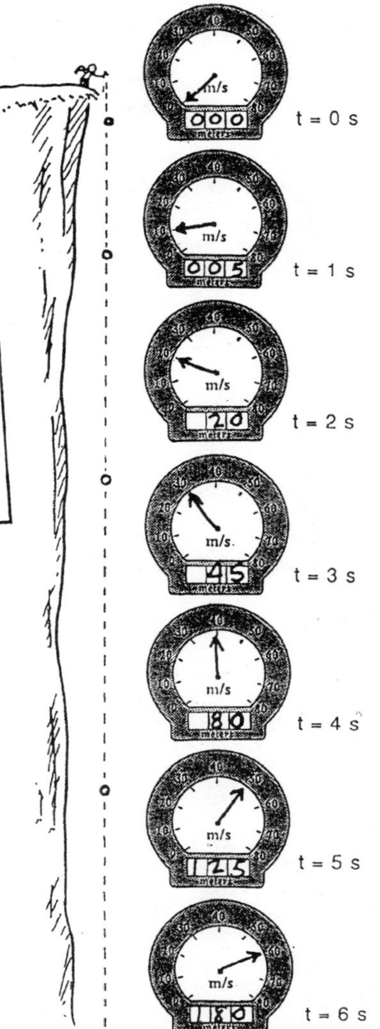

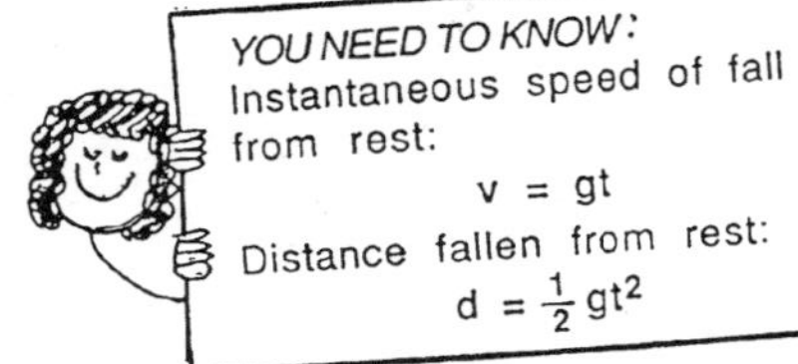

YOU NEED TO KNOW:
Instantaneous speed of fall from rest:

$$v = gt$$

Distance fallen from rest:

$$d = \frac{1}{2}gt^2$$

1. The speedometer reading increases by the same amount, **10** m/s, each second. This increase in speed per second is called **ACCELERATION**.

2. The distance fallen increases as the square of the **TIME**.

3. If it takes 7 seconds to reach the ground, then its speed at impact is **70** m/s, the total distance fallen is **245** m, and its acceleration of fall just before impact is **10** m/s^2.

Name ______________________ Date ____________

CONCEPTUAL **Physics** PRACTICE PAGE

Chapter 4 Newton's Second Law of Motion
Force and Acceleration

1. Skelly the skater, total mass 25 kg, is propelled by rocket power.

   a. Complete Table I (neglect resistance)

   $a = \frac{F}{25\text{ kg}}$

TABLE I

FORCE	ACCELERATION
100 N	4 m/s²
200 N	8 m/s²
	10 m/s²

   b. Complete Table II for a constant 50-N resistance.

   $a = \frac{F - 50\text{ N}}{25\text{ kg}}$

TABLE II

FORCE	ACCELERATION
50 N	0 m/s²
100 N	2 m/s²
200 N	6 m/s²

2. Block A on a horizontal friction-free table is accelerated by a force from a string attached to Block B. B falls vertically and drags A horizontally. Both blocks have the same mass m. (Neglect the string's mass.)

   (Circle the correct answers)

   a. The *mass* of the system [A + B] is (m) **($2m$)**.

   b. The *force* that accelerates [A + B] is the weight of (A) **(B)** (A + B).

   c. The weight of B is ($mg/2$) **(mg)** ($2mg$).

   d. Acceleration of [A + B] is **(less than g)** (g) (more than g).

   e. Use $a =$ to show the acceleration of [A + B] as a fraction of g. $a = \frac{mg}{2m} = \frac{g}{2}$

If B were allowed to fall by itself, not dragging A, then wouldn't its acceleration be g?

Yes, because the force that accelerates it would only be acting on its own mass – not twice the mass!

To better understand this, consider 3 and 4 on the other side!

CONCEPTUAL **Physics** PRACTICE PAGE

Force and Acceleration continued

3. Suppose A is still a 1-kg block, but B is a low-mass feather (or a coin).

   a. Compared to the acceleration of the system in 2, previous page, the acceleration of [A + B] here is **(less)** (more) and is **(close to zero)** (close to g).

   b. In this case the acceleration of B is (practically that of free fall) **(constrained)**.

4. Suppose A is a feather or coin, and B has a mass of 1 kg.

   a. The acceleration of [A + B] here is (close to zero) **(close to g)**.

   b. In this case the acceleration of B is **(practically that of free fall)** (constrained).

5. Summarizing 2, 3, and 4, where the weight of one object causes the acceleration of two objects, we see the range of possible accelerations is **(between zero and g)** (between zero and infinity) (between g and infinity).

6. A ball rolls down a uniform-slope ramp.

   a. Acceleration is (decreasing) **(constant)** (increasing).

   b. If the ramp were steeper, acceleration would be **(more)** (the same) (less).

   c. When the ball reaches the bottom and rolls along the smooth level surface it (continues to accelerate) **(does not accelerate)**.

Name ______________________ Date ________

CONCEPTUAL **Physics** PRACTICE PAGE

Chapter 4 Newton's Second Law of Motion
Friction

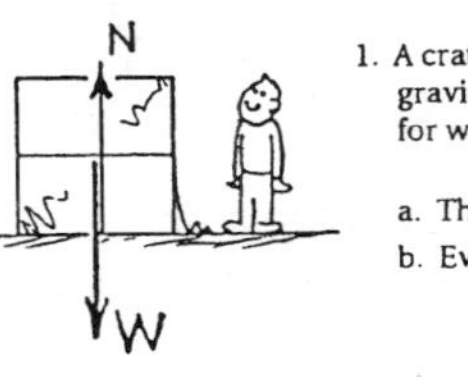

1. A crate filled with delicious junk food rests on a horizontal floor. Only gravity and the support force of the floor act on it, as shown by the vectors for weight W and normal force N.

   a. The net force on the crate is (zero) (greater than zero).

   b. Evidence for this is NO ACCELERATION

2. A slight pull P is exerted on the crate, not enough to move it. A force of friction f now acts,

   a. which is (less than) (equal to) (greater than) P.

   b. Net force on the crate is (zero) (greater than zero).

3. Pull P is increased until the crate begins to move. It is pulled so that it moves with constant velocity across the floor.

   a. Friction f is (less than) (equal to) (greater than) P.

   b. Constant velocity means acceration is (zero) (greater than zero).

   c. Net force on the crate is (less than) (equal to) (greater than) zero.

4. Pull P is further increased and is now greater than friction f.

   a. Net force on the crate is (less than) (equal to) (greater than) zero.

   b. The net force acts toward the right, so acceleration acts toward the (left) (right).

5. If the pulling force P is 150 N and the crate doesn't move, what is the magnitude of f? 150 N

6. If the pulling force P is 200 N and the crate doesn't move, what is the magnitude of f? 200 N

7. If the force of sliding friction is 250 N, what force is necessary to keep the crate sliding at constant velocity? 250 N

8. If the mass of the crate is 50 kg and sliding friction is 250 N, what is the acceleration of the crate when the pulling force is 250 N? 0 m/s² 300 N? 1 m/s² 500 N? 5 m/s²

CONCEPTUAL **Physics** PRACTICE PAGE

Falling and Air Resistance

Bronco skydives and parachutes from a stationary helicopter. Various stages of fall are shown in positions *a* through *f*. Using Newton's 2nd law,

$$a = \frac{F_{NET}}{m} = \frac{W-R}{m}$$

find Bronco's acceleration at each position (answer in the blanks to the right). You need to know that Bronco's mass *m* is 100 kg so his weight is a constant 1000 N. Air resistance *R* varies with speed and cross-sectional area as shown.

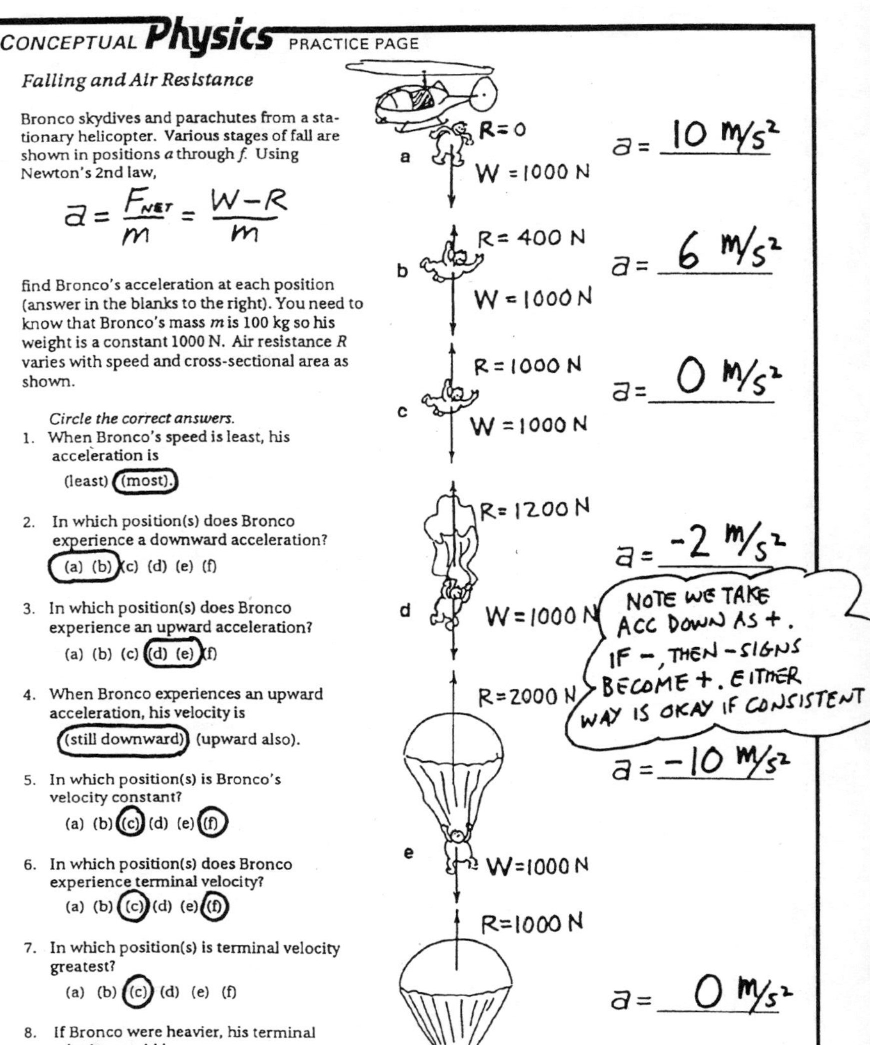

Circle the correct answers.

1. When Bronco's speed is least, his acceleration is

   (least) (most).

2. In which position(s) does Bronco experience a downward acceleration?

   (a) (b) (c) (d) (e) (f)

3. In which position(s) does Bronco experience an upward acceleration?

   (a) (b) (c) (d) (e) (f)

4. When Bronco experiences an upward acceleration, his velocity is

   (still downward) (upward also).

5. In which position(s) is Bronco's velocity constant?

   (a) (b) (c) (d) (e) (f)

6. In which position(s) does Bronco experience terminal velocity?

   (a) (b) (c) (d) (e) (f)

7. In which position(s) is terminal velocity greatest?

   (a) (b) (c) (d) (e) (f)

8. If Bronco were heavier, his terminal velocity would be

   (greater) (less) (the same).

Name ______________________ Date ________

CONCEPTUAL Physics PRACTICE PAGE

Chapter 5 Newton's Third Law of Motion

Action and Reaction Pairs

1. In the example below, the action-reaction pair is shown by the arrows (vectors), and the action-reaction described in words. In (*a*) through (*g*) draw the other arrow (vector) and state the reaction to the given action. Then make up your own example in (*h*).

Example:

Fist hits wall.

Wall hits fist.

Head bumps ball.

(*a*) BALL BUMPS HEAD

Windshield hits bug.

(*b*) BUG HITS WINDSHIELD

Bat hits ball.

(*c*) BALL HITS BAT

Hand touches nose.

(*d*) NOSE TOUCHES HAND

Hand pulls on flower.

(*e*) FLOWER PULLS ON HAND

Athlete pushes bar upward.

(*f*) BAR PUSHES ATHLETE DOWNWARD

Compressed air pushes balloon surface outward.

(*g*) BALLOON SURFACE PUSHES COMPRESSED AIR INWARD

STUDENT DRAWING (OPEN)

(*h*) THING A ACTS ON THING B — THING B ACTS ON THING A

2. Draw arrows to show the chain of at least six pairs of action-reaction forces below.

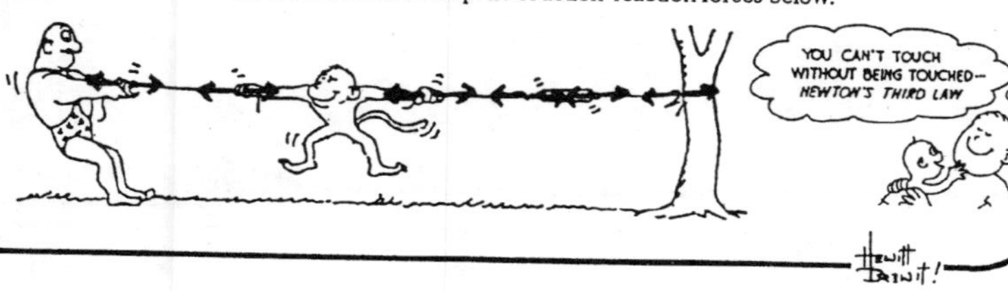

YOU CAN'T TOUCH WITHOUT BEING TOUCHED— NEWTON'S THIRD LAW

CONCEPTUAL Physics PRACTICE PAGE

Interactions

3. Nellie Newton holds an apple weighing 1 newton at rest on the palm of her hand. The force vectors shown are the forces that act on the apple.

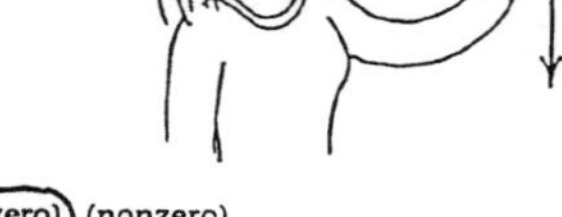

a. To say the weight of the apple is 1 N is to say that a downward gravitational force of 1 N is exerted on the apple by **(the earth)** (her hand).

b. Nellie's hand supports the apple with normal force N, which acts in a direction opposite to W. We can say N (equals W) **(has the same magnitude as W)**.

c. Since the apple is at rest, the net force on the apple is **(zero)** (nonzero).

d. Since N is equal and opposite to W, we (can) **(cannot)** say that N and W comprise an action-reaction pair. The reason is because action and reaction always (act on the same object) **(act on different objects)**, and here we see N and W **(both acting on the apple)** (acting on different objects).

e. In accord with the rule, "If ACTION is A acting on B, then REACTION is B acting on A," if we say *action* is the earth pulling down on the apple, *reaction* is **(the apple pulling up on the earth)** (N, Nellie's hand pushing up on the apple).

f. To repeat for emphasis, we see that N and W are equal and opposite to each other (and comprise an action-reaction pair) **(but do *not* comprise an action-reaction pair)**.

To identify a pair of action-reaction forces in any situation, first identify the pair of interacting objects involved. Something is interacting with something else. In this case the whole earth is interacting (gravitationally) with the apple. So the earth pulls downward on the apple (call it action), while the apple pulls upward on the earth (reaction).

Simply put, earth pulls on apple (action); apple pulls on earth (reaction).

Better put, apple and earth pull on each other with equal and opposite forces that comprise a *single* interaction.

g. Another pair of forces is N [shown] and the downward force of the apple against Nellie's hand [not shown]. This force pair **(is)** (isn't) an action-reaction pair.

h. Suppose Nellie now pushes upward on the apple with a force of 2 N. The apple (is still in equilibrium) **(accelerates upward)** and compared to W, the magnitude of N is (the same) **(twice)** (not the same, and not twice).

i. Once the apple leaves Nellie's hand, N is **(zero)** (still twice the magnitude of W), and the net force on the apple is (zero) **(only W)** (still W - N, a negative force).

Name ______________________ Date __________

CONCEPTUAL Physics PRACTICE PAGE

Chapter 5 Newton's Third Law of Motion
Vectors and the Parallelogram Rule

1. When vectors A and B are at an angle to each other, they add to produce the resultant C by the *parallelogram rule.* Note that C is the diagonal of a parallelogram where A and B are adjacent sides. Resultant C is shown in the first two diagrams, *a* and *b*. Construct the resultant C in diagrams *c* and *d*. Note that in diagram *d* you form a rectangle (a special case of a parallelogram).

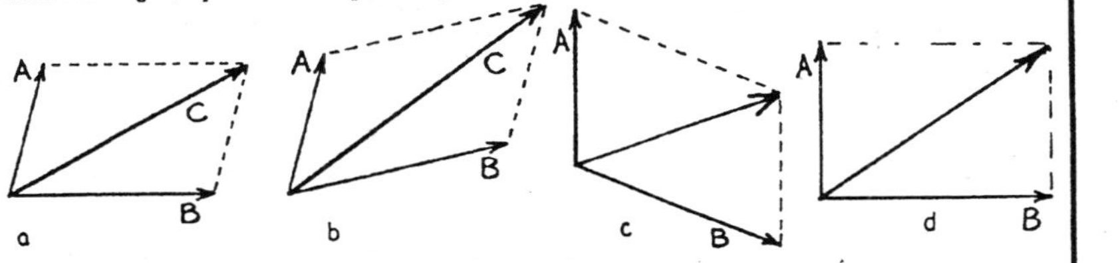

2. Below we see a top view of an airplane being blown offcourse by wind in various directions. Use the parallelogram rule to show the resulting speed and direction of travel for each case. In which case does the airplane travel fastest across the ground? d Slowest? c

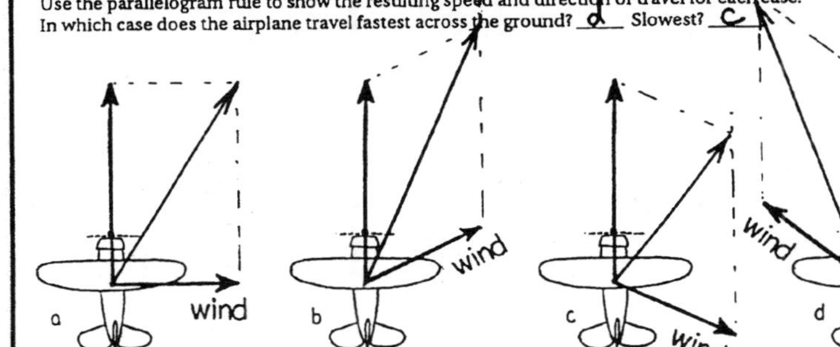

3. To the right we see top views of 3 motorboats crossing a river. All have the same speed relative to the water, and all experience the same water flow.

Construct resultant vectors showing the speed and direction of the boats.

a. Which boat takes the shortest path to the opposite shore? a

b. Which boat reaches the opposite shore first? b

c. Which boat provides the fastest ride? c

CONCEPTUAL Physics PRACTICE PAGE

Velocity Vectors and Components

1. Draw the resultants of the four sets of vectors below.

2. Draw the horizontal and vertical components of the four vectors below.

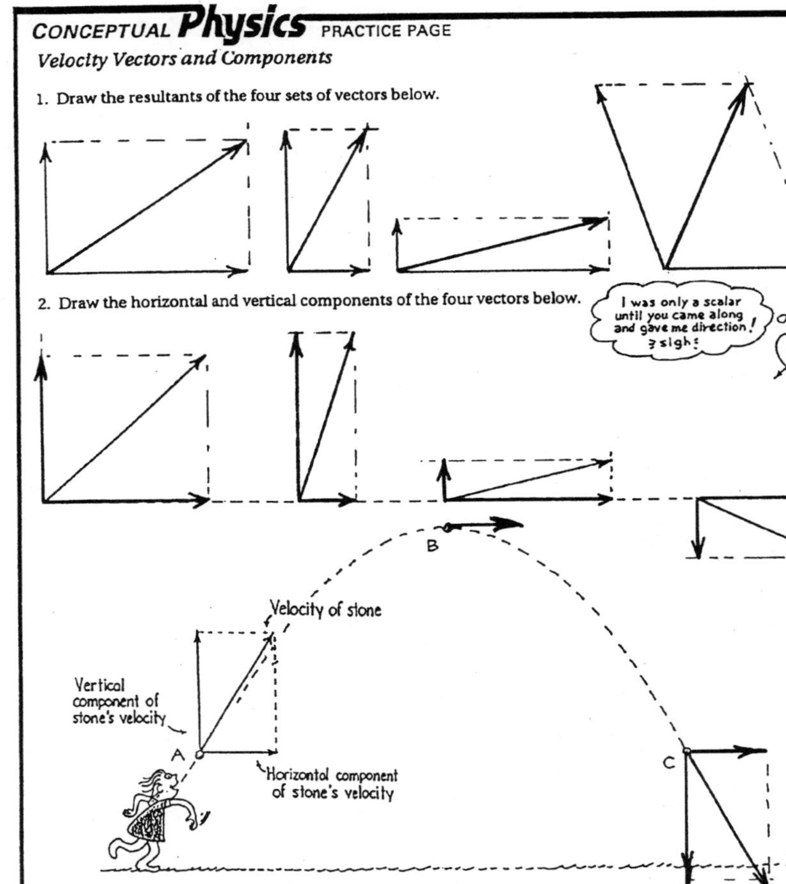

3. She tosses the ball along the dashed path. The velocity vector, complete with its horizontal and vertical components is shown at position A. Carefully sketch the appropriate velocity vectors with appropriate components for positions B and C.

a. Since there is no acceleration in the horizontal direction, how does the horizontal component of velocity compare for positions A, B, and C? SAME

b. What is the value of the vertical component of velocity at position B? 0 m/s

c. How does the vertical component of velocity at position C compare with that of position A? EQUAL AND OPPOSITE

Name ______________________ Date ________

CONCEPTUAL **Physics** PRACTICE PAGE

Chapter 5 Newton's Third Law of Motion
Force and Velocity Vectors

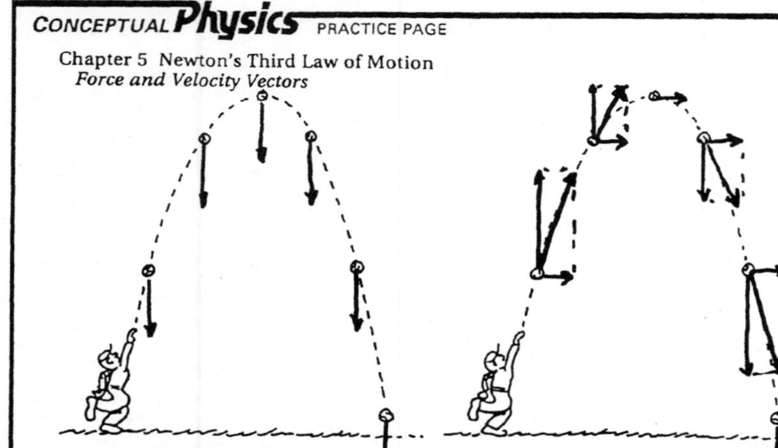

1. Draw sample vectors to represent the force of gravity on the ball in the positions shown above (after it leaves the thrower's hand). Neglect air drag.

2. Draw sample bold vectors to represent the velocity of the ball in the positions shown above. With lighter vectors, show the horizontal and vertical components of velocity for each position.

3. (a) Which velocity component in the previous question remains constant ? Why?

HORIZONTAL COMPONENT CONSTANT – NO HORIZ COMP ON BALL

(b) Which velocity component changes along the path? Why?

VERTICAL COMPONENT CHANGES DUE TO GRAVITY IN VERTICAL DIRECTION

4. It is important to distinguish between force and velocity vectors. Force vectors combine with other force vectors, and velocity vectors combine with other velocity vectors. Do velocity vectors combine with force vectors? NO

5. All forces on the bowling ball, weight down and support of alley up, are shown by vectors at its center before it strikes the pin (*a*). Draw vectors of all the forces that act on the ball (*b*) when it strikes the pin, and (*c*) after it strikes the pin.

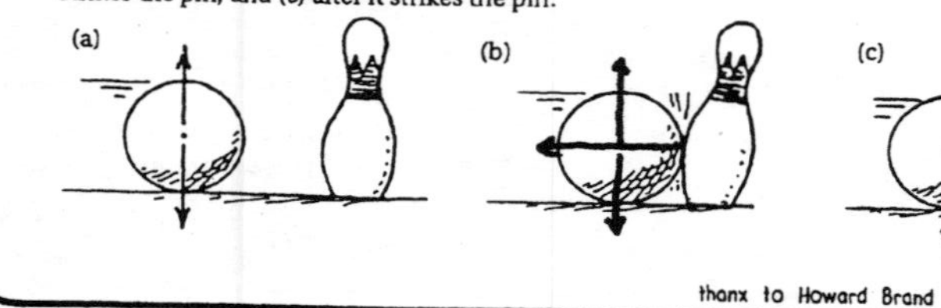

thanx to Howard Brand

CONCEPTUAL **Physics** PRACTICE PAGE

Force Vectors and the Parallelogram Rule

1. The heavy ball is supported in each case by two strands of rope. The tension in each strand is shown by the vectors. Use the paralellogram rule to find the resultant of each vector pair.

Note it's the angle, not the length of the rope, that affects tension!

a. Is your resultant vector the same for each case? YES

b. How do you think the resultant vector compares to the weight of the ball? SAME (BUT OPPOSITE DIRECTION)

2. Now let's do the opposite of what we've done above. More often, we know the weight of the suspended object but we don't know the rope tensions. In each case below, the weight of the ball is shown by the vector W. Each dashed vector represents the resultant of the pair of rope tensions. Note that each is equal and opposite to vectors W (they must be; otherwise the ball wouldn't be at rest).

a. Construct parallelograms wherein the ropes define adjacent sides and the dashed vectors are the diagonals.

b. How do the relative lengths of the sides of each parallelogram compare to rope tensions? SAME

c. Draw rope-tension vectors.

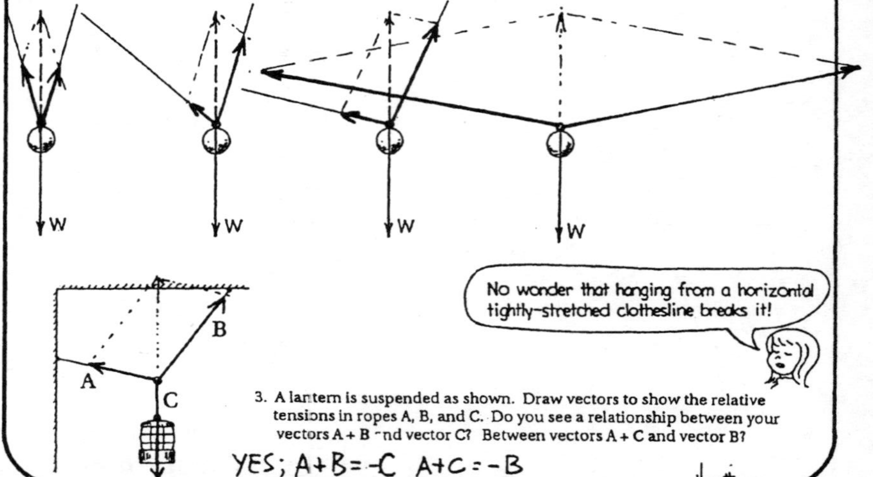

No wonder that hanging from a horizontal tightly-stretched clothesline breaks it!

3. A lantern is suspended as shown. Draw vectors to show the relative tensions in ropes A, B, and C. Do you see a relationship between your vectors A + B and vector C? Between vectors A + C and vector B?

YES; A+B=-C A+C=-B

Name ____________________ Date ________

CONCEPTUAL **Physics** PRACTICE PAGE

Chapter 5 Newton's Third Law of Motion

Force-Vector Diagrams

In each case, a rock is acted on by one or more forces. Draw an accurate vector diagram showing all forces acting on the rock, and no other forces. Use a ruler, and do it in pencil so you can correct mistakes. The first two are done as examples. Show by the parallelogram rule in 2 that the vector sum of A + B is equal and opposite to W (that is, A + B = -W). Do the same for 3 and 4. Draw and label vectors for the weight and normal support forces in 5 to 10, and for the appropriate forces in 11 and 12.

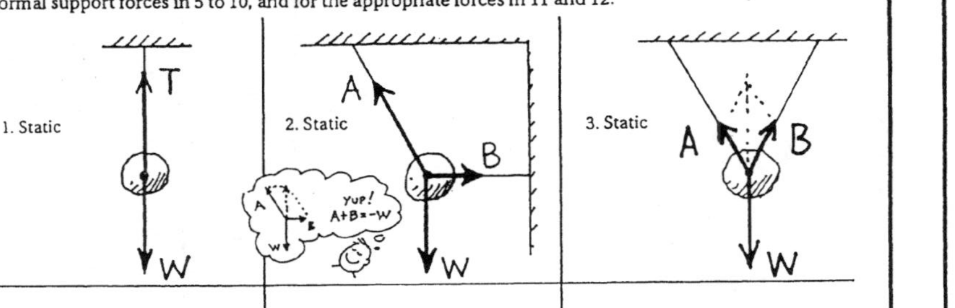

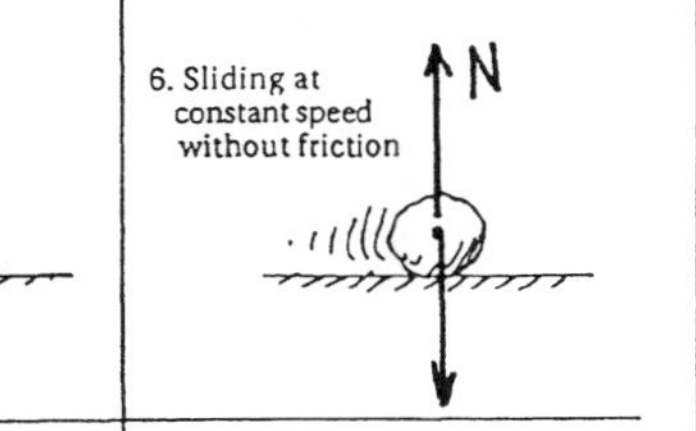

1. Static
2. Static
3. Static
4. Static
5. Static
6. Sliding at constant speed without friction
7. Decelerating due to friction
8. Static (Friction prevents sliding
9. Rock slides (No friction)
10. Static
11. Rock in free fall
12. Falling at terminal velocity

thanx to Jim Court

Name ____________________ Date ________

CONCEPTUAL **Physics** PRACTICE PAGE

Appendix D More About Vectors

Vectors and Sailboats

(Do not attempt this until you have studied Appendix D!)

1. The sketch shows a top view of a small railroad car pulled by a rope. The force *F* that the rope exerts on the car has one component along the track, and another component perpendicular to the track.

   a. Draw these components on the sketch. Which component is larger? PERPENDICULAR COMP.

   b. Which component produces acceleration? COMP PARALLEL TO TRACK

   c. What would be the effect of pulling on the rope if it were perpendicular to the track? NO ACCELERATION

2. The sketches below represent simplified top views of sailboats in a cross-wind direction. The impact of the wind produces a FORCE vector on each as shown. (We do NOT consider *velocity* vectors here!)

   WIND

   a. Why is the position of the sail above useless for propelling the boat along its forward direction? (Relate this to Question 1c. above. Where the train is constrained by tracks to move in one direction, the boat is similarly constrained to move along one direction by its deep vertical fin — the *keel*.)
   AS IN 1C ABOVE, THERE'S NO COMP PARALLEL TO DIRECTION of MOTION.

   b. Sketch the component of force parallel to the direction of the boat's motion (along its keel), and the component perpendicular to its motion. Will the boat move in a forward direction? (Relate this to Question 1b. above.)
   YES, AS IN 1b ABOVE, THERE IS A COMP PARALLEL TO DIRECTION OF MOTION.

CONCEPTUAL **Physics** PRACTICE PAGE

3. The boat to the right is oriented at an angle into the wind. Draw the force vector and its forward and perpendicular components.

   a. Will the boat move in a forward direction and tack into the wind? Why or why not?
   YES, BECAUSE THERE IS A COMPONENT OF FORCE PARALLEL TO THE DIRECTION OF MOTION.

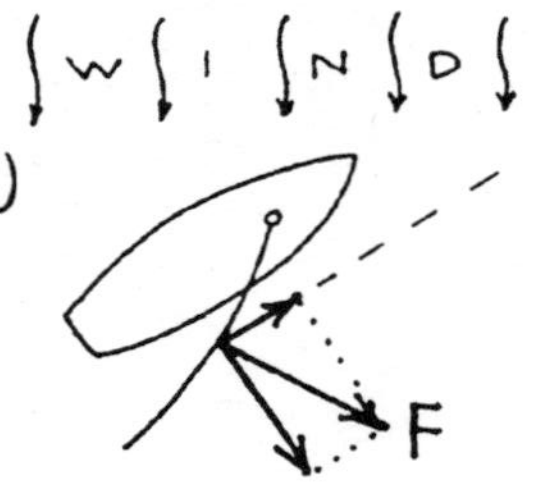

4. The sketch below is a top view of five identical sailboats. Where they exist, draw force vectors to represent wind impact on the sails. Then draw components parallel and perpendicular to the keels of each boat.

   a. Which boat will sail the fastest in a forward direction?
   BOAT 4 (WILL USUALLY EXCEED BOAT 1)

   b. Which will respond least to the wind?
   BOAT 2 (OR BOAT 3)*

   c. Which will move in a backward direction?
   BOAT 5

   d. Which will experience less and less wind impact with increasing speed?
   BOAT 1 (NO IMPACT AT WIND SPEED)

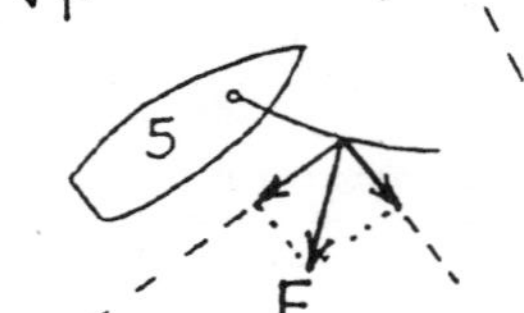
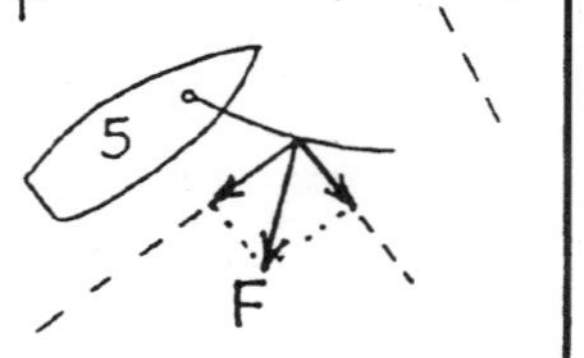
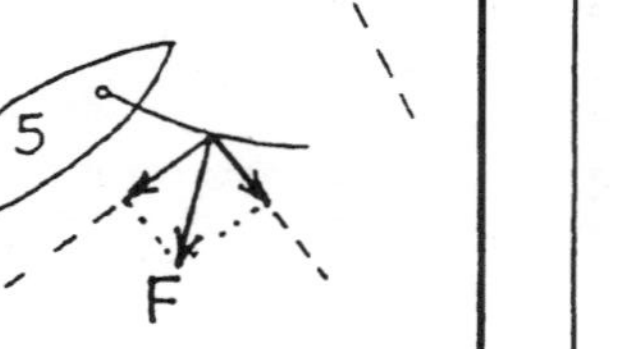
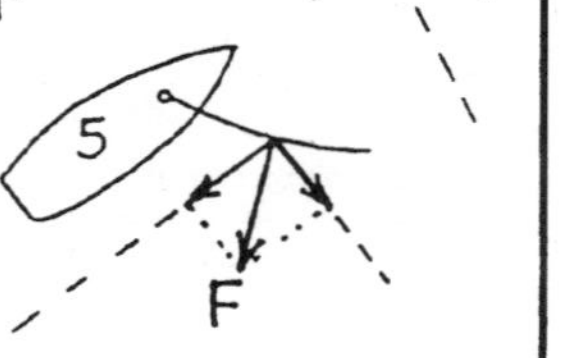
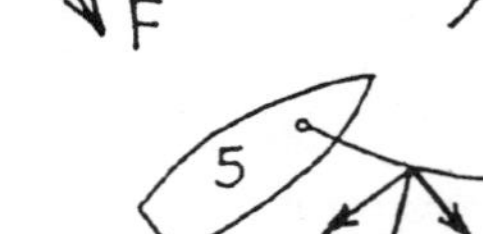

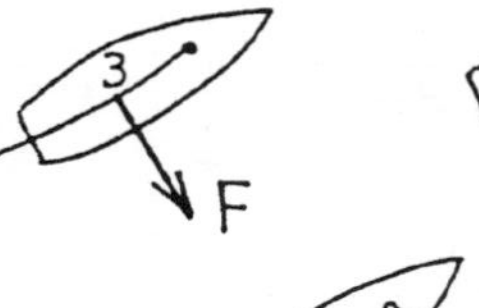

* THE WIND MISSES THE SAIL OF BOAT 2, AND THERE'S NO COMPONENT PARALLEL TO THE KEEL FOR BOAT 3.

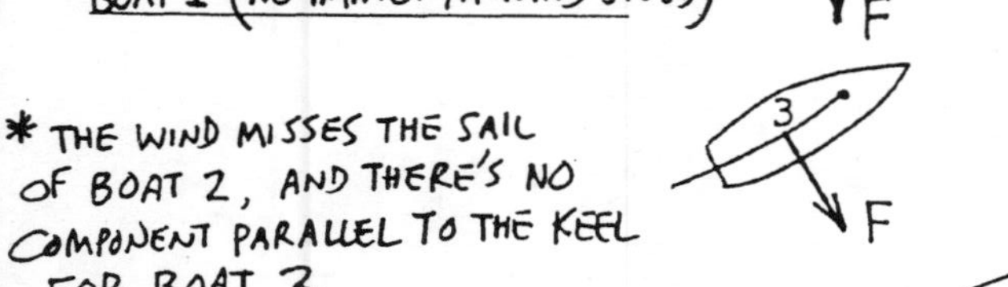

Name ____________ Date ____________

CONCEPTUAL **Physics** PRACTICE PAGE

Chapter 6 Momentum
Impulse and Momentum

1. A moving car has momentum. If it moves twice as fast, its momentum is TWICE as much.

2. Two cars, one twice as heavy as the other, move down a hill at the same speed. Compared to the lighter car, the momentum of the heavier car is TWICE as much.

3. The recoil momentum of a gun that kicks is
   (more than) (less than) **(the same as)**
   the momentum of the bullet it fires.
   ALTHO SPEED AND ACCELERATION OF BULLET GREATER

4. If a man firmly holds a gun when fired, then the momentum of the bullet is equal to the recoil momentum of the
   (gun alone) **(gun-man system)** (man alone)

5. Suppose you are traveling in a bus at highway speed on a nice summer day and the momentum of an unlucky bug is suddenly changed as it splatters onto the front window.

   a. Compared to the force that acts on the bug, how much force acts on the bus?
   (more) **(the same)** (less)

   b. The time of impact is the same for both the bug and the bus. Compared to the impulse on the bug, this means the impulse on the bus is
   (more) **(the same)** (less)

   c. Although the momentum of the bus is very large compared to the momentum of the bug, the change in momentum of the bus, compared to the *change* of momentum of the bug is
   (more) **(the same)** (less)

   d. Which undergoes the greater acceleration?
   (bus) (both the same) **(bug)**

   e. Which therefore, suffers the greater damage?
   (bus) (both the same) **(the bug of course!)**

CONCEPTUAL **Physics** PRACTICE PAGE

Systems

1. When the compressed spring is released, Blocks A and B will slide apart. There are 3 systems to consider here, indicated by the closed dashed lines below — System A, System B, and System A+B. Ignore the vertical forces of gravity and the support force of the table.

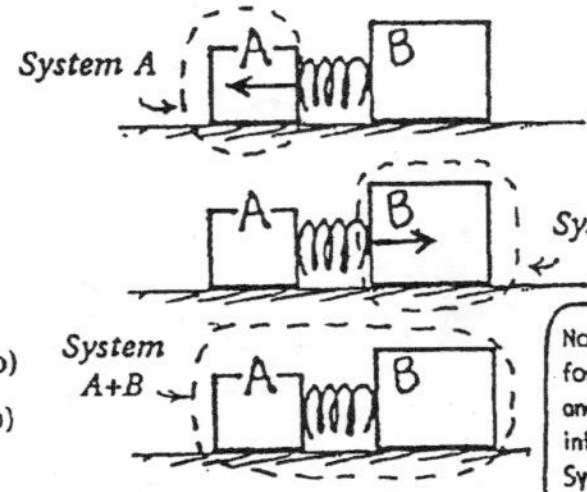

   a. Does an external force act on System A? (yes) (no)
   Will the momentum of System A change? (yes) (no)

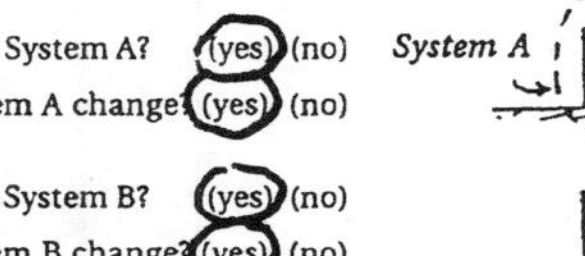

   b. Does an external force act on System B? (yes) (no)
   Will the momentum of System B change? (yes) (no)

   System B

   c. Does an external force act on System A+B? (yes) (no)
   Will the momentum of System A+B change? (yes) (no)

   System A+B

   Note that external forces on System A and System B are internal to System A+B, so they cancel!

2. Billiard ball A collides with billiard ball B at rest. Isolate each system with a closed dashed line. Draw only the external force vectors that act on each system.

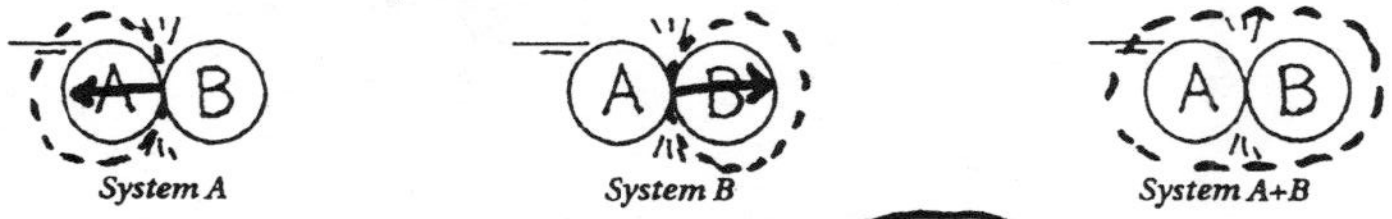

   System A — System B — System A+B

   a. Upon collision, the momentum of System A (increases) (decreases) (remains unchanged).
   b. Upon collision, the momentum of System B (increases) (decreases) (remains unchanged).
   c. Upon collision, the momentum of System A+B (increases) (decreases) (remains unchanged).

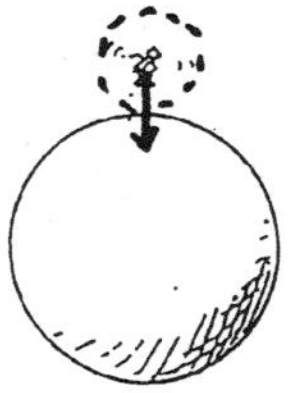

3. A girl jumps upward. In the sketch to the left, draw a closed dashed line to indicate the system of the girl.

   a. Is there an external force acting on her? (yes) (no)
   Does her momentum change? (yes) (no)
   Is the girl's momentum conserved? (yes) (no)

   b. In the sketch to the right, draw a closed dashed line to indicate the system [girl + earth]. Is there an external force due to the interaction between the girl and the earth that acts on the system? (yes) (no)
   Is the momentum of the system conserved? (yes) (no)

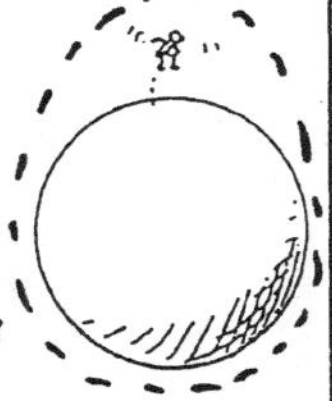

4. A block strikes a blob of jelly. Isolate 3 systems with a closed dashed line and show the external force on each. In which system is momentum conserved?

   ONE ON RIGHT

5. A truck crashes into a wall. Isolate 3 systems with a closed dashed line and show the the external force on each. In which system is momentum conserved?

   ONE ON RIGHT

   MOMENTUM TRANSFER

thanx to Cedric Linder

Hewitt Drew it!

Name ____________________ Date ________

CONCEPTUAL **Physics** PRACTICE PAGE

Chapter 6 Momentum
Momentum Conservation

In the tables below, fill in the numerical values for total momentum before and after collisions of the two-body systems. Show how you calculate the momentum before. Also fill in the velocity blanks.

1. Bumper cars are fun. Assume each car with its occupant has a mass of 200 kg.

Before	Momentum BEFORE	Momentum AFTER	After
$v = 2$ m/s, $v = 0$	$(200)2 + 0 =$ 400	$0 + (200)2$ 400	$v = 0$, $v =$ 2 m/s
$v = 2$ m/s, $v = -1$ m/s	$(200)2 + 200(-1)$ 200	$200(-1) + (200)2$ 200	$v = -1$ m/s, $v =$ 2 m/s
$v = 2$ m/s, $v = -1$ m/s	200	$(400)\frac{1}{2} =$ 200	$v = \frac{1}{2}$ m/s

Sticky goo!

This time they stick!

2. Granny whizzes around the rink and is suddenly confronted with Ambrose at rest directly in her path. Rather than knock him over, she picks him up and continues in motion without "braking."

DATA
Granny's mass; 50 kg
Granny's initial speed; 3 m/s
Ambrose's mass; 25 kg
Ambrose's initial speed; 0 m/s

Momentum BEFORE	Momentum AFTER
150	150

$v =$ 2 m/s

GRANNY'S MMT + 0 = GRANNY AND AMBROSE MMT

$50(3) + 0 = (50+25)v$

$150 = 75v$ $\quad v = 2$ m/s

Hewitt Drew it!

Name ____________________ Date ________

CONCEPTUAL Physics PRACTICE PAGE

Chapter 7 Energy
Work and Energy

1. How much work (energy) is needed to lift an object that weighs 200 N to a height of 4 m?
 800 J

2. How much power is needed to lift the 200-N object to a height of 4 m in 4 s?
 200 W

3. What is the power output of an engine that does 60 000 J of work in 10 s?
 6 kW

4. The block of ice weighs 500 newtons.

 a. How much force is needed to push it up the incline (neglect friction)?
 250 N

 b. How much work is required to push it up the incline compared with lifting the block vertically 3 m?
 SAME (250×6 = 500×3)

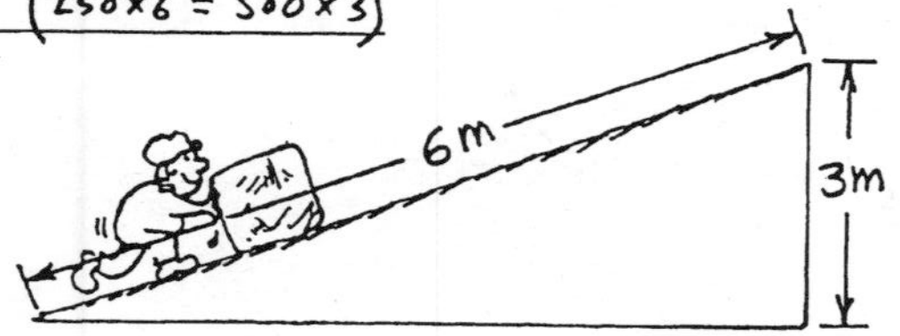
6m
3m

5. All the ramps are 5 m high. We know that the KE of the block at the bottom of the ramp will be equal to the loss of PE (conservation of energy). Find the speed of the block at ground level in each case. [Hint: Do you recall from earlier chapters how long it takes something to fall a vertical distance of 5 m from a positon of rest (assume $g = 10\ m/s^2$)? And how much speed a falling object acquires in this time? This gives you the answer to Case 1. Discuss with your classmates how energy conservation gives you the answers to Cases 2 and 3.]

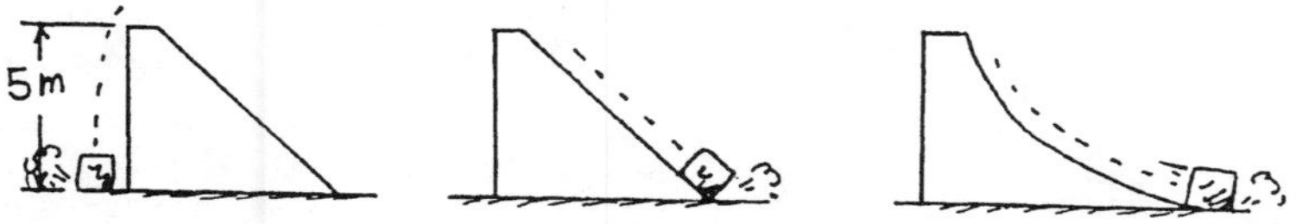
5m

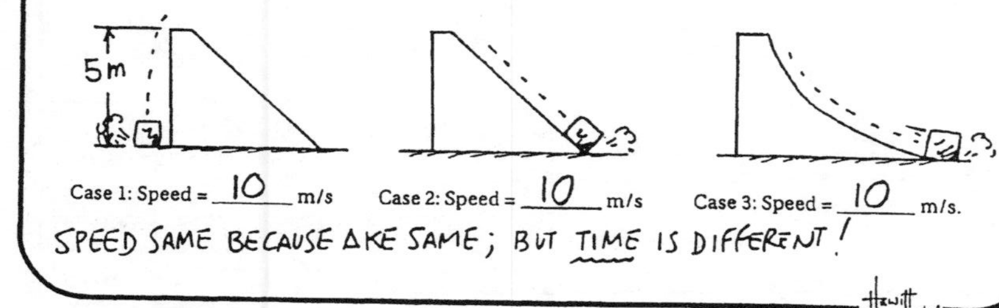

Case 1: Speed = 10 m/s   Case 2: Speed = 10 m/s   Case 3: Speed = 10 m/s.

SPEED SAME BECAUSE ΔKE SAME; BUT TIME IS DIFFERENT!

CONCEPTUAL Physics PRACTICE PAGE

6. Which block gets to the bottom of the incline first? Assume no friction. (Be careful!) Explain your answer.
 BLOCK A BECAUSE GREATER ACCELERATION AND LESS RAMP DISTANCE. SO A HAS SHORTER SLIDING TIME --- BUT SAME SPEED

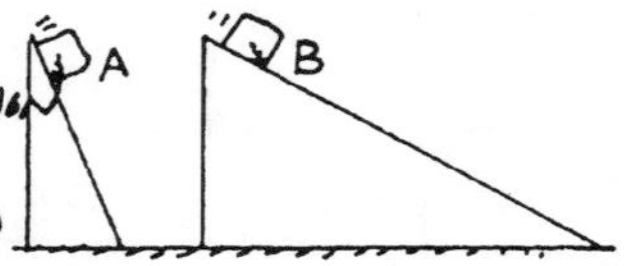
A
B

7. The KE and PE of a block freely sliding down a ramp are shown in only one place in the sketch. Fill in the missing values.

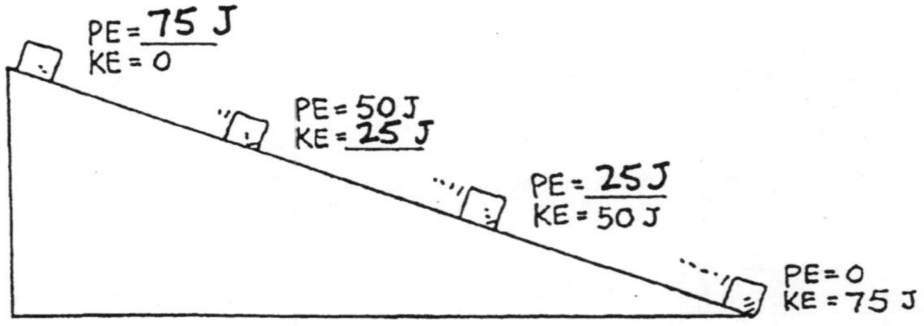
PE = 75 J
KE = 0
PE = 50 J
KE = 25 J
PE = 25 J
KE = 50 J
PE = 0
KE = 75 J

8. A big metal bead slides due to gravity along an upright friction-free wire. It starts from rest at the top of the wire as shown in the sketch. How fast is it traveling as it passes

 Point B? 10 m/s

 Point D? 10 m/s

 Point E? 10 m/s

 At what point does it have the maximum speed? C

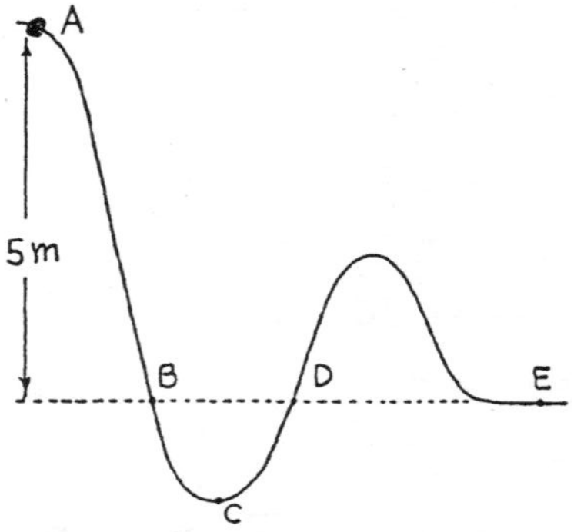
A
5m
B
C
D
E

THINK ENERGY CONSERVATION!

9. Rows of wind-powered generators are used in various windy locations to generate electric power. Does the power generated affect the speed of the wind? Would locations behind the 'windmills' be windier if they weren't there? Discuss this in terms of energy conservation with your classmates.
 YES, BY CONS OF ENERGY, ENERGY GAINED BY WINDMILLS IS TAKEN FROM WIND KE, SO WIND MUST SLOW DOWN. LOCATIONS BEHIND WOULD BE A BIT WINDIER WITHOUT THE WINDMILLS!

Name ______________________ Date __________

CONCEPTUAL Physics PRACTICE PAGE

Chapter 7 Energy
Conservation of Energy

1. Fill in the blanks for the six systems shown.

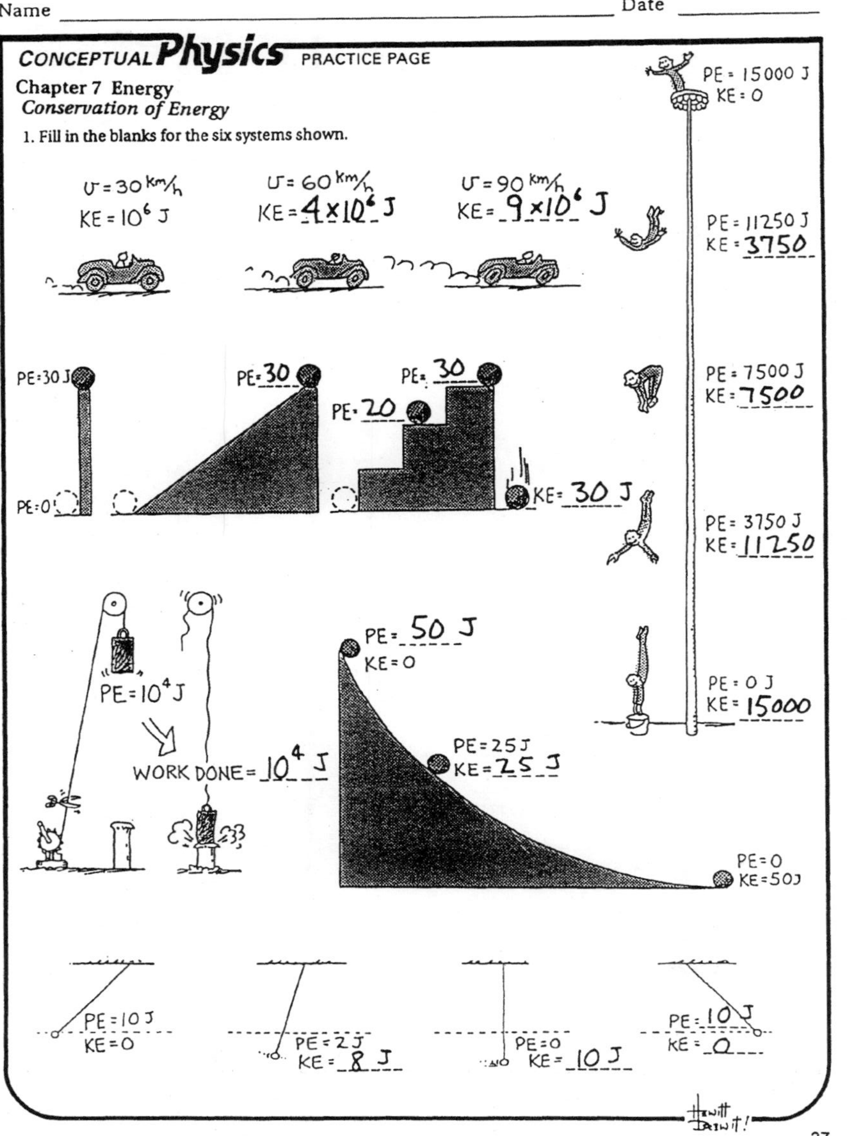

CONCEPTUAL Physics PRACTICE PAGE

2. The woman supports a 100-N load with the friction-free pulley systems shown below. Fill in the spring-scale readings that show how much force she must exert.

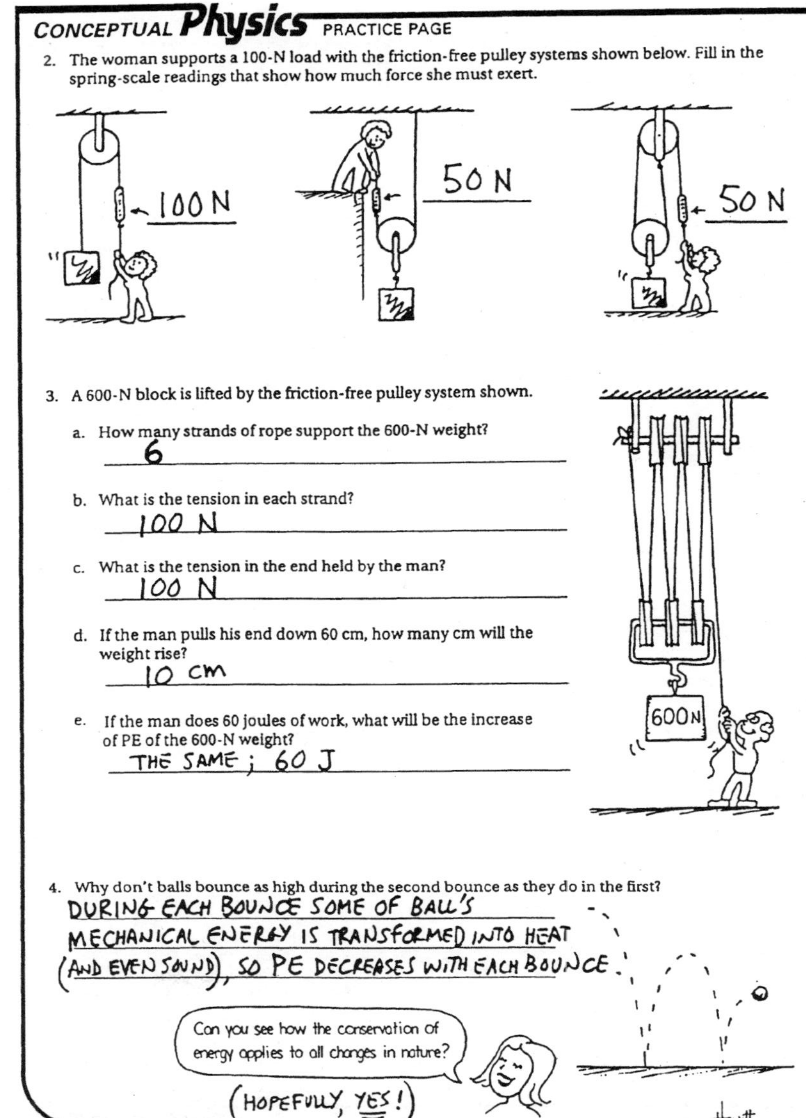

3. A 600-N block is lifted by the friction-free pulley system shown.

a. How many strands of rope support the 600-N weight?
6

b. What is the tension in each strand?
100 N

c. What is the tension in the end held by the man?
100 N

d. If the man pulls his end down 60 cm, how many cm will the weight rise?
10 cm

e. If the man does 60 joules of work, what will be the increase of PE of the 600-N weight?
THE SAME ; 60 J

4. Why don't balls bounce as high during the second bounce as they do in the first?
DURING EACH BOUNCE SOME OF BALL'S MECHANICAL ENERGY IS TRANSFORMED INTO HEAT (AND EVEN SOUND), SO PE DECREASES WITH EACH BOUNCE.

Can you see how the conservation of energy applies to all changes in nature?

(HOPEFULLY, YES!)

Name ______________________ Date ____________

CONCEPTUAL Physics PRACTICE PAGE

Chapter 7 Energy
Momentum and Energy

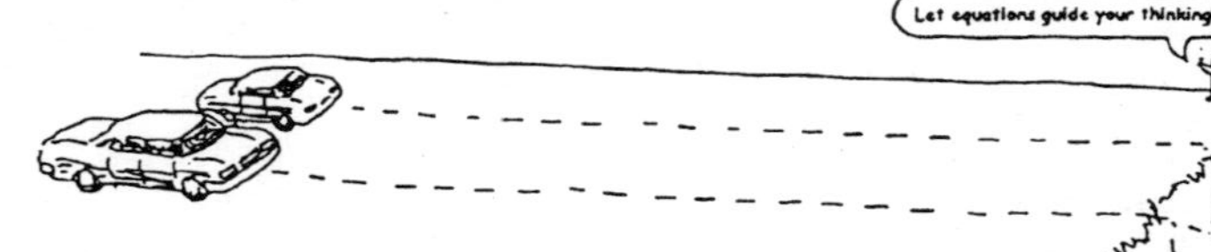

Bronco Brown wants to put $Ft = \Delta mv$ to the test and try bungee jumping. Bronco leaps from a high cliff and experiences free fall for 3 seconds. Then the bungee cord begins to stretch, reducing his speed to zero in 2 seconds. Fortunately, the cord stretches to its maximum length just short of the ground below.

Fill in the blanks. Bronco's mass is 100 kg. Acceleration of free fall is 10 m/s².

Express values in SI units (*distance* in m, *velocity* in m/s, *momentum* in kg-m/s, *impulse* in N-s, and *deceleration* in m/s²).

$t = 0$ s $v =$ 0 *momentum* = 0

$t = 1$ s $v =$ 10 m/s *momentum* = 1000 kg m/s

$t = 2$ s $v =$ 20 m/s *momentum* = 2000 kg m/s

$t = 3$ s $v =$ 30 m/s *momentum* = 3000 kg m/s

$t = 5$ s $v =$ 0 *momentum* = 0

The 3-s free-fall distance of Bronco just before the bungee cord begins to stretch = 45 m

Δmv during the 3-s interval of free fall = 3000 kg m/s

Δmv during the 2-s interval of slowing down = 3000 kg m/s

Impulse during the 2-s interval of slowing down = 3000 N·s

Average force exerted by the cord during the 2-s interval of slowing down = 1500 N

How about *work* and *energy*? How much KE does Bonco have 3 s after his jump? 45 000 J

How much does gravitational PE decrease during this 3 s? 45 000 J

What two kinds of PE are changing during the slowing-down interval? GRAVITATIONAL AND ELASTIC

Hewitt Drew it!

CONCEPTUAL Physics PRACTICE PAGE

Energy and Momentum

A Honda Civic and a Lincoln Town Car are initially at rest on a horizontal parking lot at the edge of a steep cliff. For simplicity, we assume that the Town Car has twice as much mass as the Civic. Equal constant forces are applied to each car and they accelerate across equal distances (we ignore the effects of friction). When they reach the far end of the lot the force is suddenly removed, whereupon they sail through the air and crash to the ground below. (The cars are beat up to begin with, and this is a scientific experiment!)

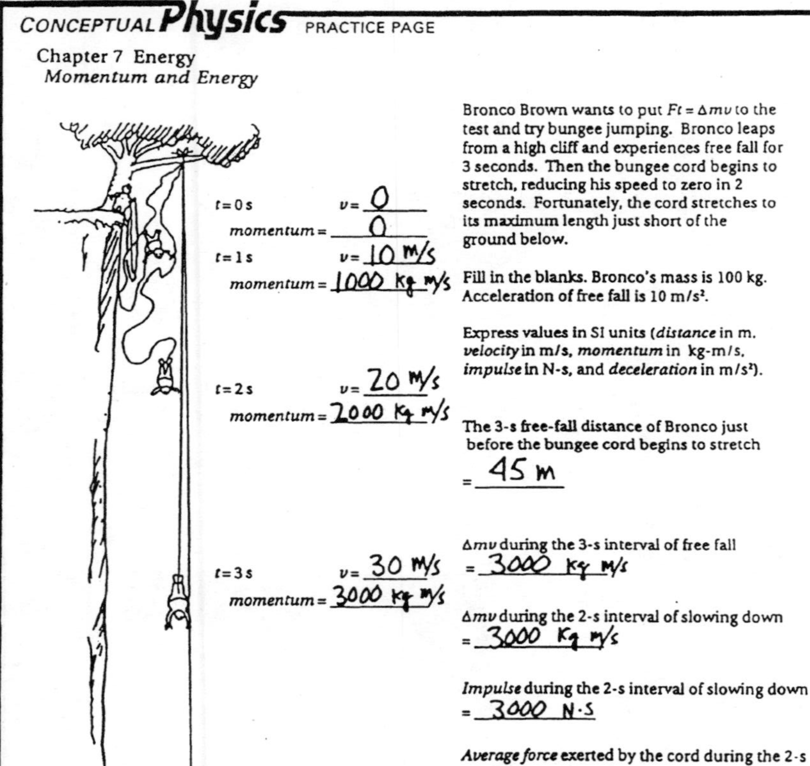

Let equations guide your thinking!
Impulse = Δmomentum
Ft = Δmv
Work = Fd = ΔKE = Δ½mv²
Making the distinction between momentum and kinetic energy is high-level physics.

1. Which car has the greater acceleration ? (Think $a = F/m$)
CIVIC (LESS MASS ACTED ON BY THE SAME FORCE)

2. Which car spends more time along the surface of the lot? (The faster or slower one?)
TOWN CAR (SLOWER DUE TO LESS ACCELERATION)

3. Which car has the larger impulse imparted to it by the applied force? (Think Impulse = Ft)
Defend your answer.
TOWN CAR. SAME FORCE IS APPLIED OVER A LONGER TIME.

4. Which car has the greater momentum at the cliff's edge? (Think $Ft = \Delta mv$) Defend your answer.
TOWN CAR. MORE IMPULSE PRODUCES MORE MOMENTUM CHANGE

5. Which car has the greater work done on it by the applied force? (Think $W = Fd$) Defend your answer in terms of the distance traveled.
SAME ON EACH. FORCE x DISTANCE IS SAME FOR EACH.

6. Which car has the greater kinetic energy at the edge of the cliff? (Think $W = \Delta KE$)
Does your answer follow from your explanation of 5? YES
Does it contradict your answer to 3? Why or why not? NO
SAME, BECAUSE OF SAME WORK. NO CONTRADICTION BECAUSE GREATER IMPULSE DOESN'T MEAN GREATER WORK

7. Which car spends more time in the air, from the edge of the cliff to the ground below?
BOTH THE SAME! (LOOK AHEAD TO CHAP 10 ON INDEPENDENCE OF HORIZ AND VERTICAL MOTION)

8. Which car lands farthest horizontally from the edge of the cliff onto the ground below?
THE CIVIC, WHICH IS MOVING FASTER.

Challenge: Suppose the slower car crashes a horizontal distance of 10 m from the ledge. Then at what horizontal distance does the faster car hit?
14.1 m (CIVIC MOVES $\sqrt{2}$ TIMES FASTER DUE TO EQUAL KE AT CLIFF EDGE
$\frac{1}{2}(2m)v^2 = \frac{1}{2}mV^2$, WHERE $V = \sqrt{2}v$. SO $\sqrt{2}$ FASTER MEANS $\sqrt{2}$ FARTHER IN SAME TIME!

Hewitt Drew it!

Name ______________________ Date __________

CONCEPTUAL **Physics** PRACTICE PAGE

Chapter 8 Rotational Motion
Torques

1. Apply what you know about torques by making a mobile. Shown below are five horizontal arms with fixed 1- and 2-kg masses attached, and four hangers with ends that fit in the loops of the arms, lettered A through R. You are to figure where the loops should be attached so that when the whole system is suspended from the spring scale at the top, it will hang as a proper mobile, with its arms suspended horizontally. This is best done by working from the bottom upward. Circle the loops where the hangers should be attached. When the mobile is complete, how many kilograms will be indicated on the scale? (Assume the horizontal struts and connecting hooks are practically massless compared to the 1- and 2-kg masses.) On a separate sheet of paper, make a sketch of your completed mobile.

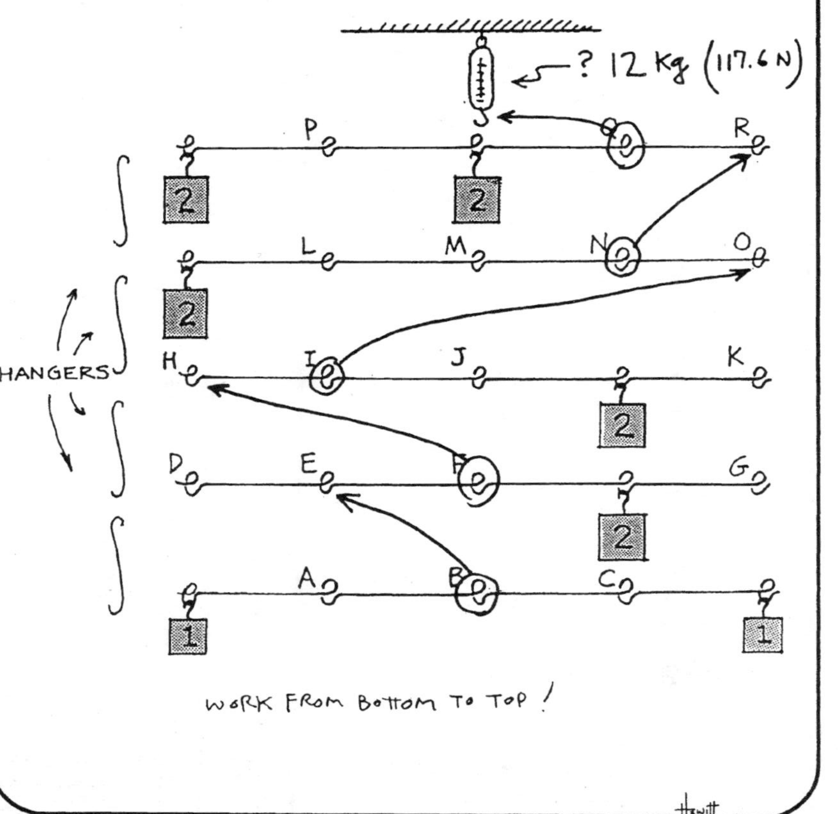

CONCEPTUAL **Physics** PRACTICE PAGE

2. Complete the data for the three seesaws in equilibrium.

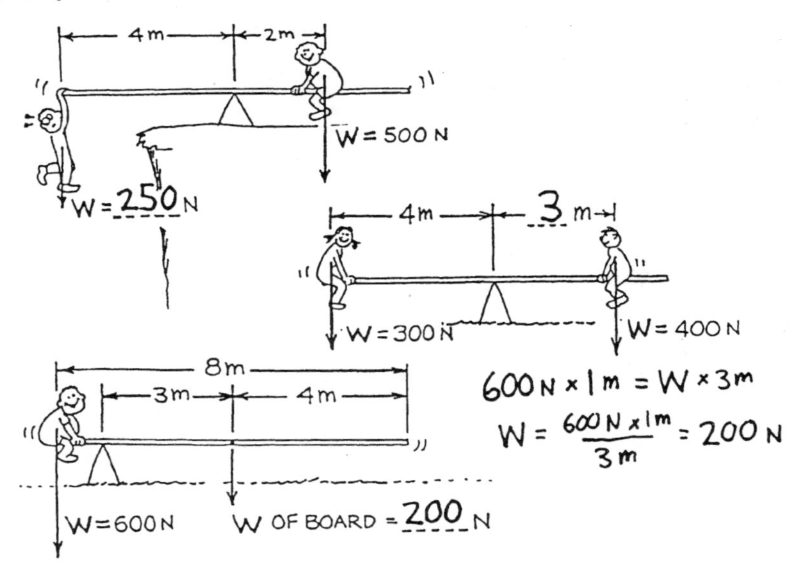

$600\,N \times 1\,m = W \times 3\,m$

$W = \frac{600\,N \times 1\,m}{3\,m} = 200\,N$

3. The broom balances at its CG. If you cut the broom in half at the CG and weigh each part of the broom, which end would weigh more?

PIECE WITH BRUSH WEIGHS MORE

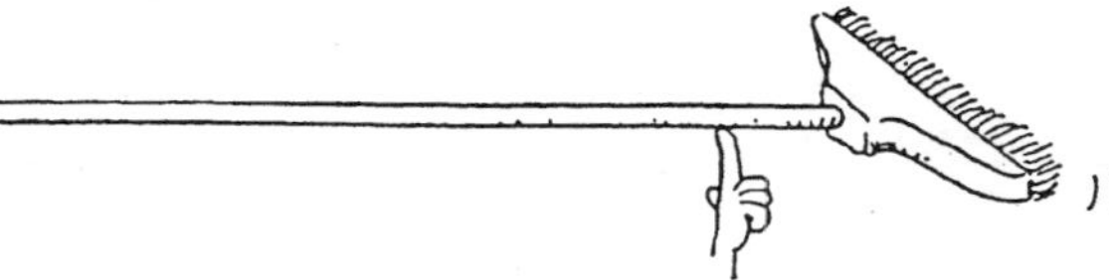

Explain why each end has or does not have the same weight? (Hint: Compare this to one of the seesaw systems above.)

WEIGHT ON EITHER SIDE ISN'T SAME, BUT TORQUE IS! LIKE SEESAWS ABOVE, SHORTER LEVER ARM HAS MORE WEIGHT

Name _______________ Date _______

CONCEPTUAL **Physics** PRACTICE PAGE

Chapter 8 Rotational Motion
Torques and Rotation

1. Pull the string gently and the spool rolls. The direction of roll depends on the way the torque is applied.

   In (1) and (2) below, the force and lever arm are shown for the torque about the point where surface contact is made (shown by the triangular "fulcrum"). The lever arm is the heavy dashed line, which is different for each different pulling position.

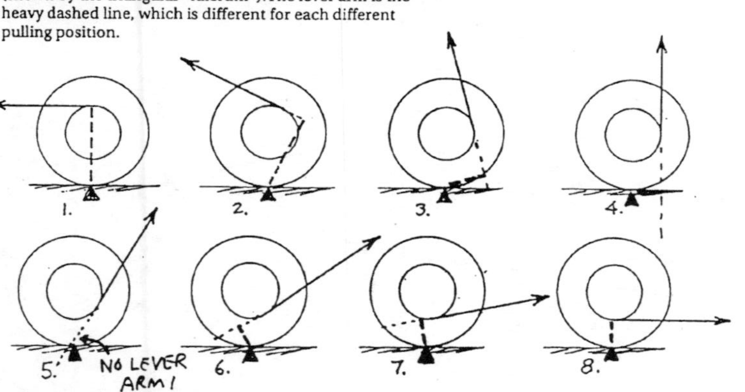

a. Construct the lever arm for the other positions.

b. Lever arm is longer when the string is on the **(top)** (bottom) of the spool spindle.

c. For a given pull, the torque is greater when the string is on the **(top)** (bottom).

d. For the same pull, rotational acceleration is greater when the string is on **(top)** (bottom) (makes no difference).

e. At which positions does the spool roll to the left? 1, 2, 3, 4

f. At which positions does the spool roll to the right? 6, 7, 8

g. At which position does the spool not roll at all? 5

h. Why does the spool slide rather than roll at this position?
LINE OF ACTION EXTENDS TO FULCRUM; NO LEVER ARM, NO TORQUE

Be sure your right angle is between the force's *line of action* and the lever arm.

2. We all know that a ball rolls down an incline. But relatively few people know that the reason the ball picks up rotational speed is because of a torque. In Sketch A, we see the ingredients of the torque acting on the ball—the force due to gravity and the lever arm to the point where surface contact is made.

   a. Constuct the lever arms for positions B and C.

   b. As the incline becomes steeper, the torque **(increases)** (decreases)

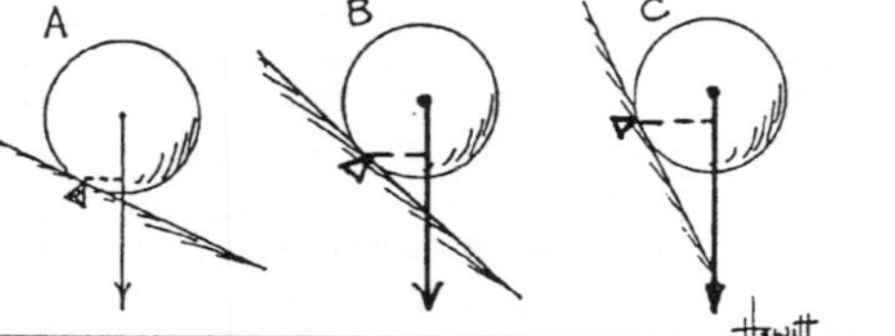

Hewitt Drew it!

CONCEPTUAL **Physics** PRACTICE PAGE

Acceleration and Circular Motion

Newton's 2nd law, $a = F/m$, tells us that net force and its corresponding acceleration are always in the same direction. (Both force and acceleration are vector quantities). But force and acceleration are not always in the direction of velocity (another vector).

1. You're in a car at a traffic light. The light turns green and the driver "steps on the gas."
   a. Your body lurches (forward) (not at all) **(backward)**.
   b. The car accelerates **(forward)** (not at all) (backward).
   c. The force on the car acts **(forward)** (not at all) (backward).

   The sketch shows the top view of the car. Note the directions of the velocity and acceleration vectors.

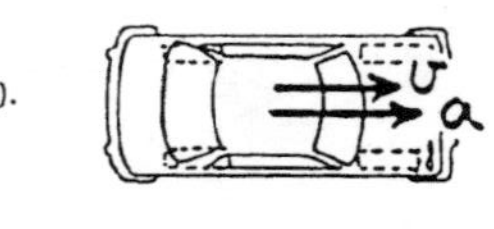

2. You're driving along and approach a stop sign. The driver steps on the brakes.
   a. Your body lurches **(forward)** (not at all) (backward).
   b. The car accelerates (forward) (not at all) **(backward)**.
   c. The force on the car acts (forward) (not at all) **(backward)**.

   The sketch shows the top view of the car. Draw vectors for velocity and acceleration.

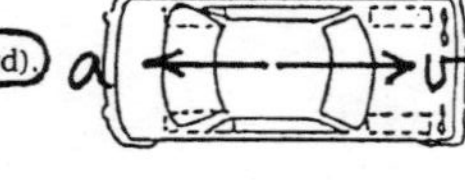

3. You continue driving, and round a sharp curve to the left at constant speed.
   a. Your body leans (inward) (not at all) **(outward)**.
   b. The direction of the car's acceleration is **(inward)** (not at all) (outward).
   c. The force on the car acts **(inward)** (not at all) (outward).

   Draw vectors for velocity and acceleration of the car.

4. In general, the directions of lurch and acceleration, and therefore the directions of lurch and force, are (the same) (not related) **(opposite)**.

5. The whirling stone's direction of motion keeps changing.
   a. If it moves faster, its direction changes **(faster)** (slower).
   b. This indicates that as speed increases, acceleration **(increases)** (decreases) (stays the same).

6. Consider whirling the stone on a shorter string—that is, of smaller radius.
   a. For a given speed, the rate that the stone changes direction is (less) **(more)** (the same).
   b. This indicates that as the radius decreases, acceleration **(increases)** (decreases) (stays the same).

thanx to Jim Harper

Hewitt Drew it!

Name ________________________ Date ____________

CONCEPTUAL **Physics** PRACTICE PAGE

Chapter 8 Rotational Motion
Simulated Gravity and Frames of Reference

Susie Spacewalker and Bob Biker are in outer space. Bob experiences earth-normal gravity in a rotating habitat, where centripetal force on his feet provides a normal support force that feels like weight. Susie hovers outside in a weightless condition, motionless relative to the stars and the center of mass of the habitat.

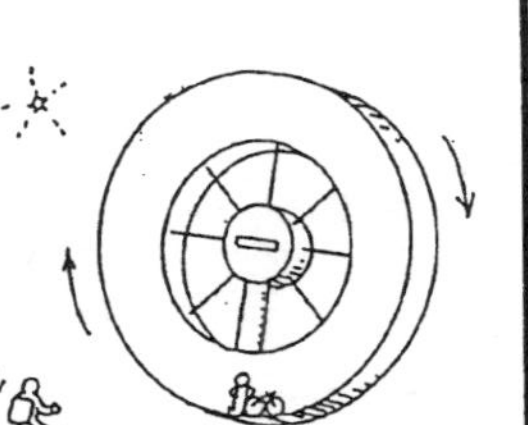

1. Susie sees Bob rotating clockwise in a circular path at a linear speed of 30 km/h. Suzie and Bob are facing each other, and from Bob's point of view, he is at rest and he sees Suzie moving

(clockwise) (counter clockwise).

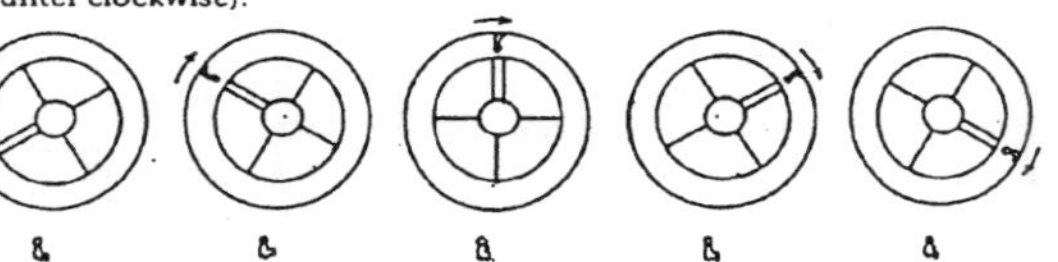

Bob at rest on the floor
Suzie hovering in space

2. The rotating habitat seems like home to Bob—until he rides his bicycle. When he rides in the opposite direction as the habitat rotates, Suzie sees him moving (faster) (slower).

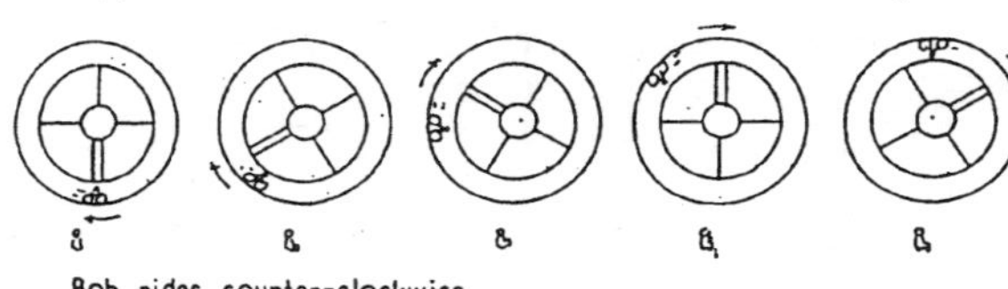

Bob rides counter-clockwise

3. As Bob's bicycle speedometer reading increases, his rotational speed

(decreases) (remains unchanged) (increases) and the normal force that feels like weight

(decreases) (remains unchanged) (increases). So friction between the tires and the floor

(decreases) (remains unchanged) (increases).

4. When Bob nevertheless gets his speed up to 30 km/h, as read on his bicycle speedometer, Suzie sees him

(moving at 30 km/h) (motionless) (moving at 60 km/h)

thanx to Bob Becker

CONCEPTUAL **Physics** PRACTICE PAGE

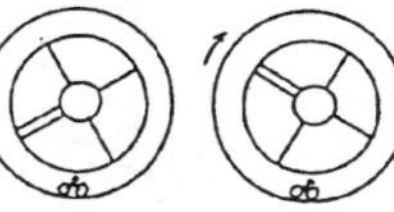
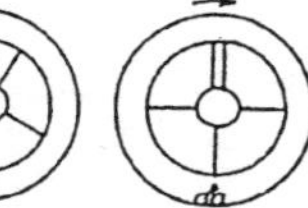
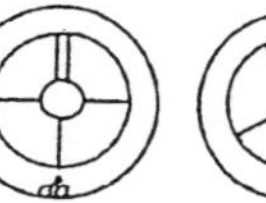
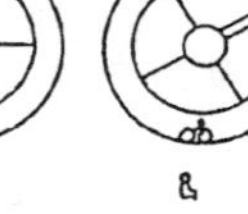
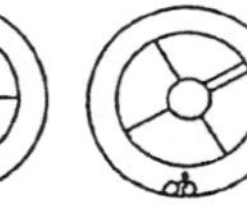
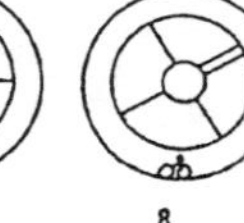
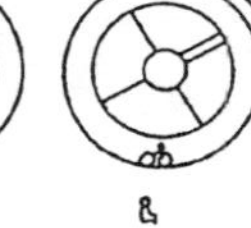

Bob rides at 30 km/h with respect to the floor

5. Bounding off the floor a bit while riding at 30 km/h, and neglecting wind effects, Bob

(drifts toward the ceiling in midspace as the floor whizzes by him at 30 km/h)
(falls as he would on earth)
(slams onto the floor with increased force)

and he finds himself

(in the same frame of reference as Suzie)
(as if he rode at 30 km/h on the earth's surface)
(pressed harder against the bicycle seat).

6. Bob manuevers back to his initial condition, whirling at rest with the habitat, standing beside his bicycle. But not for long. Urged by Suzie, he rides in the opposite direction, clockwise with the rotation of the habitat.

Now Suzie sees him moving (faster) (slower).

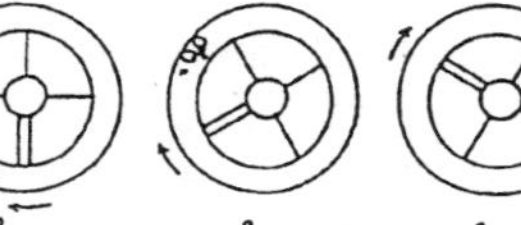

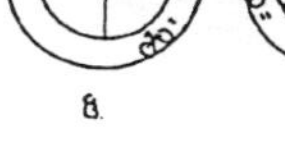

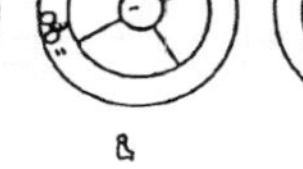

Bob rides clockwise

7. As Bob gains speed, the normal support force that feels like weight

(decreases) (remains unchanged) (increases).

8. When Bob's speedometer reading gets up to 30 km/h, Suzie sees him moving

(30 km/h) (not at all) (60 km/h)

and Bob finds himself

(weightless like Suzie)
(just as if he rode at 30 km/h on the earth's surface)
(pressed harder against the bicycle seat).

Next, Bob goes bowling. You decide whether the game depends on which direction the ball is rolled!

Name ______________________ Date ____________

CONCEPTUAL **Physics** PRACTICE PAGE

Chapter 9 Gravity
Inverse-Square Law

1. Paint spray travels radially away from the nozzle of the can in straight lines. Like gravity, the strength (intensity) of the spray obeys an inverse-square law. Complete the diagram by filling in the blank spaces.

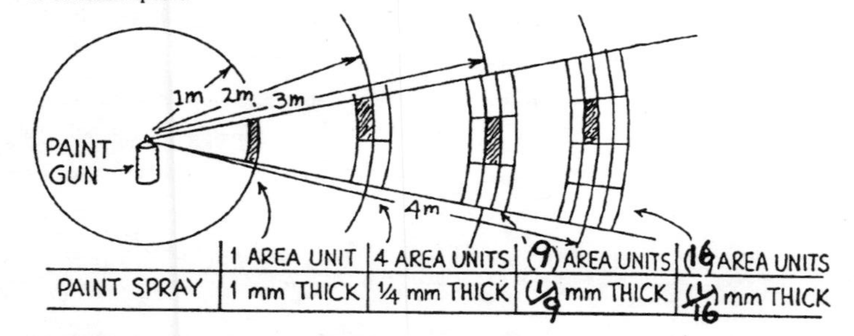

	1 AREA UNIT	4 AREA UNITS	(9) AREA UNITS	(16) AREA UNITS
PAINT SPRAY	1 mm THICK	¼ mm THICK	(1/9) mm THICK	(1/16) mm THICK

2. A small light source located 1 m in front of an opening of area 1 m² illuminates a wall behind. If the wall is 1 m behind the opening (2 m from the light source), the illuminated area covers 4 m². How many square meters will be illuminated if the wall is

5 m from the source? 25 m²

10 m from the source? 100 m²

1 m² OPENING
LIGHT SOURCE
1m
2m
4 m² OF ILLUMINATION

3. If we stand on a weighing scale and find that we are pulled toward the earth with a force of 500 N, then we weigh 500 N. Strictly speaking, we weigh 500 N relative to the earth. How much does the earth weigh? If we tip the scale upside down and repeat the weighing process, we can say that we and the earth are still pulled together with a force of 500 N, and therefore, relative to us, the whole 6 000 000 000 000 000 000 000 000-kg earth weighs 500 N! Weight, unlike mass, is a relative quantity.

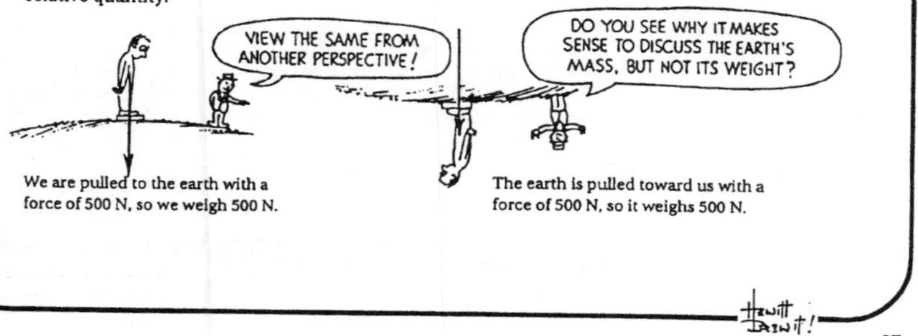

VIEW THE SAME FROM ANOTHER PERSPECTIVE!
DO YOU SEE WHY IT MAKES SENSE TO DISCUSS THE EARTH'S MASS, BUT NOT ITS WEIGHT?

We are pulled to the earth with a force of 500 N, so we weigh 500 N.

The earth is pulled toward us with a force of 500 N, so it weighs 500 N.

CONCEPTUAL **Physics** PRACTICE PAGE

4. The spaceship is attracted to both the planet and the planet's moon. The planet has four times the mass of its moon. The force of attraction of the spaceship to the planet is shown by the vector.

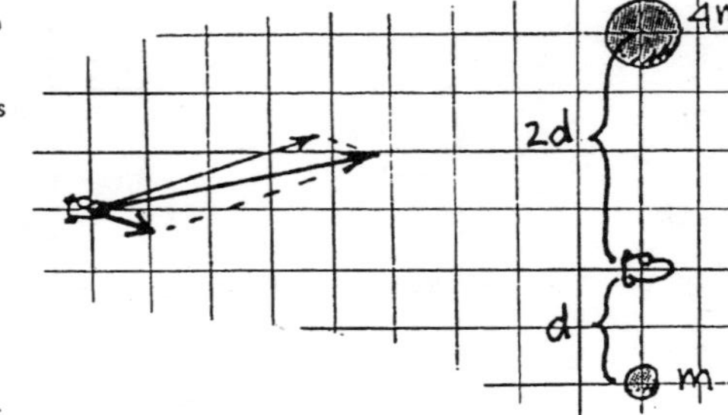

a. Carefully sketch another vector to show the spaceship's attraction to the moon. Then use the parallelogram method of Chapter 3 and sketch the resultant force.

b. Determine the location between the planet and its moon (along the dotted line) where gravitational forces cancel. Make a sketch of the spaceship there.

5. Consider a planet of uniform density that has a straight tunnel from the north pole through the center to the south pole. At the surface of the planet, an object weighs 1 ton.

a. Fill in the gravitational force on the object when it is half way to the center, then at the center.

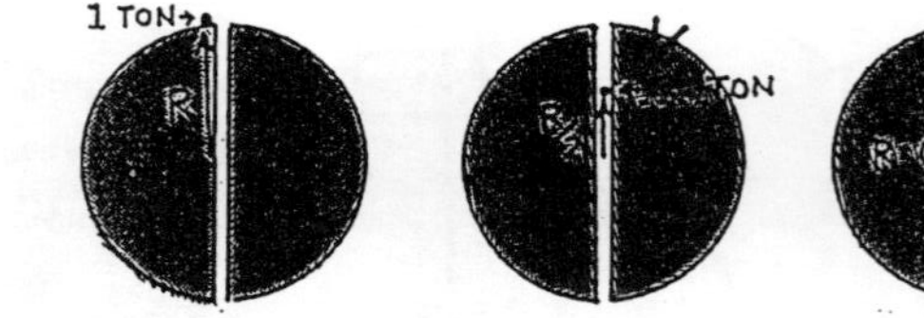

b. Describe the motion you would experience if you fell into the tunnel.
TO AND FRO (IN SIMPLE HARMONIC MOTION)

6. Consider an object that weighs 1 ton at the surface of a planet, just before the planet gravitationally collapses.

a. Fill in the weights of the object on the planet's shrinking surface at the radial values shown.

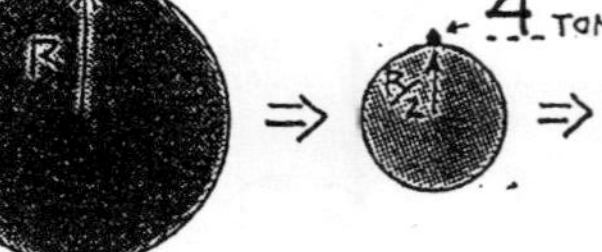

b. When the planet has collapsed to 1/10 of its initial radius, a ladder is erected that puts the object as far from its center as the object was originally. Fill in its weight at this position.

Name ______________________ Date __________

CONCEPTUAL Physics PRACTICE PAGE

Chapter 9 Gravity

Our Ocean Tides

1. Consider two equal-mass blobs of water, A and B, initially at rest in the moon's gravitational field. The vector shows the gravitational force of the moon on A.

Moon   A   B

a. Draw a force vector on B due to the moon's gravity.

b. Is the force on B more or less than the force on A? LESS

c. Why? FARTHER AWAY

d. The blobs accelerate toward the moon. Which has the greater acceleration? (A) (B)

e. Because of the different accelerations, with time (A gets farther ahead of B) (A and B gain identical speeds) and the distance between A and B (increases) (stays the same) (decreases).

f. If A and B were connected by a rubber band, with time the rubber band would (stretch) (not stretch).

g. This (stretching) (non-stretching) is due to the (difference) (non-difference) in the moon's gravitational pulls.

h. The two blobs will eventually crash into the moon. To orbit around the moon instead of crashing into it, the blobs should move (away from the moon) (tangentially). Then their accelerations will consist of changes in (speed) (direction).

2. Now consider the same two blobs located on opposite sides of the earth.

Moon   A Earth B

a. Because of differences in the moon's pull on the blobs, they tend to (spread away from each other) (approach each other).

b. Does this spreading produce ocean tides? (Yes) (No)

c. If earth and moon were closer, gravitational force between them would be (more) (the same) (less), and the difference in gravitational forces on the near and far parts of the ocean would be (more) (the same) (less).

d. Because the earth's orbit about the sun is slightly elliptical, earth and sun are closer in December than in June. Taking the sun's tidal force into account, on a world average, ocean tides are greater in (December) (June) (no difference).

Hewitt Drew it!

Name ______________________ Date __________

CONCEPTUAL Physics PRACTICE PAGE

Chapter 10 Projectile and Satellite Motion

Independence of Horizontal and Vertical Components of Motion

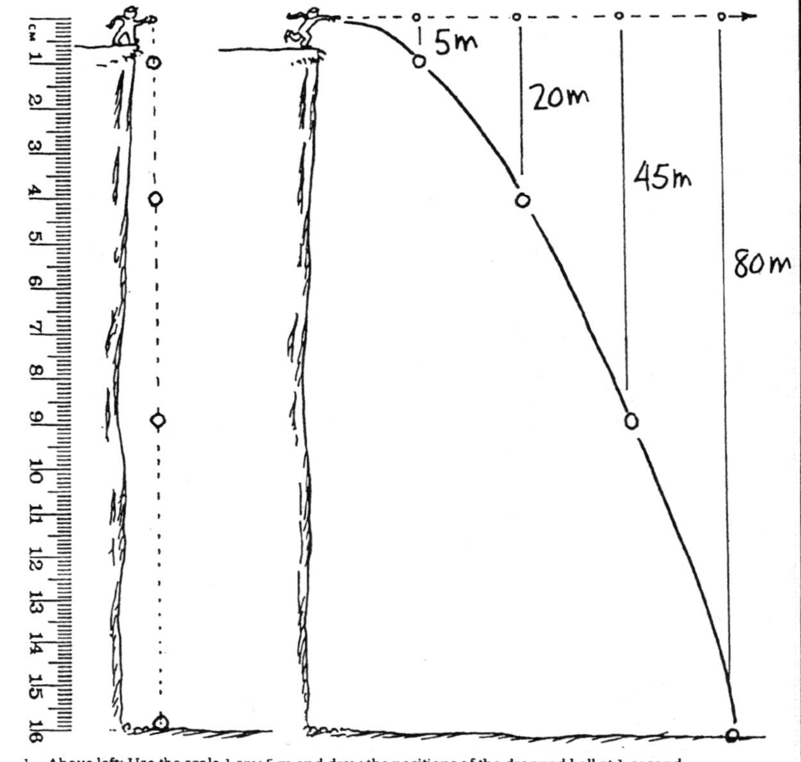

1. Above left: Use the scale 1 cm: 5 m and draw the positions of the dropped ball at 1-second intervals. Neglect air drag and assume $g = 10\ m/s^2$. Estimate the number of seconds the ball is in the air.

   4 seconds.

2. Above right: The four positions of the thrown ball with *no gravity* are at 1-second intervals. At 1 cm: 5 m, carefully draw the positions of the ball *with* gravity. Neglect air drag and assume $g = 10\ m/s^2$. Connect your positions with a smooth curve to show the path of the ball. How is the motion in the vertical direction affected by motion in the horizontal direction?

   ONLY VERT MOTION AFFECTED BY GRAVITY; HORIZ MOTION INDEPENDENT

Hewitt Drew it!

CONCEPTUAL **Physics** PRACTICE PAGE

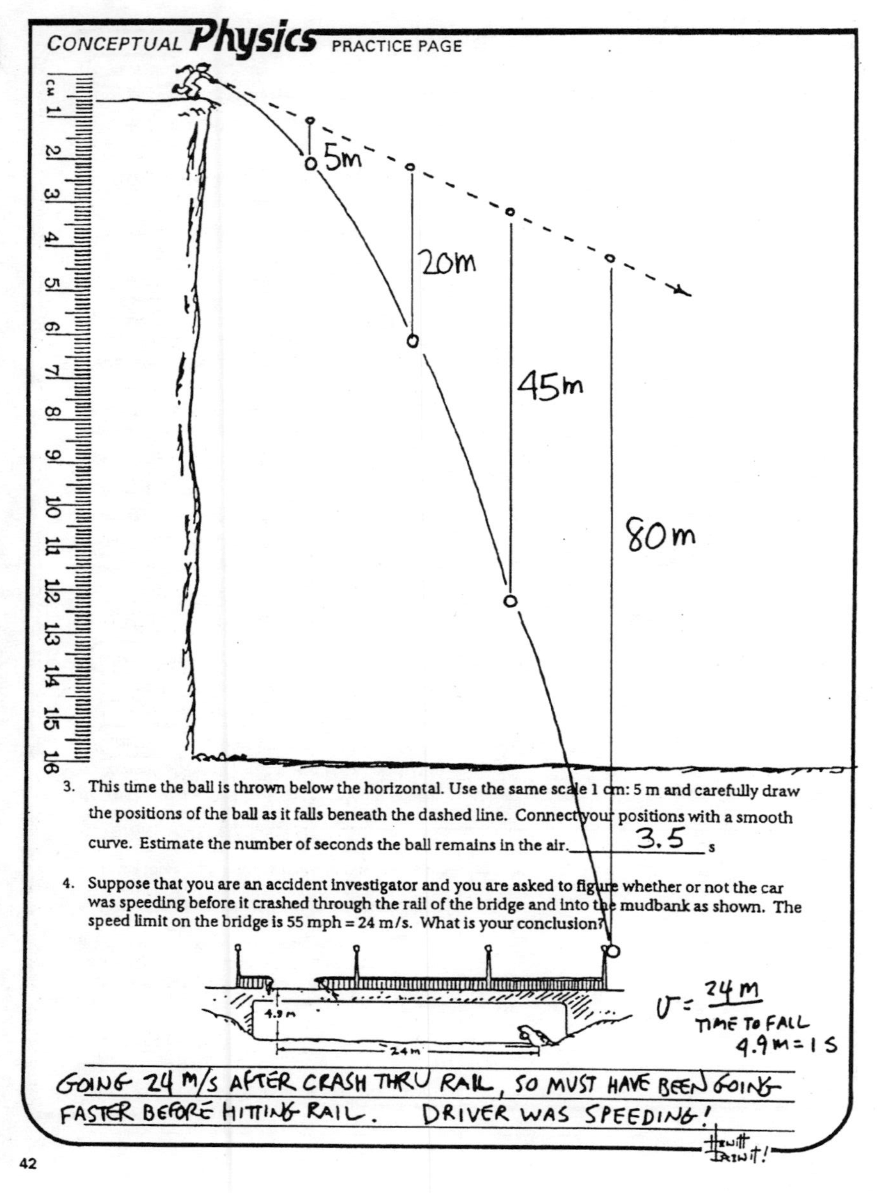

3. This time the ball is thrown below the horizontal. Use the same scale 1 cm: 5 m and carefully draw the positions of the ball as it falls beneath the dashed line. Connect your positions with a smooth curve. Estimate the number of seconds the ball remains in the air. 3.5 s

4. Suppose that you are an accident investigator and you are asked to figure whether or not the car was speeding before it crashed through the rail of the bridge and into the mudbank as shown. The speed limit on the bridge is 55 mph = 24 m/s. What is your conclusion?

$v = \frac{24\ m}{\text{TIME TO FALL } 4.9\ m = 1\ s}$

GOING 24 M/S AFTER CRASH THRU RAIL, SO MUST HAVE BEEN GOING FASTER BEFORE HITTING RAIL. DRIVER WAS SPEEDING!

Hewitt Drew it!

Name ______________________ Date __________

CONCEPTUAL **Physics** PRACTICE PAGE

Chapter 10 Projectile and Satellite Motion

Tossed Ball

A ball tossed upward has initial velocity components 30 m/s vertical, and 5 m/s horizontal. The position of the ball is shown at 1-second intervals. Air resistance is negligible, and $g = 10\ m/s^2$. Fill in the boxes, writing in the values of velocity ***components*** ascending, and your calculated ***resultant velocities*** descending.

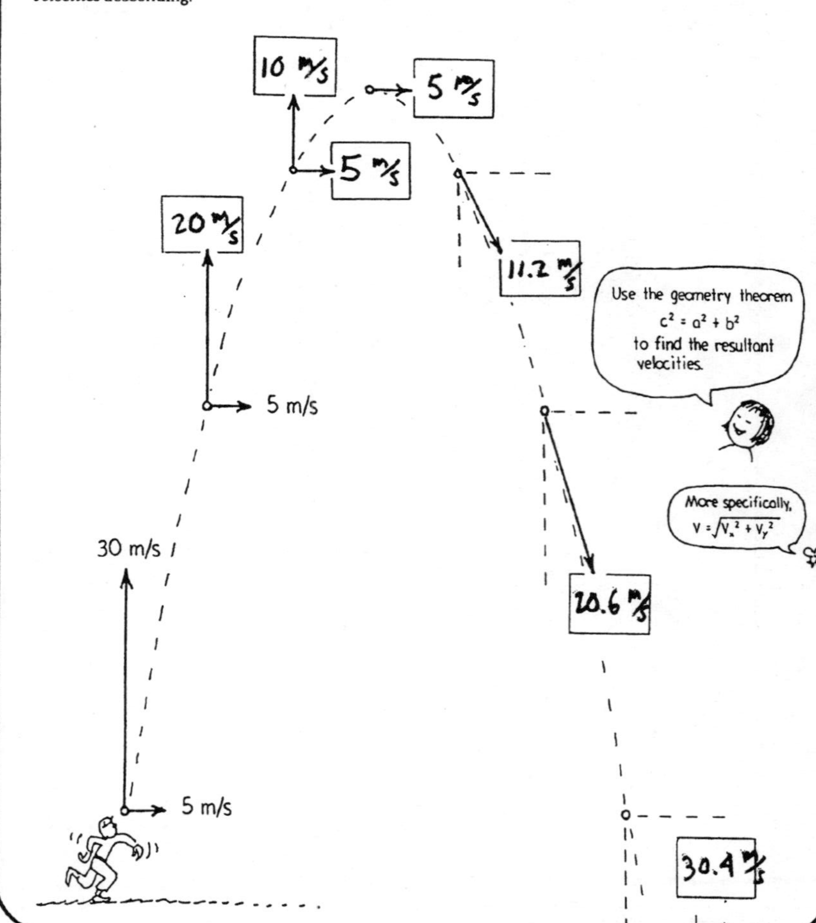

Hewitt Drew it!

Name ______________________ Date __________

CONCEPTUAL **Physics** PRACTICE PAGE

Chapter 10 Projectile and Satellite Motion
Satellite in Circular Orbit

1. Figure A shows "Newton's Mountain," so high that its top is above the drag of the atmosphere. The cannonball is fired and hits the ground as shown.

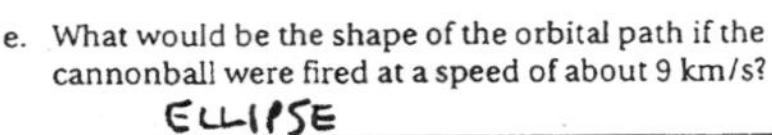

Figure A

   a. You draw the path the cannonball might take if it were fired a little bit faster.

   b. Repeat for a still greater speed, but still less than 8 km/s.

   c. Then draw the orbital path it would take if its speed were 8 km/s.

   d. What is the shape of the 8 km/s curve? CIRCLE

   e. What would be the shape of the orbital path if the cannonball were fired at a speed of about 9 km/s? ELLIPSE

2. Figure B shows a satellite in circular orbit.

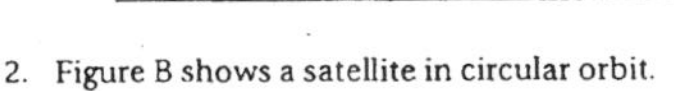

Figure B

   a. At each of the four positions draw a vector that represents the gravitational *force* exerted on the satellite.

   b. Label the force vectors *F*.

   c. Then draw at each position a vector to represent the *velocity* of the satellite at that position, and label it *V*.

   d. Are all four *F* vectors the same length? Why or why not? YES; SATELLITE IS AT SAME DISTANCE, SAME FORCE

   e. Are all four *V* vectors the same length? Why or why not? YES; IN CIRCULAR ORBIT F ⊥ V SO THERE'S NO COMPONENT OF FORCE ALONG V TO CHANGE SPEED V.

   f. What is the angle between your *F* and *V* vectors? 90°

   g. Is there any component of *F* along *V*? NO (F ⊥ V)

   h. What does this tell you about the work the force of gravity does on the satellite? NO WORK BECAUSE THERE'S NO COMP OF FORCE ALONG PATH

   i. Does the KE of the satellite in Figure B remain constant, or does it vary? CONSTANT

   j. Does the PE of the satellite remain constant, or does it vary? REMAINS CONSTANT

CONCEPTUAL **Physics** PRACTICE PAGE

Satellite in Elliptical Orbit

a. Repeat the procedure you used for the circular orbit, drawing vectors *F* and *V* for each position, including proper labeling. Show equal magnitudes with equal lengths, and greater magnitudes with greater lengths, but don't bother making the scale accurate.

Figure C

b. Are your vectors *F* all the same magnitude? Why or why not? NO, FORCE DECREASES WHEN DISTANCE FROM EARTH INCREASES

c. Are your vectors *V* all the same magnitude? Why or why not? NO. WHEN KE DECREASES (AS SATELLITE MOVES FARTHER FROM EARTH) SPEED DECREASES. WHEN KE INCREASES (CLOSER TO EARTH) SPEED INCREASES

d. Is the angle between vectors *F* and *V* everywhere the same, or does it vary? IT VARIES

e. Are there places where there is a component of *F* along *V*? YES (EVERYWHERE EXCEPT AT THE APOGEE AND PERIGEE)

f. Is work done on the satellite when there is a component of *F* along and in the same direction of *V* and if so, does this increase or decrease the KE of the satellite? YES. THIS INCREASES KE OF SATELLITE

g. When there is a component of *F* along and opposite to the direction of *V*, does this increase or decrease the KE of the satellite? THIS DECREASES KE OF SATELLITE

h. What can you say about the sum KE + PE along the orbit? CONSTANT (IN ACCORD WITH CONSERVATION OF ENERGY)

Be very very careful when placing both velocity and force vectors on the same diagram. Not a good practice, for one may construct the resultant of the vectors -- ouch!

Name ______________________ Date __________

CONCEPTUAL **Physics** PRACTICE PAGE

Mechanics Overview

1. The sketch shows the elliptical path described by a satellite about the earth. In which of the marked positions, A - D, (put S for "same everywhere") does the satellite experience the maximum
   a. gravitational force? A
   b. speed? A
   c. momentum? A
   d. kinetic energy? A
   e. gravitational potential energy? C
   f. total energy (KE + PE)? S
   g. acceleration? A ← $a = \frac{F}{m}$
   h. angular momentum? S

2. Answer the above questions for a satellite in circular orbit.
   a. S b. S c. S d. S e. S f. S g. S h. S

3. In which position(s) is there momentarily no work being done on the satellite by the force of gravity? Why?
   A AND C BECAUSE NO FORCE COMPONENTS IN DIRECTION OF MOTION

4. Work changes energy. Let the equation for work, $W = Fd$, guide your thinking on these: Defend your answers in terms of $W = Fd$.
   a. In which position will a several-minutes thrust of rocket engines pushing the satellite forward do the most work on the satellite and give it the greatest change in kinetic energy? (Hint: think about where the most distance will be traveled during the application of a several-minutes thrust?)
   A, WHERE FORCE ACTS OVER LONGEST DISTANCE
   b. In which position will a several-minutes thrust of rocket engines pushing the satellite forward do the least work on the satellite and give it the least boost in kinetic energy?
   C, WHERE FORCE ACTS OVER SHORTEST DISTANCE
   c. In which positon will a several-minutes thrust of a retro-rocket (pushing opposite to the satellite's direction of motion) do the most work on the satellite and change its kinetic energy the most?
   A, MOST "NEGATIVE WORK" + MOST KE OCCURS WHERE FORCE ACTS OVER THE LONGEST DISTANCE.

Name ______________________ Date __________

CONCEPTUAL **Physics** PRACTICE PAGE

Chapter 11 The Atomic Nature of Matter

ATOMS ARE CLASSIFIED BY THEIR ATOMIC NUMBER, WHICH IS THE SAME AS THE NUMBER OF PROTONS IN THE NUCLEUS.

TO CHANGE THE ATOMS OF ONE ELEMENT INTO THOSE OF ANOTHER, PROTONS MUST BE ADDED OR SUBTRACTED!

Use the periodic table foldout in your text to help you answer the following questions.

1. When the atomic nuclei of hydrogen and lithium are squashed together (nuclear fusion) the element that is produced is
   BERYLLIUM
2. When the atomic nuclei of a pair of lithium nuclei are fused, the element produced is
   CARBON
3. When the atomic nuclei of a pair of aluminum nuclei are fused, the element produced is
   IRON
4. When the nucleus of a nitrogen atom absorbs a proton, the resulting element is
   OXYGEN
5. What element is produced when a gold nucleus gains a proton?
   MERCURY
6. Which results in the more valuable product – *adding* or *subtracting* protons from gold nuclei?
   SUBTRACTING PRODUCES MORE VALUABLE PLATINUM (ADDING A PROTON PRODUCES MERCURY)
7. What element is produced when a uranium nucleus ejects an elementary particle composed of two protons and two neutrons?
   THORIUM
8. If a uranium nucleus breaks into two pieces (nuclear fission) and one of the pieces is zirconium (atomic number 40), the other piece is the element
   TELLURIUM (ATOMIC NUMBER 52)
9. Which has more mass, a nitrogen molecule (N_2) or an oxygen molecule (O_2)?
   AN OXYGEN MOLECULE
10. Which has the greater number of atoms, a gram of helium or a gram of neon?
    A GRAM OF HELIUM

Name ______________________ Date __________

CONCEPTUAL Physics PRACTICE PAGE

Chapter 12 Solids
Scaling

1. Consider a cube 1 cm x 1 cm x 1 cm (about the size of a sugar cube). Its volume is 1 cm^3. The surface area of one of its faces is 1 cm^2. The total surface area of the cube is 6 cm^2 because it has 6 sides. Now consider a second cube, scaled up by a factor of 2 so it is 2 cm x 2 cm x 2 cm.

a. What is the total surface area of each cube?

1st cube __6__ cm^2; 2nd cube __24__ cm^2

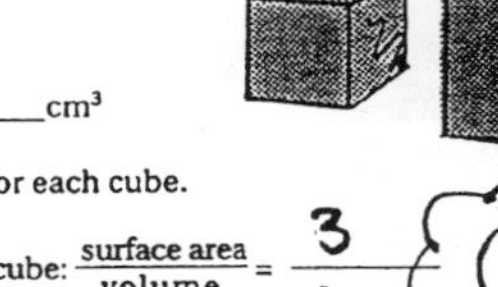

b. What is the volume of each cube?

1st cube __1__ cm^3; 2nd cube __8__ cm^3

c. Compare the ratio of surface area to volume for each cube.

1st cube: $\frac{\text{surface area}}{\text{volume}} = \frac{6}{1}$; 2nd cube: $\frac{\text{surface area}}{\text{volume}} = \frac{3}{1}$ $\left(\frac{24}{8}\right)$

2. Now consider a third cube, scaled up by a factor of 3 so it is 3 cm x 3 cm x 3 cm.

a. What is its total surface area? __54__ cm^2

b. What is its volume? __27__ cm^3

c. What is its ratio of surface area to volume?

$\frac{\text{surface area}}{\text{volume}} = \frac{2}{1}$ $\left(\frac{54}{27}\right)$

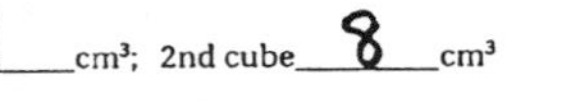

3. When the size of a cube is scaled up by a certain factor (2 and then 3 for the above examples), the area increases as the __SQUARE__ of the factor, and the volume increases as the __CUBE__ of the factor.

4. Does the ratio of surface area to volume increase or decrease as things are scaled up?

__RATIO DECREASES__

5. Does the rule for the scaling up of cubes apply also to other shapes? __YES__
Would your answers have been different if we started with a sphere of diameter 1 cm and scaled it up to a sphere of diameter 2 cm, and then 3 cm? __NO; SAME RATIOS__

6. The effects of scaling are beneficial to some creatures and detrimental to others. Check either beneficial (B) or detrimental (D) for each of the following:

a. an insect falling from a tree __B__ b. an elephant falling from the same tree __D__
c. a small fish trying to flee a big fish __D__ d. a big fish chasing a small fish __B__
e. a hungry mouse __D__ f. an insect that falls in the water __D__

Hewitt Drew it!

CONCEPTUAL Physics PRACTICE PAGE

Scaling Circles

1. Complete the table.

CIRCLES		
RADIUS	CIRCUMFERENCE	AREA
1 cm	$2\pi(1 cm) = 2\pi$ cm	$\pi(1 cm)^2 = \pi cm^2$
2 cm	$2\pi(2 cm) = 4\pi$ cm	$\pi(2 cm)^2 = 4\pi cm^2$
3 cm	$2\pi(3 cm) = 6\pi$ cm	$\pi(3 cm)^2 = 9\pi cm^2$
10 cm	$2\pi(10 cm) = 20\pi$ cm	$\pi(10 cm)^2 = 100\pi cm^2$

FOR THE CIRCUMFERENCE OF A CIRCLE, $C = 2\pi r$

AND FOR THE AREA OF A CIRCLE, $A = \pi r^2$

2. From your completed table, when the radius of a circle is doubled, its area increases by a factor of __4__. When the radius is increased by a factor of 10, the area increases by a factor of __100__.

3. Consider a round pizza that costs \$5.00. Another pizza of the same thickness has twice the diameter. How much should the larger pizza cost?

__\$25.00__

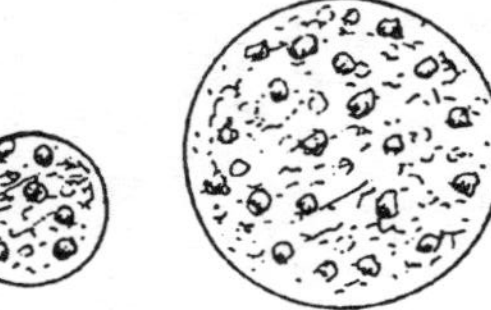

4. *True or false:* If the radius of a circle is increased by a certain factor, say 5, then the area increases by the *square* of the factor, in this case 5^2 or 25. __TRUE__

So if you scale up the radius of a circle by a factor of 10, its area will increase by a factor of __100__.

5. *(Application:)* Suppose you raise chickens and spend \$50 to buy wire for a chicken pen. To hold the most chickens inside, you should make the shape of the pen

(square) **(circular)** (either, for both provide the same area)

Hewitt Drew it!

Name ______________ Date ______________

CONCEPTUAL Physics PRACTICE PAGE

Chapter 13 Liquids

Archimedes' Principle I

1. Consider a balloon filled with 1 liter of water (1000 cm³) in equilibrium in a container of water, as shown in Figure 1.

   a. What is the mass of the 1 liter of water?
   1 kg

WATER DOES NOT SINK IN WATER!

   b. What is the weight of the 1 liter of water?
   9.8 N (OR 10 N)

   c. What is the weight of water displaced by the balloon?
   9.8 N "

   d. What is the buoyant force on the balloon?
   9.8 N "

   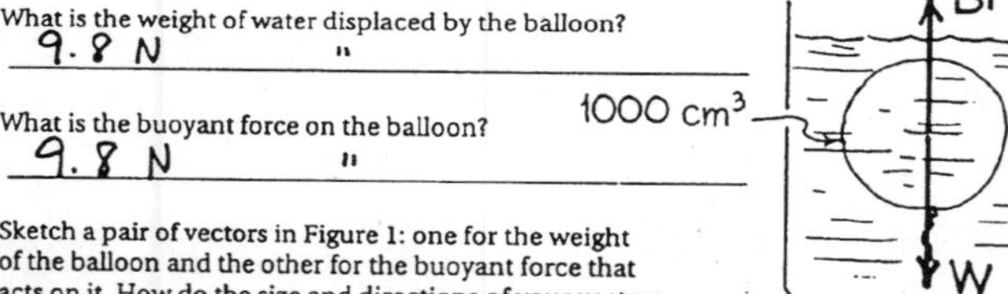

   Figure 1

   e. Sketch a pair of vectors in Figure 1: one for the weight of the balloon and the other for the buoyant force that acts on it. How do the size and directions of your vectors compare?
   VECTORS EQUAL IN MAGNITUDE, OPPOSITE IN DIRECTION

2. As a thought experiment, pretend we could remove the water from the balloon but still have it remain the same size of 1 liter. Then inside the balloon is a vacuum.

   a. What is the mass of the liter of nothing?
   0 kg

   b. What is the weight of the liter of nothing?
   0 N

   ANYTHING THAT DISPLACES 9.8 N OF WATER EXPERIENCES 9.8 N OF BUOYANT FORCE.

   c. What is the weight of water displaced by the massless balloon?
   9.8 N

   d. What is the buoyant force on the massless balloon?
   9.8 N

   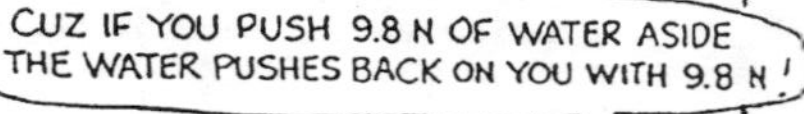

CUZ IF YOU PUSH 9.8 N OF WATER ASIDE THE WATER PUSHES BACK ON YOU WITH 9.8 N!

   e. In which direction would the massless balloon be accelerated?
   UPWARD

CONCEPTUAL Physics PRACTICE PAGE

3. Assume the balloon is replaced by a 0.5-kilogram piece of wood that has exactly the same volume (1000 cm³), as shown in Figure 2. The wood is held in the same submerged position beneath the surface of the water.

   1000 cm³

   Figure 2

   a. What volume of water is displaced by the wood?
   1000 cm³ = 1 L

   b. What is the mass of the water displaced by the wood?
   1 kg

   c. What is the weight of the water displaced by the wood?
   9.8 N

   d. How much buoyant force does the surrounding water exert on the wood?
   9.8 N

   e. When the hand is removed, what is the net force on the wood?
   NET FORCE = BUOYANT FORCE − WEIGHT of WOOD = 9.8 N − 4.9 N = 4.9 N (UPWARD)

   f. In which direction does the wood accelerate when released? UPWARD

THE BUOYANT FORCE ON A SUBMERGED OBJECT EQUALS THE WEIGHT OF WATER DISPLACED

... NOT THE WEIGHT OF THE OBJECT ITSELF!

... UNLESS IT IS FLOATING!

4. Repeat parts *a* through *f* in the previous question for a 5-kg rock that has the same volume (1000 cm³), as shown in Figure 3. Assume the rock is suspended by a string in the container of water.

   a. 1000 cm³ (SAME)
   b. 1 kg (SAME)
   c. 9.8 N (SAME)
   d. 9.8 N (SAME)
   e. 39 N DOWNWARD *
   f. DOWNWARD

   1000 cm³

   Figure 3

   WHEN THE WEIGHT OF AN OBJECT IS GREATER THAN THE BUOYANT FORCE EXERTED ON IT, IT SINKS!

* NET FORCE = BUOYANT FORCE − WT ROCK = 9.8 N − 49 N = −39 N (DOWNWARD)

Name ______________________ Date ____________

CONCEPTUAL Physics PRACTICE PAGE

Chapter 13 Liquids
Archimedes' Principle II

1. The water lines for the first three cases are shown. Sketch in the appropriate water lines for cases *d* and *e*, and make up your own for case *f*.

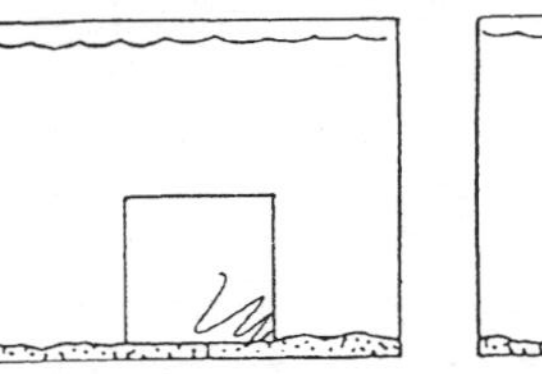

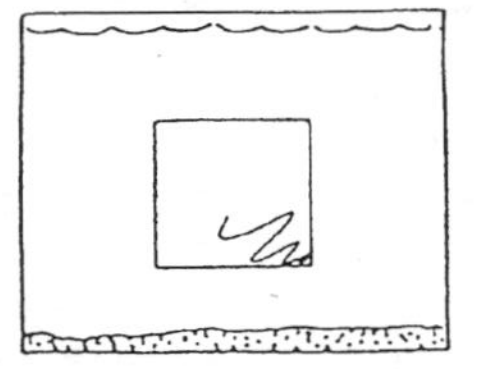

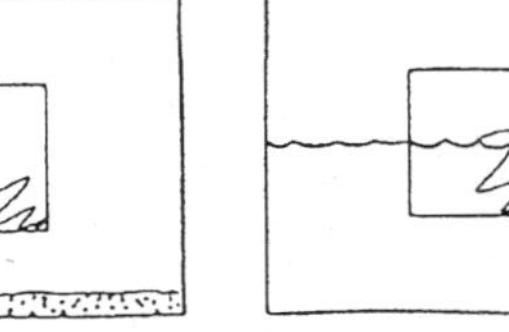

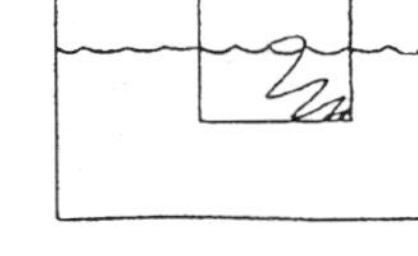

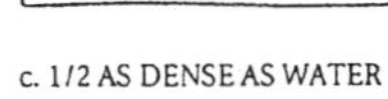

a. DENSER THAN WATER

b. SAME DENSITY AS WATER

c. 1/2 AS DENSE AS WATER

d. 1/4 AS DENSE AS WATER

e. 3/4 AS DENSE AS WATER

f. ____ AS DENSE AS WATER

2. If the weight of a ship is 100 million N, then the water it displaces weighs **100 MILLION N**. If cargo weighing 1000 N is put on board then the ship will sink down until an extra **1000 N** of water is displaced.

3. The first two sketches below show the water line for an empty and a loaded ship. Draw in the appropriate water line for the third sketch.

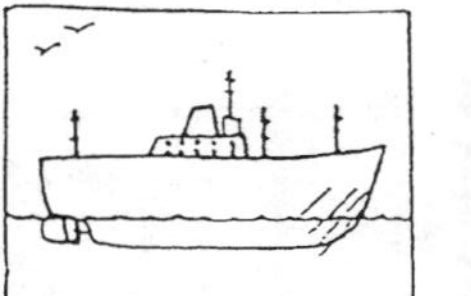

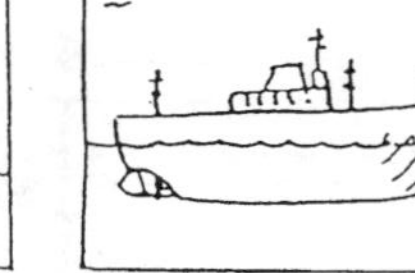

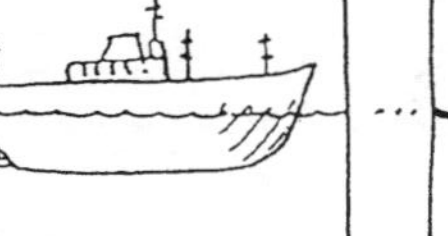

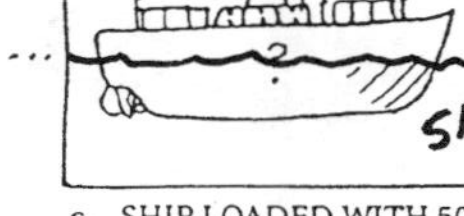

a. SHIP EMPTY

b. SHIP LOADED WITH 50 TONS OF IRON

c. SHIP LOADED WITH 50 TONS OF STYROFOAM

CONCEPTUAL Physics PRACTICE PAGE

4. Here is a glass of ice water with an ice cube floating in it. Draw the water line after the ice cube melts. (Will the water line rise, fall, or remain the same?)
REMAINS SAME. VOL OF WATER WITH SAME WT OF ICE CUBE EQUALS VOL OF SUBMERGED PORTION OF ICE CUBE. THIS IS ALSO THE VOL OF WATER FROM MELTED ICE CUBE.

5. The air-filled balloon is weighted so it sinks in water. Near the surface, the balloon has a certain volume. Draw the balloon at the bottom (inside the dashed square) and show whether it is bigger, smaller, or the same size.

a. Since the weighted balloon sinks, how does its overall density compare to the density of water?
THE DENSITY OF BALLOON IS GREATER

b. As the weighted balloon sinks, does its density increase, decrease, or remain the same?
DENSITY INCREASES (BECAUSE VOL DECREASES)

c. Since the weighted balloon sinks, how does the buoyant force on it compare to its weight?
BF IS LESS THAN ITS WEIGHT

d. As the weighted balloon sinks deeper, does the buoyant force on it increase, decrease, or remain the same?
BF DECREASES (BECAUSE VOL DECREASES)

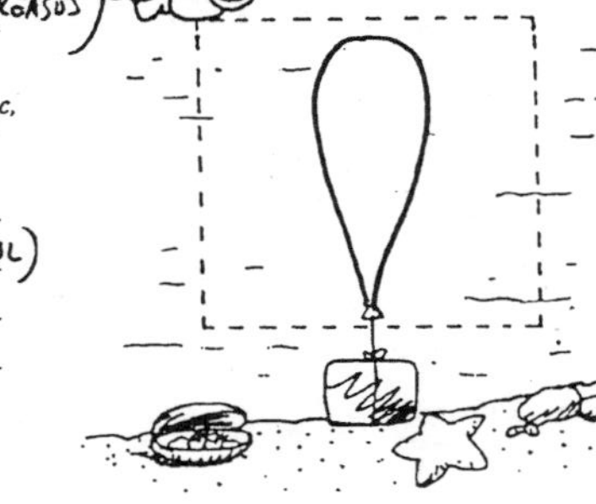

5. What would be your answers to Questions *a*, *b*, *c*, and *d* for a rock instead of an air-filled balloon?

a. **DENSITY OF ROCK IS GREATER**
b. **DENSITY REMAINS SAME (SAME VOL)**
c. **BF IS LESS THAN ITS WEIGHT**
d. **BF STAYS SAME (VOL STAYS SAME)**

Name ______________________ Date __________

CONCEPTUAL Physics PRACTICE PAGE

Chapter 14 Gases

1. A principle difference between a liquid and a gas is that when a liquid is under pressure, its volume
   (increases) (decreases) **(doesn't change noticeably)**
   and its density
   (increases) (decreases) **(doesn't change noticeably)**
   When a gas is under pressure, its volume
   (increases) **(decreases)** (doesn't change noticeably)
   and its density
   (increases) (decreases) (doesn't change noticeably)

PHYSICS SIGHS

2. The sketch shows the launching of a weather balloon at sea level. Make a sketch of the same weather balloon when it is high in the atmosphere. In words, what is different about its size and why?
   BALLOON GROWS AS IT RISES. ATM PRESSURE TENDS TO COMPRESS THINGS—EVEN BALLOONS. MORE PRESSURE AT GROUND LEVEL, & MORE COMPRESSION. LESS COMPRESSION AT HIGH ALTITUDES, & BIGGER BALLOON

HIGH-ALTITUDE SIZE

GROUND-LEVEL SIZE

3. A hydrogen-filled balloon that weighs 10 N must displace 10 N of air in order to float in air.
   If it displaces less than 10 N it will be buoyed up with less than 10 N and sink.
   If it displaces more than 10 N of air it will move upward.

4. Why is the cartoon more humorous to physics types than to non-physics types? What physics has occurred?
   IN ACCORD WITH BERNOULLI'S PRINCIPLE, MOVEMENT OF AIR OVER CURVED TOP OF UMBRELLA CAUSES A REDUCTION OF AIR PRESSURE (LIKE AIRPLANE WING). THIS LIKELY PRODUCED A NET UPWARD FORCE THAT TURNED THE UMBRELLA INSIDE OUT.

RATS TO YOU TOO, DANIEL BERNOULLI!

Hewitt Drew it!

Name ______________________ Date __________

CONCEPTUAL Physics PRACTICE PAGE

Chapter 15 Temperature, Heat, and Expansion

1. Complete the table:

TEMPERATURE OF MELTING ICE	0 °C	32 °F	273 K
TEMPERATURE OF BOILING WATER	100 °C	212 °F	373 K

2. Suppose you apply a flame and heat one liter of water, raising its temperature 10°C. If you transfer the same heat energy to two liters, how much will the temperature rise? For three liters? *Record your answers on the blanks in the drawing at the right.*

1L: $\Delta T = 10°C$ 2L: $\Delta T = 5°C$ 3L: $\Delta T = 3.3°C$

3. A thermometer is in a container half-filled with 20°C water.
   a. When an equal volume of 20°C water is added, the temperature of the mixture is
      (10°C) **(20°C)** (40°C)
   b. When instead an equal volume of 40°C water is added, the temperature of the mixture will be
      (20°C) **(30°C)** (40°C)
   c. When instead a small amount of 40°C water is added, the temperature of the mixture will be
      (20°C) **(between 20°C and 30°C)** (30°C) (more than 30°C)

20°C 20°C 40°C 20°C

4. A red-hot piece of iron is put into a bucket of cool water. *Mark the following statements true (T) or false (F).* (Ignore heat transfer to the bucket.)
   a. The decrease in iron temperature equals the increase in the water temperature. F
   b. The quantity of heat lost by the iron equals the quantity of heat gained by the water. T
   c. The iron and water both will reach the same temperature. T
   d. The final temperature of the iron and water is halfway between the initial temperatures of each. F

CAN COMMON ICE BE COLDER THAN 0°C? YES

Hewitt Drew it!

CONCEPTUAL **Physics** PRACTICE PAGE

Thermal Expansion

1. The weight hangs above the floor from the copper wire. When a candle is moved along the wire and heats it, what happens to the height of the weight above the floor? Why?
   HEIGHT DECREASES AS WIRE LENGTHENS

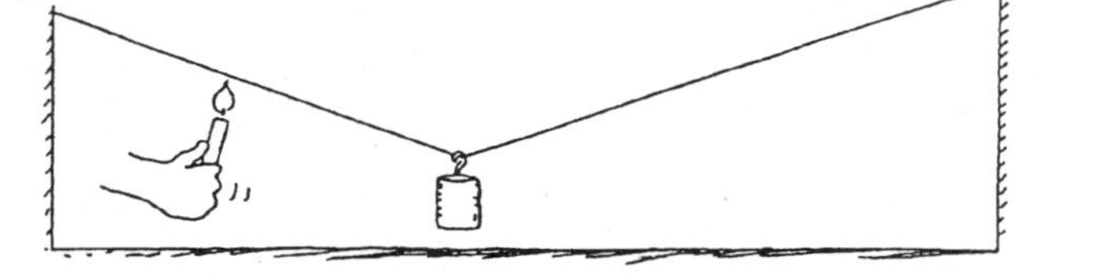

2. The levels of water at 0°C and 1°C are shown below in the first two flasks. At these temperatures there is microscopic slush in the water. There is slightly more slush at 0°C than at 1°C. As the water is heated, some of the slush collapses as it melts, and the level of the water falls in the tube. That's why the level of water is slightly lower in the 1°C-tube. Make rough estimates and sketch in the appropriate levels of water at the other temperatures shown. What is important about the level when the water reaches 4°C?
   SINCE WATER IS MOST DENSE AT 4°C, WATER LEVEL IS LOWEST AT 4°C

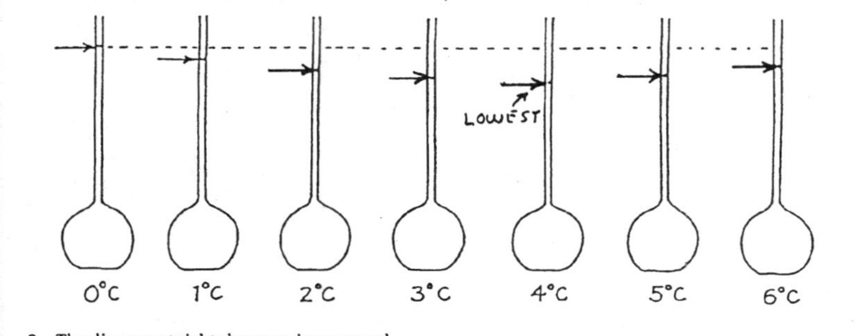

LOWEST
0°C
1°C
2°C
3°C
4°C
5°C
6°C

3. The diagram at right shows an ice-covered pond. Mark the probable temperatures of water at the top and bottom of the pond.

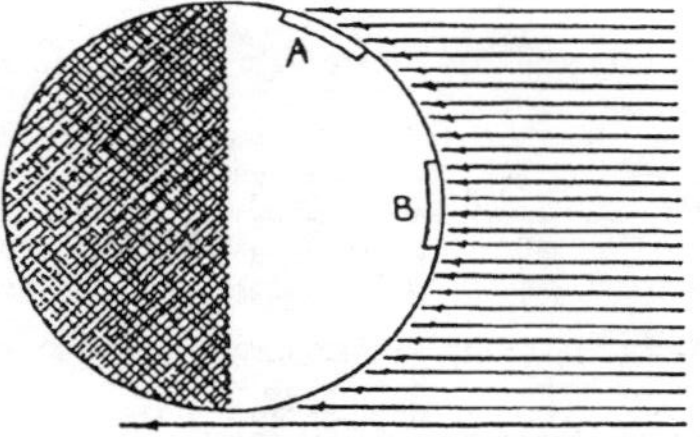

ICE
0 °C
4 °C

I CAN'T GET THIS METAL LID OFF THE JAR··· SHOULD I HEAT THE LID OR COOL IT? WHY? HEAT IT SO IT WILL EXPAND!
WHICH WILL WEIGH MORE, 1 LITER OF ICE OR 1 LITER OF WATER?
WATER (MORE DENSE)

Name ______ Date ______

CONCEPTUAL **Physics** PRACTICE PAGE

Chapter 16 Heat Transfer

Transmission of Heat

1. The tips of both brass rods are held in the gas flame. *Mark the following true (T) or false (F).*
   a. Heat is conducted only along Rod A. F
   b. Heat is conducted only along Rod B. F
   c. Heat is conducted equally along both Rod A and Rod B. T
   d. The idea that "heat rises" applies to heat transfer by *convection*, not by *conduction*. T

2. Why does a bird fluff its feathers to keep warm on a cold day?
   FLUFFED FEATHERS TRAP AIR THAT ACTS AS AN INSULATOR

3. Why does a down-filled sleeping bag keep you warm on a cold night? Why is it useless if the down is wet?
   AS IN #2. WHEN WATER TAKES PLACE OF TRAPPED AIR, INSULATION IS REDUCED

4. What does *convection* have to do with the holes in the shade of the desk lamp?
   WARMED AIR RISES THRU HOLES INSTEAD OF BEING TRAPPED

5. The warmth of equatorial regions and coldness of polar regions on the earth can be understood by considering light from a flashlight striking a surface. If it strikes perpendicularly, light energy is more concentrated as it covers a smaller area; if it strikes at an angle, the energy spreads over a larger area. So the energy per unit area is less.

   The arrows represent rays of light from the distant sun incident upon the earth. Two areas of equal size are shown, Area A near the north pole and Area B near the equator. Count the rays that reach each area, and explain why B is warmer than A.
   3 ON A; 6 ON B. AREA B GETS TWICE SOLAR ENERGY AS AREA A SO IS WARMER

CONCEPTUAL Physics PRACTICE PAGE

6. The earth's seasons result from the 23.5-degree tilt of the earth's daily spin axis as it orbits the sun. When the earth is at the point shown on the right in the sketch below (not to scale), the Northern Hemisphere tilts toward the sun, and sunlight striking it is strong (more rays per area). Sunlight striking the Southern Hemisphere is weak (fewer rays per area). Days in the north are warmer, and daylight is longer. You can see this by imagining the earth making its complete daily 24-hour spin.

   Do two things on the sketch: (1) Shade the part of the earth in nighttime darkness for all positions, as is already done in the left position. (2) Label each position with the proper month — March, June, September, or December.

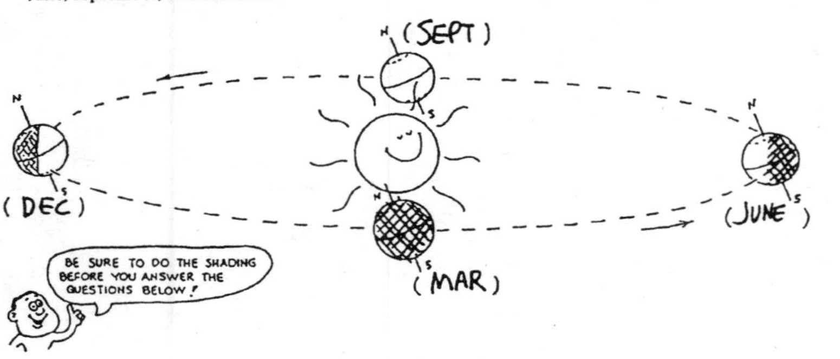
(SEPT)
(DEC)
(JUNE)
(MAR)
N
S
BE SURE TO DO THE SHADING BEFORE YOU ANSWER THE QUESTIONS BELOW!

a. When the earth is in any of the four positions shown, during one 24-hour spin a location at the equator receives sunlight half the time and is in darkness the other half the time.

   This means that regions at the equator always get about 12 hours of sunlight and 12 hours of darkness.

b. Can you see that in the June position regions farther north have longer daylight hours and shorter nights? Locations north of the Arctic Circle (dotted line in Northern Hemisphere) always face toward the sun as the earth spins, so they get daylight 24 hours a day.

c. How many hours of light and darkness are there in June at regions south of the Antarctic Circle (dotted line in Southern Hemisphere)?

   ZERO HOURS OF LIGHT, OR 24 HOURS OF DARKNESS PER DAY

d. Six months later, when the earth is at the December position, is the situation in the Antarctic the same or is it the reverse?

   REVERSE; MORE SUNLIGHT PER AREA IN DEC IN SOUTHERN HEMISPHERE

e. Why do South America and Australia enjoy warm weather in December instead of June?

   IN DEC THE SOUTHERN HEMISPHERE IS TILTED TOWARD THE SUN AND GETS MORE SUNLIGHT PER AREA THAN IN JUNE

PHYSICS YEA!

Hewitt Drew it!

Name ______________________ Date __________

CONCEPTUAL Physics PRACTICE PAGE

Chapter 17 Change of Phase

All matter can exist in the solid, liquid, or gaseous phases. The solid phase exists at relatively low temperatures, the liquid phase at higher temperatures, and the gaseous phase at still higher temperatures. Water is the most common example, not only because of its abundance but also because the temperatures for all three phases are common. Study "Energy and Changes of Phase" in your textbook and then answer the following:

1. How many calories are needed to change 1 gram of 0°C ice to water?

   80

   ICE 0°C ⟹ WATER 0°C

2. How many calories are needed to change the temperature of 1 gram of water by 1°C?

   1

   T ⟹ T+1°C

3. How many calories are needed to melt 1 gram of 0°C ice and turn it to water at a room temperature of 23°C?

   80 CAL + 23 CAL = 103 CAL

   0°C ⟹ 0°C ⟹ 23°C

4. A 50-gram sample of ice at 0°C is placed in a glass beaker that contains 200 g of water at 20°C.

   a. How much heat is needed to melt the ice? 4000 CAL

      SINCE THERE'S 50 g OF ICE, AND 80 CAL IS REQUIRED PER GRAM, HEAT REQUIRED IS $50\,g \times (80\ cal/g) = 4000\ CAL$

   b. By how much would the temperature of the water change if it gave up this much heat to the ice? BY 20°C

      200 g OF WATER GIVES OFF 200 CAL FOR EACH 1°C DROP IN TEMP.

      SO $4000\ CAL / 200\ CAL/°C = 20°C$

   c. What will be the final temperature of the mixture? (Disregard any heat absorbed by the glass or given off by the surrounding air.) 0°C

5. How many calories are needed to change 1 gram of 100°C boiling water to 100°C steam?

   540 CAL

   100°C ⟹ 100°C

6. Fill in the number of calories at each step below for changing the state of 1 gram of 0°C ice to 100°C steam.

   1 GRAM ICE 0°C —CHANGE OF PHASE⟹ 1 GRAM WATER 0°C —TEMP. RISE⟹ 1 GRAM WATER 100°C —CHANGE OF PHASE⟹ 1 GRAM STEAM 100°C

   HEAT NEEDED = 80 CAL + 100 CAL + 540 CAL = 720 CAL

Hewitt Drew it!

CONCEPTUAL Physics PRACTICE PAGE

7. One gram of steam at 100°C condenses, and the water cools to 22°C.

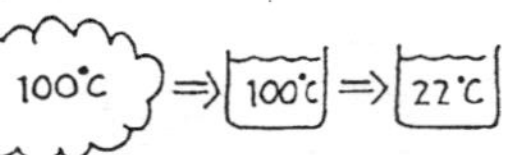

   a. How much heat is released when the steam condenses? 540 CAL

   b. How much heat is released when the water cools from 100°C to 22°C?
   78 CAL (SINCE WATER COOLS BY 100° - 22° = 78°C)

   c. How much heat is released altogether? 618 CAL

8. In a household radiator 1000 g of steam at 100°C condenses, and the water cools to 90°C.

   a. How much heat is released when the steam condenses?
   540,000 CAL

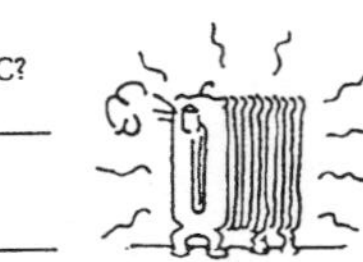

   b. How much heat is released when the water cools from 100°C to 90°C?
   10,000 CAL

   c. How much heat is released altogether?
   550,000 CAL

9. Why is it difficult to make tea on the top of a high mountain?
WATER BOILS AT A LOWER TEMP, AND GETS NO HOTTER THAN THIS TEMP.

10. How many calories are given up by 1 gram of 100°C steam that condenses to 100°C water?
540 CAL

11. How many calories are given up by 1 gram of 100°C steam that condenses and drops in temperature to 22°C water?
$540 + (100 - 22) = 618$ CAL

12. How many calories are given to a household radiator when 1000 grams of 100°C steam condenses, and drops in temperature to 90°C water?
$1000(540 + [100 - 90]) = 550{,}000$ CAL

13. To get water from the ground, even in the hot desert, dig a hole about a half meter wide and a half meter deep. Place a cup at the bottom. Spread a sheet of plastic wrap over the hole and place stones along the edge to hold it secure. Weight the center of the plastic with a stone so it forms a cone shape. Why will water collect in the cup? (Physics can save your life if you're ever stranded in a desert!)

EVAPORATED WATER FROM GROUND IS TRAPPED, AND CONDENSES ON UNDER-SIDE OF PLASTIC AND RUNS INTO THE CUP. (AT NIGHT, CONDENSATION FROM AIR COLLECTS ON TOP OF THE PLASTIC.)

Name ______________________ Date __________

CONCEPTUAL Physics PRACTICE PAGE

Chapter 17 Change of Phase

Evaporation

1. Why does it feel colder when you swim at a pool on a windy day?
WATER EVAPORATES FROM YOUR BODY FASTER AND COOLS YOU.

2. Why does your skin feel cold when a little rubbing alcohol is applied to it?
ALCOHOL RAPIDLY EVAPORATES AND COOLS YOU IN PROCESS.

3. Briefly explain from a molecular point of view why evaporation is a cooling process.
THE MORE ENERGETIC AND FASTER MOLECULES ESCAPE INTO THE AIR. ENERGY TAKEN WITH THEM REDUCES AVERAGE KE OF REMAINING MOLECULES.

4. When hot water rapidly evaporates, the result can be dramatic. Consider 4 g of boiling water spread over a large surface so that 1 g rapidly evaporates. Suppose further that the surface and surroundings are very cold so that all 540 calories for evaporation come from the remaining 3 g of water.

   a. How many calories are taken from each gram of water?
   $\frac{540 \text{ CAL}}{3} = 180$ CAL

   b. How many calories are released when 1 g of 100°C water cools to 0°C?
   100 CAL

   c. How many calories are released when 1 g of 0°C water changes to 0°C ice?
   80 CAL

   d. What happens in this case to the remaining 3 g of boiling water when 1 g rapidly evaporates?
   THE REMAINING WATER FREEZES! (EACH GRAM OF WATER RELEASES 180 CAL IN COOLING AND FREEZING)

Name ____________________ Date __________

CONCEPTUAL **Physics** PRACTICE PAGE

Chapter 18 Thermodynamics
Absolute Zero

A mass of air is contained so that the volume can change but the pressure remains constant. Table I shows air volumes at various temperatures when the air is heated slowly.

1. Plot the data in Table I on the graph, and connect the points.

Table I

TEMP. (°C)	VOLUME (mL)
0	50
25	55
50	60
75	65
100	70

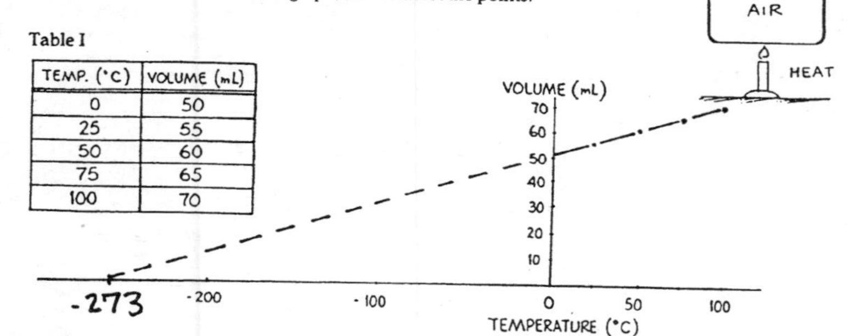

2. The graph shows how the volume of air varies with temperature at constant pressure. The straightness of the line means that the air expands uniformly with temperature. From your graph, you can predict what will happen to the volume of air when it is cooled.

   Extrapolate (extend) the straight line of your graph to find the temperature at which the volume of the air would become zero. Mark this point on your graph. Estimate this temperature: **-273°C**

3. Although air would liquify before cooling to this temperature, the procedure suggests that there is a lower limit to how cold something can be. This is the absolute zero of temperature. Careful experiments show that absolute zero is **-273** °C.

4. Scientists measure temperature in *kelvins* instead of degrees Celsius, where the absolute zero of temperature is 0 kelvins. If you relabeled the temperature axis on the graph in Question 1 so that it shows temperature in kelvins, would your graph look like the one below? **YES**

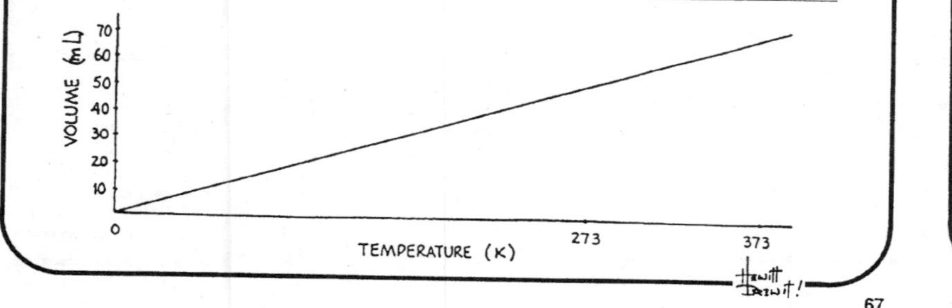

CONCEPTUAL **Physics** PRACTICE PAGE

Our Earth's Hot Interior

A major puzzle faced scientists in the 19th Century. Volcanoes showed that the Earth is molten beneath its crust. Penetration into the crust by bore holes and mines showed that the Earth's temperature increases with depth. Scientists knew that heat flows from the interior to the surface. They assumed that the source of the Earth's internal heat was primordial, the afterglow of its fiery birth. Measurements of cooling rates indicated a relatively young Earth—some 25 to 30 millions years in age. But geological evidence indicated an older Earth. This puzzle wasn't solved until the discovery of radioactivity. Then it was learned that the interior is kept hot by the energy of radioactive decay. We now know the age of the Earth is some 4.5 billions years—a much older Earth.

All rock contains trace amounts of radioactive minerals. Those in common granite release energy at the rate 0.03 J/kg y. Granite at the Earth's surface transfers this energy to the surroundings practically as fast as it is generated, so we don't find granite warm to the touch. But what if a sample of granite were thermally insulated? That is, suppose the increase of internal energy due to radioactivity were contained. Then it would get hotter. How much? Let's figure it out, using 790 joule/kilogram kelvin as the specific heat of granite.

Calculations to make:

1. How many joules are required to increase the temperature of 1 kg of granite by 1000 K?

   $Q = cm\Delta T = (790\ \frac{J}{kg\,C°})(1\ kg)(1000\ C°) = 790{,}000\ J$

2. How many years would it take radioactive decay in a kilogram of granite to produce this many joules?

   $\frac{790{,}000\ J}{0.03\ \frac{J}{kg\,yr}} \times 1\ kg \approx$ 26.3 MILLION YEARS

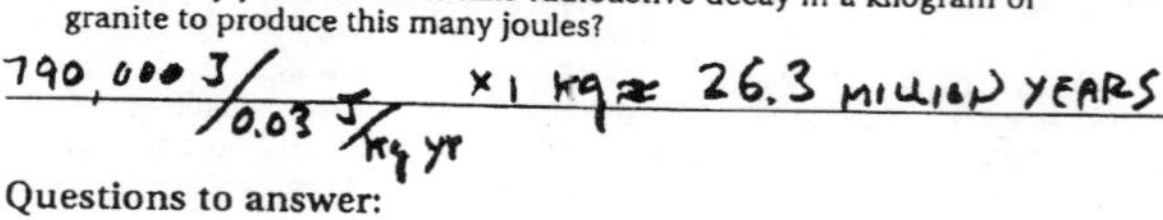

Questions to answer:

1. How many years would it take a thermally insulated 1-kg chunk of granite to undergo a 1000 K increase in temperature?

   Same 26.3 million years

2. How many years would it take a thermally insulated one-million-kilogram chunk of granite to undergo a 1000 K increase in temperature?

   SAME (DUE TO CORRESPONDINGLY MORE RADIATION)

3. Why does the Earth's interior remain molten hot?

   BECAUSE OF RADIOACTIVITY

4. Rock has a higher melting temperature deep in the interior. Why?

   GREATER PRESSURE (LIKE WATER IN A PRESSURE COOKER)

5. Why doesn't the Earth just keep getting hotter until it all melts?

   INTERIOR IS NOT PERFECTLY INSULATED - HEAT MIGRATES TO THE SURFACE.

6. **True** or false: The energy produced by Earth radioactivity ultimately becomes terrestrial radiation.

An electric toaster stays hot while electric energy is supplied, and doesn't cool until switched off. Similarly, do you think the energy source now keeping the earth hot will one day suddenly switch off like a disconnected toaster — or gradually decrease over a long time?

Name ______________________ Date __________

CONCEPTUAL **Physics** PRACTICE PAGE

Chapter 19 Vibrations and Waves

1. A sine curve that represents a transverse wave is drawn below. With a ruler, measure the wavelength and amplitude of the wave.

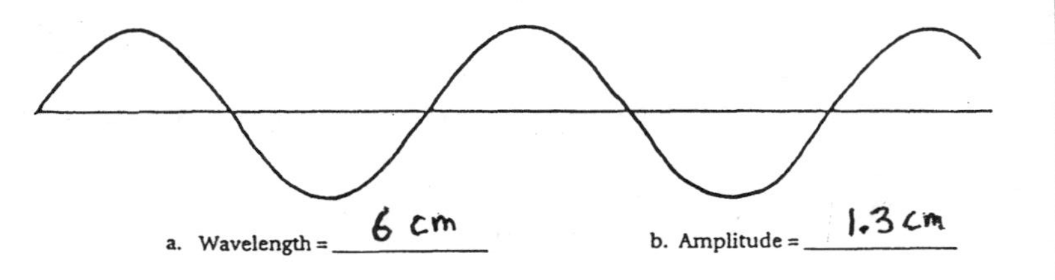

a. Wavelength = 6 cm b. Amplitude = 1.3 cm

2. A kid on a playground swing makes a complete to-and-fro swing each 2 seconds. The frequency of swing is

(0.5 hertz) (1 hertz) (2 hertz)

and the period is

(0.5 second) (1 second) (2 seconds)

3. *Complete the statements.*

THE PERIOD OF A 440-HERTZ SOUND WAVE IS 1/440 SECOND.

A MARINE WEATHER STATION REPORTS WAVES ALONG THE SHORE THAT ARE 8 SECONDS APART. THE FREQUENCY OF THE WAVES IS THEREFORE 1/8 HERTZ.

4. The annoying sound from a mosquito is produced when it beats its wings at the average rate of 600 wingbeats per second.

a. What is the frequency of the soundwaves? 600 Hz

b. What is the wavelength? (Assume the speed of sound is 340 m/s.) 0.57 m

$\lambda = \frac{340 \text{ m/s}}{600 \text{ Hz}}$

CONCEPTUAL **Physics** PRACTICE PAGE

5. A machine gun fires 10 rounds per second. The speed of the bullets is 300 m/s.

a. What is the distance in the air between the flying bullets? 30 m

b. What happens to the distance between the bullets if the rate of fire is increased?
DISTANCE BETWEEN BULLETS DECREASES

6. Consider a wave generator that produces 10 pulses per second. The speed of the waves is 300 cm/s.

a. What is the wavelength of the waves? 30 cm

b. What happens to the wavelength if the frequency of pulses is increased?
λ DECREASES, JUST AS DISTANCE BETWEEN BULLETS IN #5 DECREASES

7. The bird at the right watches the waves. If the portion of a wave between 2 crests passes the pole each second, what is the speed of the wave?
$v = f\lambda = 2 \times 1\text{ m} = 2 \text{ m/s}$

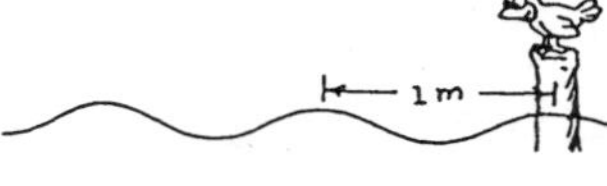

What is its period?
$T = \frac{1}{f} = \frac{1}{2} = 0.5 \text{ s}$

8. If the distance between crests in the above question were 1.5 meters apart, and 2 crests pass the pole each second, what would be the speed of the wave?
$v = f\lambda = 2 \times 1.5 = 3 \text{ m/s}$

What would be its period?
SAME (0.5 s)

9. When an automobile moves toward a listener, the sound of its horn seems relatively

(low pitched) (normal)
(high pitched)

and when moving away from the listener, its horn seems

(low pitched) (normal)
(high pitched)

10. The changed pitch of the Doppler effect is due to changes in

(wave speed) (wave frequency)

Name ______________________ Date __________

CONCEPTUAL **Physics** PRACTICE PAGE

Chapter 19 Vibrations and Waves
Shock Waves

The cone-shaped shock wave produced by a supersonic aircraft is actually the result of overlapping spherical waves of sound, as indicated by the overlapping circles in Figure 18.19 in your textbook. Sketches *a, b, c, d,* and *e,* at the left show the "animated" growth of only one of the many spherical sound waves (shown as an expanding circle in the two-dimensional sketch). The circle originates when the aircraft is in the position shown in *a.* Sketch *b* shows both the growth of the circle and position of the aircraft at a later time. Still later times are shown in *c, d,* and *e.* Note that the circle grows and the aircraft moves farther to the right. Note also that the aircraft is moving farther than the sound wave. This is because the aircraft is moving faster than sound.

Careful examination will reveal how fast the aircraft is moving compared to the speed of sound. Sketch *e* shows that in the same time the sound travels from O to A, the aircraft has traveled from O to B — twice as far. You can check this with a ruler.

Circle the answer.

1. Inspect sketches *b* and *d.* Has the aircraft traveled twice as far as sound in the same time in these postions also?

   (yes) (no) — *(yes) circled*

2. For greater speeds, the angle of the shock wave would be

   (wider) (the same) (narrower) — *(narrower) circled*

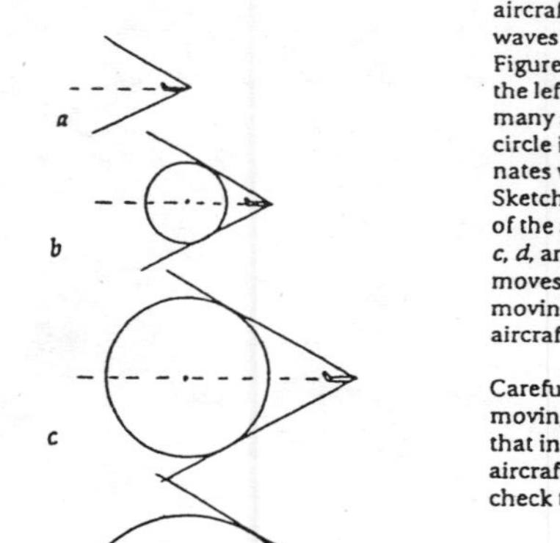
DURING THE TIME THAT SOUND TRAVELS FROM O TO A, THE PLANE TRAVELS TWICE AS FAR --- FROM O TO B.
SO IT'S FLYING AT TWICE THE SPEED OF SOUND!

CONCEPTUAL **Physics** PRACTICE PAGE

3. Use a ruler to estimate the speeds of the aircraft that produce the shock waves in the two sketches below.

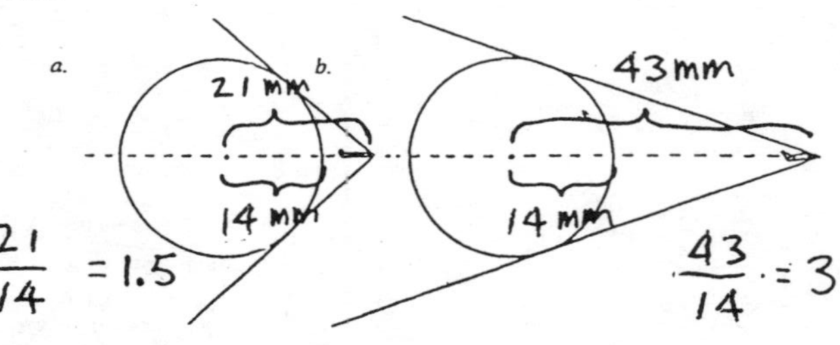
a.
b.
21 mm
14 mm
43mm
14 mm

$\frac{21}{14} = 1.5$ $\qquad$ $\frac{43}{14} = 3$

Aircraft *a* is traveling about 1.5 times the speed of sound.

Aircraft *b* is traveling about 3.0 times the speed of sound.

4. Draw your own circle (anywhere) and estimate the speed of the aircraft to produce the shock wave shown below.

ANY CIRCLE WILL DO. HERE WE'VE USED A QUARTER AND FOUND 2 ADDITIONAL ONES REACH THE APEX

FOR ANY CIRCLE, THE DISTANCE TO THE APEX WILL BE 5 TIMES GREATER THAN RADIUS OF THE CIRCLE.

The speed is about 5 times the speed of sound.

5. In the space below, draw the shock wave made by a supersonic missile that travels at four times the speed of sound.

HERE WE USE A QUARTER AGAIN (THO A BIGGER CIRCLE IS EASIER)

Name ______________________ Date ________

CONCEPTUAL Physics PRACTICE PAGE

Chapter 20 Sound

Wave Superposition

A pair of pulses travel toward each at equal speeds. The composite waveforms as they pass through each other and interfere are shown at 1-second intervals. In the left column note how the pulses interfere to produce the composite waveform (solid line). Make a similar construction for the two wave pulses in the right column. Like the pulses in the first column, they each travel at 1 space per second.

t = 0 s

t = 1 s

t = 2 s

t = 3 s

Add vertical displacements at each point, + up and – down.

t = 4 s

t = 5 s

t = 6 s

t = 7 s

thanx to Marshall Ellenstein

Hewitt Drew it!

CONCEPTUAL Physics PRACTICE PAGE

Construct the composite waveforms at 1-second intervals for the two waves traveling toward each other at equal speed.

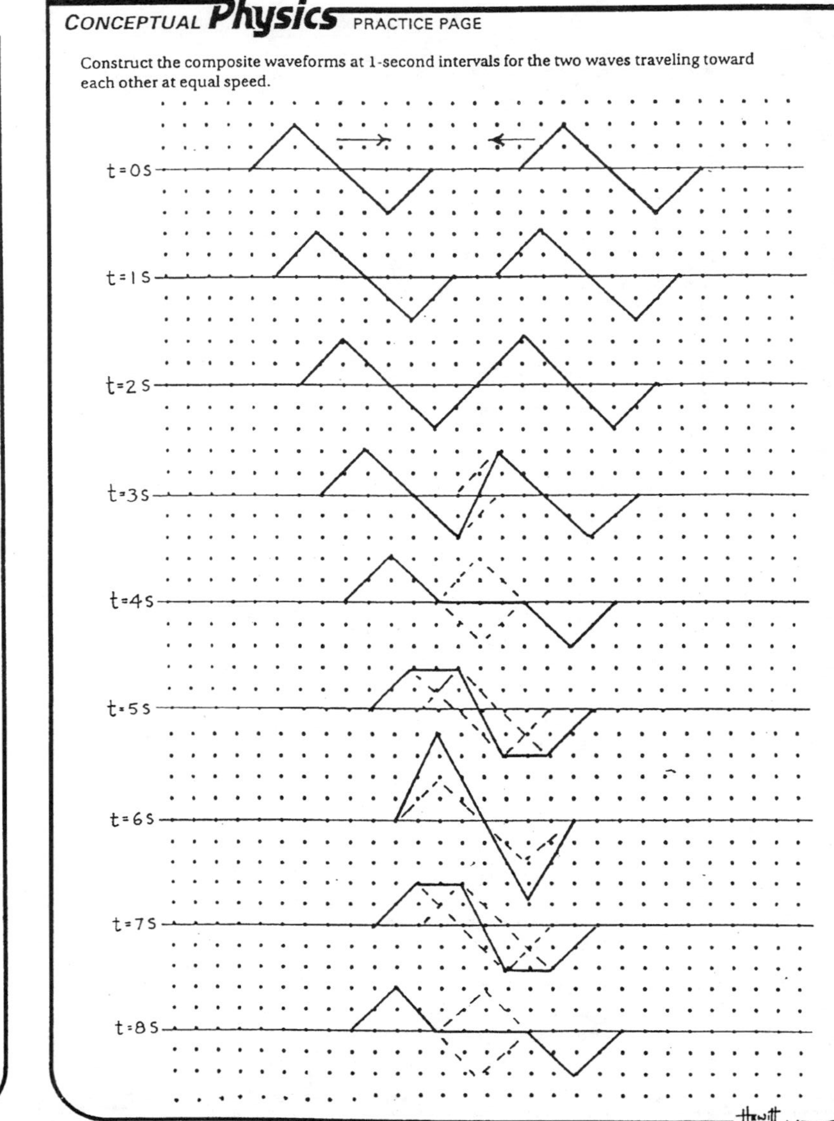

Name ______________________ Date __________

CONCEPTUAL Physics PRACTICE PAGE

Chapter 22 Electrostatics

1. Consider the diagrams below. (a) A pair of insulated metal spheres, A and B, touch each other, so in effect they form a single uncharged conductor. (b) A positively charged rod is brought near A, but not touching, and electrons in the metal sphere are attracted toward the rod. Charges in the spheres have redistributed, and the negative charge is labeled. Draw the appropriate + signs that are repelled to the far side of B. (c) Draw the signs of charge in (c), when the spheres are separated while the rod is still present, and in (d) after the rod has been removed. Your completed work should be similar to Figure 21.7 in the textbook. The spheres have been charged by *induction*.

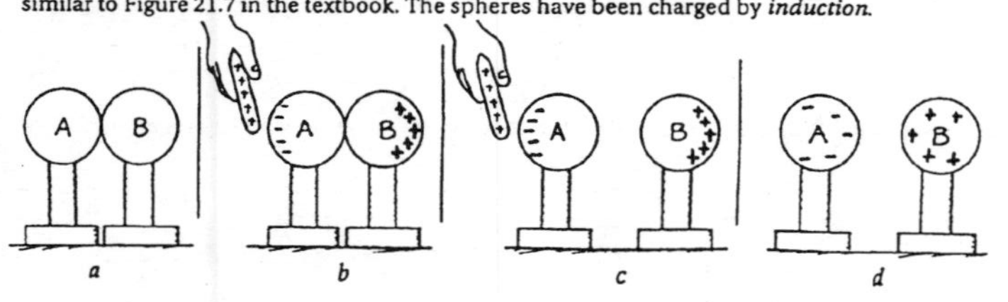

2. Consider below a single metal insulated sphere, (a) initially uncharged. When a negatively charged rod is nearby, (b), charges in the metal are separated. Electrons are repelled to the far side. When the sphere is touched with your finger, (c), electrons flow out to the sphere to the earth through the hand. The sphere is "grounded." Note the positive charge left (d) while the rod is still present and your finger removed, and (e) when the rod is removed. This is an example of *charge induction by grounding*. In this procedure the negative rod "gives" a positive charge to the sphere.

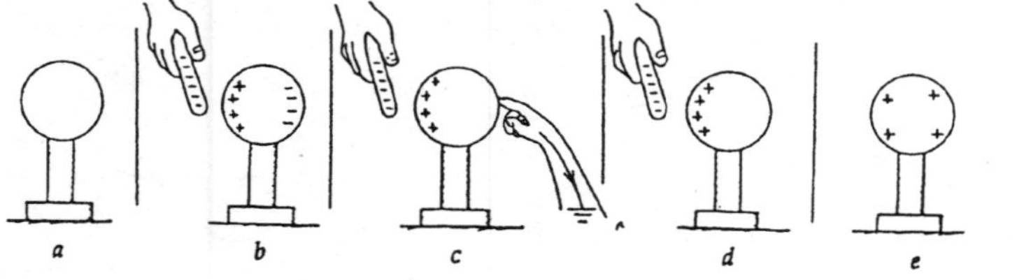

The diagrams below show a similar procedure with a positive rod. Draw the correct charges in the diagrams.

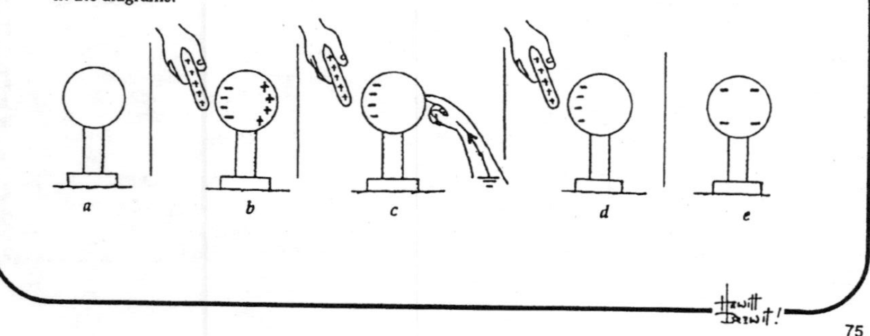

CONCEPTUAL Physics PRACTICE PAGE

Electric Potential

1. 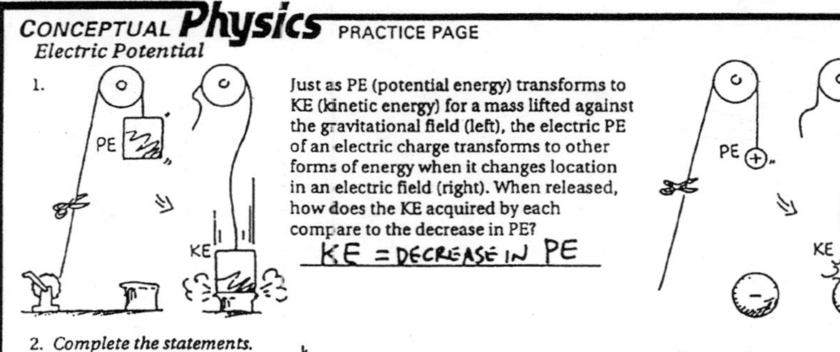

Just as PE (potential energy) transforms to KE (kinetic energy) for a mass lifted against the gravitational field (left), the electric PE of an electric charge transforms to other forms of energy when it changes location in an electric field (right). When released, how does the KE acquired by each compare to the decrease in PE?
KE = DECREASE IN PE

2. *Complete the statements.*

A force compresses the spring. The work done in compression is the product of the average force and the distance moved. W = Fd. This work increases the PE of the spring.

Similarly, a force pushes the charge (call it a test charge) closer to the charged sphere. The work done in moving the test charge is the product of the average FORCE and the DISTANCE moved. W = Fd. This work INCREASES the PE of the test charge.

If the test charge is released, it will be repelled and fly past the starting point. Its gain in KE at this point is EQUAL to its decrease in PE.

At any point, a greater quantity of test charge means a greater amount of PE, but not a greater amount of PE *per quantity* of charge. The quantities PE (measured in joules) and PE/charge (measured in volts) are different concepts.

By definition: **Electric Potential = PE/charge.** 1 volt = 1 joule/1coulomb.

3. *Complete the statements.*

ELECTRIC PE/CHARGE HAS THE SPECIAL NAME ELECTRIC POTENTIAL

SINCE IT IS MEASURED IN VOLTS IT IS COMMONLY CALLED VOLTAGE

4. If a conductor connected to the terminal of a battery has a potential of 12 volts, then each coulomb of charge on the conductor has a PE of 12 J.

5. Some people get mixed up between force and pressure. Recall that pressure is force *per area.* Similarly, some people get mixed up between electric PE and voltage. According to this chapter, voltage is electric PE *per* CHARGE.

Name ______________________ Date __________

CONCEPTUAL Physics PRACTICE PAGE

Chapter 23 Electric Current

1. Water doesn't flow in the pipe when (a) both ends are at the same level. Another way of saying this is that water will not flow in the pipe when both ends have the same potential energy (PE). Similarly, charge will not flow in a conductor if both ends of the conductor are at the same electric potential. But tip the water pipe and increase the PE of one side so there is a difference in PE across the ends of the pipe, as in (b), and water will flow. Similarly, increase the electric potential of one end of an electric conductor so there is a potential difference across the ends, and charge will flow.

   a. The units of electric potential difference are
   (volts) (amperes) (ohms) (watts) — circled: volts

   b. It is common to call electric potential difference
   (voltage) (amperage) (wattage) — circled: voltage

   c. The flow of electric charge is called electric
   (voltage) (current) (power), — circled: current

   and is measured in
   (volts) (amperes) (ohms) (watts) — circled: amperes

A VOLT IS A UNIT OF POTENTIAL
AND AN AMPERE IS A UNIT OF CURRENT

DOES VOLTAGE CAUSE CURRENT, OR DOES CURRENT CAUSE VOLTAGE? WHICH IS THE CAUSE AND WHICH IS THE EFFECT?

VOLTAGE (THE CAUSE) PRODUCES CURRENT (THE EFFECT)

2. Complete the statements:

   a. A current of 1 ampere is a flow of charge at the rate of ONE coulomb per second.
   b. When a charge of 15 C flows through any area in a circuit each second, the current is 15 A.
   c. One volt is the potential difference between two points if 1 joule of energy is needed to move ONE coulomb of charge between the two points.
   d. When a lamp is plugged into a 120-V socket, each coulomb of charge that flows in the circuit is raised to a potential energy of 120 joules.
   e. Which offers more resistance to water flow, a wide pipe or a narrow pipe? NARROW PIPE
   Similarly, which offers more resistance to the flow of charge, a thick wire or a thin wire?
   THIN WIRE

Hewitt Drew it!

CONCEPTUAL Physics PRACTICE PAGE

Ohm's Law

CURRENT = $\frac{\text{VOLTAGE}}{\text{RESISTANCE}}$ OR $I = \frac{V}{R}$

USE OHM'S LAW IN THE TRIANGLE TO FIND THE QUANTITY YOU WANT, COVER THE LETTER WITH YOUR FINGER AND THE REMAINING TWO SHOW YOU THE FORMULA!

V
I × R

1. How much current flows in a 1000-ohm resistor when 1.5 volts are impressed across it?
   0.0015 A

2. If the filament resistance in an automobile headlamp is 3 ohms, how many amps does it draw when connected to a 12-volt battery?
   4 A

3. The resistance of the side lights on an automobile are 10 ohms. How much current flows in them when connected to 12 volts?
   1.2 A

CONDUCTORS AND RESISTORS HAVE RESISTANCE TO THE CURRENT IN THEM.

4. What is the current in the 30-ohm heating coil of a coffee maker that operates on a 120-volt circuit?
   4 A

5. During a lie detector test, a voltage of 6 V is impressed across two fingers. When a certain question is asked, the resistance between the fingers drops from 400 000 ohms to 200 000 ohms. What is the current (a) initially through the fingers, and (b) when the resistance between them drops?
   (a) 0.000015 A (15 μA) (b) 0.000030 A (30 μA)

NOTE THE TRIANGLE ABOVE PROVIDES A SORT OF MATH CRUTCH

6. How much resistance allows an impressed voltage of 6 V to produce a current of 0.006 A?
   1000 Ω

7. What is the resistance of a clothes iron that draws a current of 12 A at 120 V?
   10 Ω

OHM MY GOODNESS!

8. What is the voltage across a 100-ohm circuit element that draws a current of 1 A?
   100 V

9. What voltage will produce 3 A through a 15-ohm resistor?
   45 V

10. The current in an incandescent lamp is 0.5 A when connected to a 120-V circuit, and 0.2 A when connected to a 10-V source. Does the resistance of the lamp change in these cases? Explain your answer and defend it with numerical values.
   YES, RESISTANCE INCREASES WITH HIGHER TEMP OF GREATER CURRENT.
   AT 0.2 A, $R = \frac{10\text{V}}{0.2\text{A}} = 50\,\Omega$; AT 0.5 A, $R = \frac{120\text{V}}{0.5\text{A}} = 240\,\Omega$
   (APPRECIABLY GREATER).

Hewitt Drew it!

Name ______________________ Date __________

CONCEPTUAL Physics PRACTICE PAGE

Chapter 23 Electric Current
Electric Power

Recall that the rate energy is converted from one form to another is *power.*

$$\text{power} = \frac{\text{energy converted}}{\text{time}} = \frac{\text{voltage x charge}}{\text{time}} = \text{voltage x } \frac{\text{charge}}{\text{time}} = \text{voltage x current}$$

The unit of power is the *watt* (or *kilowatt*). So in units form,

Electric power (*watts*) = current (*amperes*) x voltage (*volts*),

where 1 *watt* = *1 ampere x 1 volt.*

THAT'S RIGHT··· VOLTAGE = $\frac{\text{ENERGY}}{\text{CHARGE}}$, SO ENERGY = VOLTAGE × CHARGE··· AND $\frac{\text{CHARGE}}{\text{TIME}}$ = CURRENT ⁝ NEAT ⁝

A 100-WATT BULB CONVERTS ELECTRIC ENERGY INTO HEAT AND LIGHT MORE QUICKLY THAN A 25-WATT BULB. THAT'S WHY FOR THE SAME VOLTAGE A 100-WATT BULB GLOWS BRIGHTER THAN A 25-WATT BULB!

1. What is the power when a voltage of 120 V drives a 2-A current through a device?
   240 W

2. What is the current when a 60-W lamp is connected to 120 V?
   0.5 A

WHICH DRAWS MORE CURRENT··· THE 100-WATT OR THE 25-WATT BULB?

3. How much current does a 100-W lamp draw when connected to 120 V?
   0.83 A

4. If part of an electric circuit dissipates energy at 6 W when it draws a current of 3 A, what voltage is impressed across it?
   2 V

5. The equation

$$\text{power} = \frac{\text{energy converted}}{\text{time}}$$

WATT'S HAPPENING ?

rearranged gives

energy converted = POWER × TIME

6. Explain the difference between a kilowatt and a kilowatt-hour.
   A KILOWATT IS A UNIT OF POWER; KW-HOUR IS UNIT OF ENERGY (POWER × TIME)

7. One deterrent to burglary is to leave your front porch light on all the time. If your fixture contains a 60-W bulb at 120 V, and your local power utility sells energy at 10 cents per kilowatt-hour, how much will it cost to leave the bulb on for the whole month? Show your work on the other side of this page.
   $E = P \times t = 60\text{W} \times 1\text{ MO} \times \frac{30\text{ DAY}}{1\text{ MO}} \times \frac{24\text{ H}}{1\text{ DAY}} \times \frac{1\text{ KW}}{1000\text{ W}} = 43.2\text{ KWH}$
   MULTIPLY BY \$0.10/KWH = \$4.32

Hewitt Drew it!

Name ______________________ Date __________

CONCEPTUAL Physics PRACTICE PAGE

Chapter 23 Electric Current
Series Circuits

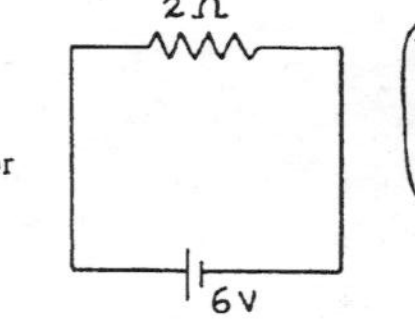

THE EQUIVALENT RESISTANCE OF RESISTORS IN SERIES IS SIMPLY THEIR SUM!

1. In the circuit shown at the right, a voltage of 6 V pushes charge through a single resistor of 2 Ω. According to Ohm's law, the current in the resistor (and therefore in the whole circuit) is
   3 A.

2. If a second identical lamp is added, as on the left, the 6-V battery must push charge through a total resistance of 6 Ω. The current in the circuit is then 1 A.

3. The equivalent resistance of three 4-Ω resistors in series is 12 Ω.

4. Does current flow *through* a resistor, or *across* a resistor? THROUGH
   Is voltage established *through* a resistor, or *across* a resistor? ACROSS

5. Does current in the lamps occur simultaneously, or does charge flow first through one lamp, then the other, and finally the last in turn?
   SIMULTANEOUSLY (SPEED OF LIGHT)

6. Circuits *a* and *b* below are identical with all bulbs rated at equal wattage (therefore equal resistance). The only difference between the circuits is that Bulb 5 has a short circuit, as shown.

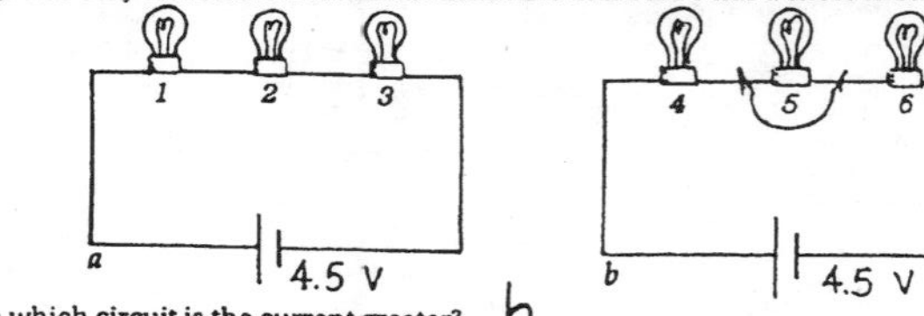

a. In which circuit is the current greater? b

b. In which circuit are all three bulbs equally bright? a

c. What bulbs are the brightest? 4 AND 6

d. What bulb is the dimmest? 5 (NOT LIT)

e. What bulbs have the largest voltage drops across them? 4 AND 6 (2.25 V EACH)

f. Which circuit dissipates more power? b (GREATER CURRENT, SAME VOLTAGE)

g. What circuit produces more light? b (MORE POWER)

Hewitt Drew it!

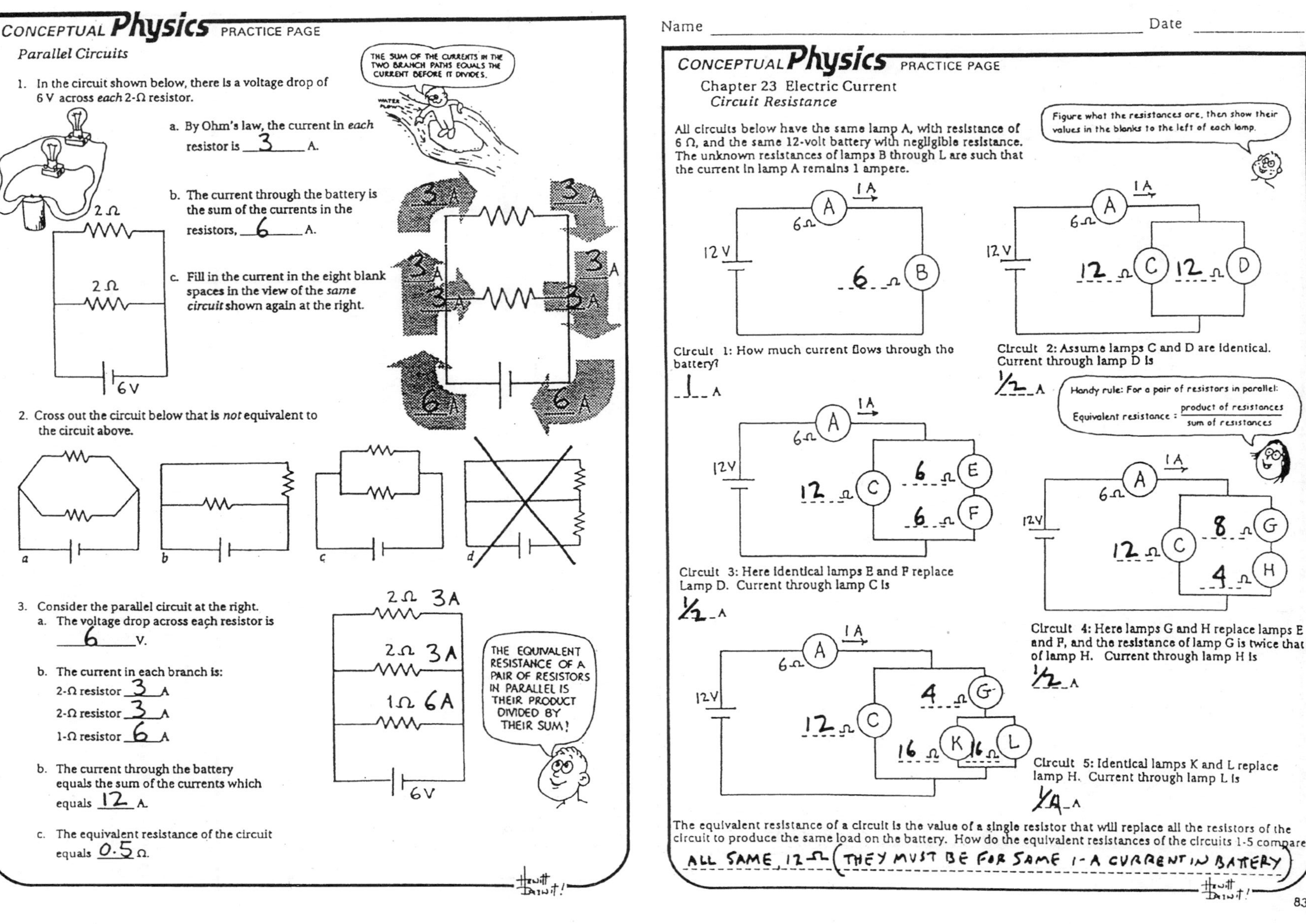

CONCEPTUAL Physics PRACTICE PAGE

Parallel Circuits

1. In the circuit shown below, there is a voltage drop of 6 V across *each* 2-Ω resistor.

   a. By Ohm's law, the current in *each* resistor is __3__ A.

   b. The current through the battery is the sum of the currents in the resistors, __6__ A.

   c. Fill in the current in the eight blank spaces in the view of the *same circuit* shown again at the right.

2. Cross out the circuit below that is *not* equivalent to the circuit above.

3. Consider the parallel circuit at the right.

   a. The voltage drop across each resistor is __6__ V.

   b. The current in each branch is:
      - 2-Ω resistor __3__ A
      - 2-Ω resistor __3__ A
      - 1-Ω resistor __6__ A

   b. The current through the battery equals the sum of the currents which equals __12__ A.

   c. The equivalent resistance of the circuit equals __0.5__ Ω.

Name ______________________ Date __________

CONCEPTUAL Physics PRACTICE PAGE

Chapter 23 Electric Current

Circuit Resistance

All circuits below have the same lamp A, with resistance of 6 Ω, and the same 12-volt battery with negligible resistance. The unknown resistances of lamps B through L are such that the current in lamp A remains 1 ampere.

Circuit 1: How much current flows through the battery?

__1__ A

Circuit 2: Assume lamps C and D are identical. Current through lamp D is

__1/2__ A

Circuit 3: Here identical lamps E and F replace Lamp D. Current through lamp C is

__1/2__ A

Circuit 4: Here lamps G and H replace lamps E and F, and the resistance of lamp G is twice that of lamp H. Current through lamp H is

__1/2__ A

Circuit 5: Identical lamps K and L replace lamp H. Current through lamp L is

__1/4__ A

The equivalent resistance of a circuit is the value of a single resistor that will replace all the resistors of the circuit to produce the same load on the battery. How do the equivalent resistances of the circuits 1-5 compare?

ALL SAME, 12 Ω (THEY MUST BE FOR SAME 1-A CURRENT IN BATTERY)

CONCEPTUAL Physics PRACTICE PAGE

The table beside circuit *a* below shows the current through each resistor, the voltage across each resistor, and the power dissipated as heat in each resistor. Find the similar correct values for circuits *b*, *c*, and *d*, and put your answers in the tables shown.

CURRENT THRU EACH BRANCH = VOLTAGE DROP ACROSS BRANCH / EQUIVALENT RESISTANCE OF BRANCH

a (2 Ω, 4 Ω, 6 Ω, 12 V)

RESISTANCE	CURRENT ×	VOLTAGE =	POWER
2 Ω	2 A	4 V	8 W
4 Ω	2 A	8 V	16 W
6 Ω	2 A	12 V	24 W

b (1 Ω, 2 Ω, 6 V)

RESISTANCE	CURRENT ×	VOLTAGE =	POWER
1 Ω	2A	2V	4W
2 Ω	2 A	4V	8W

c (6 Ω, 3 Ω, 6 V)

RESISTANCE	CURRENT ×	VOLTAGE =	POWER
6 Ω	1 A	6 V	6 W
3 Ω	2A	6 V	12 W

d (2 Ω, 2 Ω, 1 Ω, 6 V)

RESISTANCE	CURRENT ×	VOLTAGE =	POWER
2 Ω	1.5 A	3 V	4.5 W
2 Ω	1.5 A	3 V	4.5 W
1 Ω	3 A	3 V	9 W

NOTE THAT TOTAL POWER DISSIPATED BY ALL RESISTORS IN A CIRCUIT EQUALS THE POWER SUPPLIED BY THE BATTERY (VOLTAGE OF BATTERY × CURRENT THRU BATTERY)

Hewitt Drew it!

Name ______________________ Date __________

CONCEPTUAL Physics PRACTICE PAGE

Chapter 24 Magnetism

Fill in each blank with the appropriate word.

1. Attraction or repulsion of charges depends on their *signs*, positives or negatives. Attraction or repulsion of magnets depends on their magnetic POLES: NORTH or SOUTH.
2. Opposite poles attract; like poles REPEL.
3. A magnetic field is produced by the MOTION of electric charge.
4. Clusters of magnetically aligned atoms are magnetic DOMAINS.
5. A magnetic FIELD surrounds a current-carrying wire.
6. When a current-carrying wire is made to form a coil around a piece of iron, the result is an ELECTROMAGNET
7. A charged particle moving in a magnetic field experiences a deflecting FORCE that is maximum when the charge moves PERPENDICULAR to the field.
8. A current-carrying wire experiences a deflecting FORCE that is maximum when the wire and magnetic field are PERPENDICULAR to one another.
9. A simple instrument designed to detect electric current is the GALVANOMETER; when calibrated to measure current, it is an AMMETER; when calibrated to measure voltage, it is a VOLTMETER.
10. The largest size magnet in the world is the WORLD (OR EARTH) itself.

YOU HAVE A MAGNETIC PERSONALITY!
S
N
THEN TO REALLY MAKE THINGS "SIMPLE," THERE'S THE RIGHT-HAND RULE!
F
B
I

Hewitt Drew it!

CONCEPTUAL Physics PRACTICE PAGE

11. The illustration below is similar to Figure 24.2 in your textbook. Iron filings trace out patterns of magnetic field lines about a bar magnet. In the field are some magnetic compasses. The compass needle in only one compass is shown. Draw in the needles with proper orientation in the other compasses.

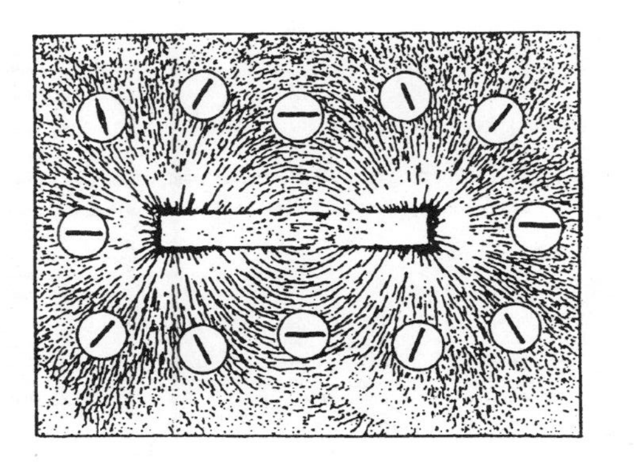

12. The illustration below is similar to Figure 24.10 (center) in your textbook. Iron filings trace out the magnetic field pattern about the loop of current-carrying wire. Draw in the compass needle orientations for all the compasses.

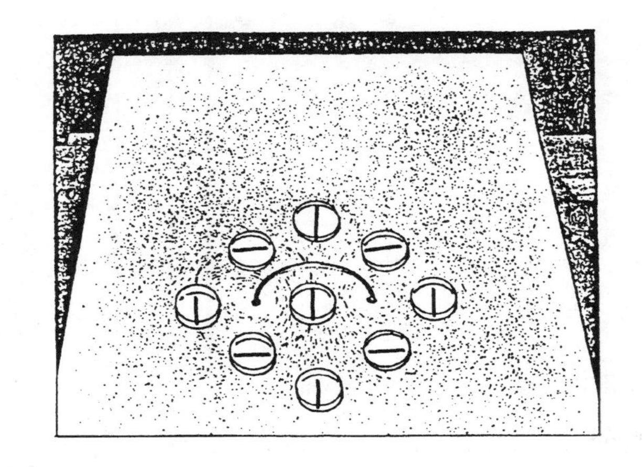

Name ______________________ Date __________

CONCEPTUAL Physics PRACTICE PAGE

Chapter 25 Electromagnetic Inducation
Faraday's Law

1. Hans Christian Oersted discovered that magnetism and electricity are

   (related) (independent of each other).

   Magnetism is produced by

   (batteries) (the motion of electric charges).

   Faraday and Henry discovered that electric current can be produced by

   (batteries) (motion of a magnet).

   More specifically, voltage is induced in a loop of wire if there is a change in the

   (batteries) (magnetic field in the loop).

   This phenomenon is called

   (electromagnetism) (electromagnetic induction).

2. When a magnet is plunged in and out of a coil of wire, voltage is induced in the coil. If the rate of the in-and-out motion of the magnet is doubled, the induced voltage

   (doubles) (halves) (remains the same).

   If instead the number of loops in the coil is doubled, the induced voltage

   (doubles) (halves) (remains the same).

3. A rapidly changing magnetic field in any region of space induces a rapidly changing

   (electric field) (magnetic field) (gravitational field)

   which in turn induces a rapidly changing

   (magnetic field) (electric field) (baseball field).

   This generation and regeneration of electric and magnetic fields makes up

   (electromagnetic waves) (sound waves) (both of these).

PHYSICS SIGH

CONCEPTUAL Physics PRACTICE PAGE

Transformers

Consider a simple transformer that has a 100-turn primary coil and a 1000-turn secondary coil. The primary is connected to a 120-V AC source and the secondary is connected to an electrical device with a resistance of 1000 ohms.

1. What will be the voltage output of the secondary? 1200 V

2. What current flows in the secondary circuit? 1.2 A

3. Now that you know the voltage and the current, what is the power in the secondary coil? 1440 W

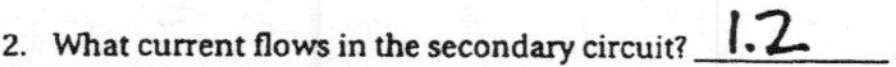

4. Neglecting small heating losses, and knowing that energy is conserved, what is the power in the primary coil? 1440 W

5. Now that you know the power and the voltage across the primary coil, what is the current drawn by the primary coil? 12 A

Circle the correct answers:

6. The results show voltage is stepped (up) (down) from primary to secondary, and that current is correspondingly stepped (up) (down).

7. For a step-up transformer, there are (more) (fewer) turns in the secondary coil than the primary. For such a transformer, there is (more) (less) current in the secondary than in the primary.

8. A transformer can step up (voltage) (energy and power), but in no way can it step up (voltage) (energy and power).

9. If 120 V is used to power a toy electric train that operates on 6 V, then a (step up) (step down) transformer should be used that has a primary to secondary turns ratio of (1/20) (20/1).

10. A transformer operates on (dc) (ac) because the magnetic field within the iron core must (continually change) (remain steady).

Electricity and magnetism connect to become light!

Name ______________________ Date __________

CONCEPTUAL Physics PRACTICE PAGE

Chapter 26 Properties of Light

1. The first investigation that led to a determination of the speed of light was performed in about 1675 by the Danish astronomer Olaus Roemer. He made careful measurements of the period of Io, a moon about the planet Jupiter, and was surprised to find an irregularity in Io's observed period. While the earth was moving away from Jupiter, the measured periods were slightly longer than average. While the earth approached Jupiter, they were shorter than average. Roemer estimated that the cumulative discrepency amounted to about 16.5 minutes. Later interpretations showed that what occurs is that light takes about 16.5 minutes to travel the extra distance across the earth's orbit. Aha! We have enough information to calculate the speed of light!

   a. What is the diameter, in kilometers, of the earth's orbit around the sun?
   300,000,000 km

   b. About how many seconds does it take light to travel across the diameter of the orbit?
   $16.5\ \text{min} \times \frac{60\ \text{s}}{1\ \text{min}} = 990\ \text{s}$

   c. How do these two quantities determine the speed of light?
   $c = \frac{d}{t} = \frac{300,000,000\ \text{km}}{990\ \text{s}} = 303,030\ \text{km/s}$

MOON Io
JUPITER
EARTH
EARTH SIX MONTHS LATER

2. Study Figure 27.5 on page 454 (Conceptual Physics, 8th Ed.) and answer the following:

   a. Which has the longer *wavelengths*, radio waves or light waves?
   RADIO WAVES

   b. Which has the longer *wavelengths*, light waves or gamma rays?
   LIGHT WAVES

   c. Which has the higher *frequencies*, ultraviolet or infrared waves?
   ULTRAVIOLET WAVES

   d. Which has the higher *frequencies*, ultraviolet waves or gamma rays?
   GAMMA RAYS

LIGHT IS THE ONLY THING WE SEE!

3. Carefully study the section "Transparent Materials" in your textbook and answer the following:

   a. Exactly what do vibrating electrons emit?
   ENERGY THAT IS CARRIED IN AN ELECTROMAGNETIC WAVE

   b. When ultraviolet light shines on glass, what does it do to electrons in the glass structure?
   UV CAUSES ELECTRONS TO VIBRATE IN RESONANCE WITH THE INCIDENT UV

   c. When energetic electrons in the glass structure vibrate against neighboring atoms, what happens to the energy of vibration?
   BECOMES THERMAL ENERGY (HEAT)

   d. What happens to the energy of a vibrating electron that does not collide with neighboring atoms?
   EMITTED AS LIGHT

e. Light in which range of frequencies, visible or ultraviolet, is absorbed in glass?

ULTRAVIOLET

f. Light in which range of frequencies, visible or ultraviolet, is transmitted through glass?

VISIBLE

g. How is the speed of light in glass affected by the succession of time delays that accompany the absorption and re-emission of light from atom to atom in the glass?

THE AVERAGE SPEED OF LIGHT IS LESS IN GLASS THAN IN AIR.

h. How does the speed of light compare in water, glass, and diamond?

SPEED OF LIGHT IS 0.75C IN WATER; 0.41 C IN A DIAMOND.

4. The sun normally shines on both the earth and on the moon. Both cast shadows. Sometimes the moon's shadow falls on the earth, and at other times the earth's shadow falls on the moon.

a. The sketch shows the sun and the earth. Draw the moon at a position for a solar eclipse.

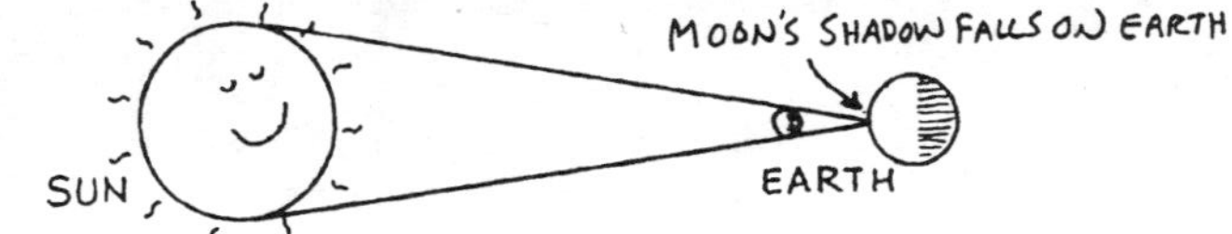

b. This sketch also shows the sun and the earth. Draw the moon at a position for a lunar eclipse.

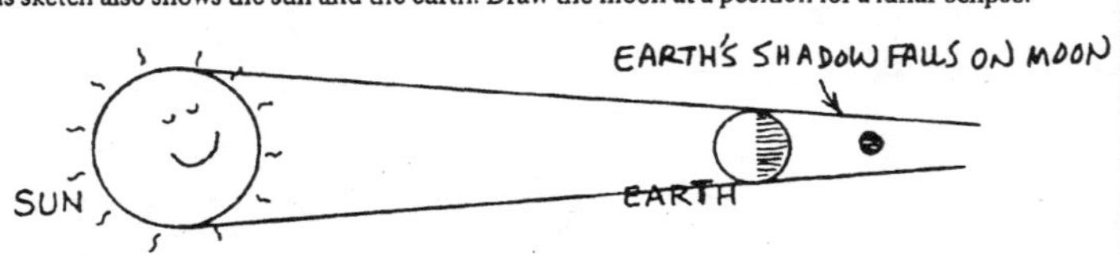

5. The diagram shows the limits of light rays when a large lamp makes a shadow of a small object on a screen. Make a sketch of the shadow on the screen, shading the umbra darker than the penumbra. In what part of the shadow could an ant see part of the lamp?

PENUMBRA

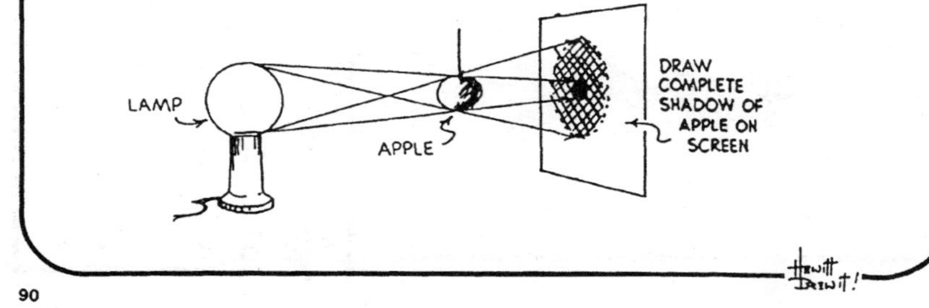

Name ______________________ Date ________

Chapter 27 Color

The sketch to the right shows the shadow of an instructor in front of a white screen in a dark room. The light source is red, so the screen looks red and the shadow looks black. Color the sketch, or label the colors with pen or pencil.

A green lamp is added and makes a second shadow. The shadow cast by the red lamp is no longer black, but is illuminated by green light. So it is green. Color or mark it green. The shadow cast by the green lamp is not black because it is illuminated by the red lamp. Indicate its color. Do the same for the background, which receives a mixture of red and green light.

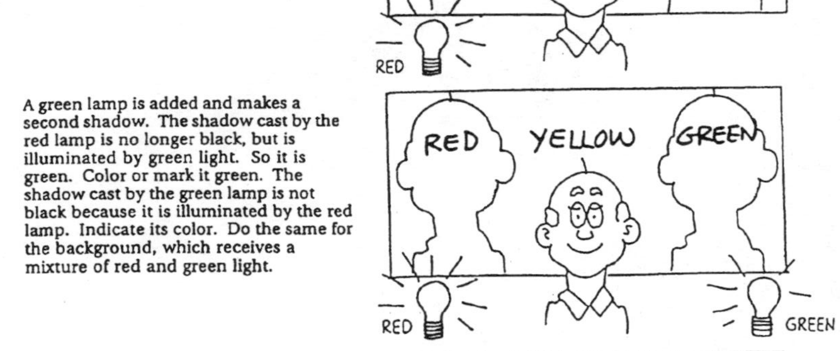

A blue lamp is added and three shadows appear. Indicate the appropriate colors of the shadows and the background.

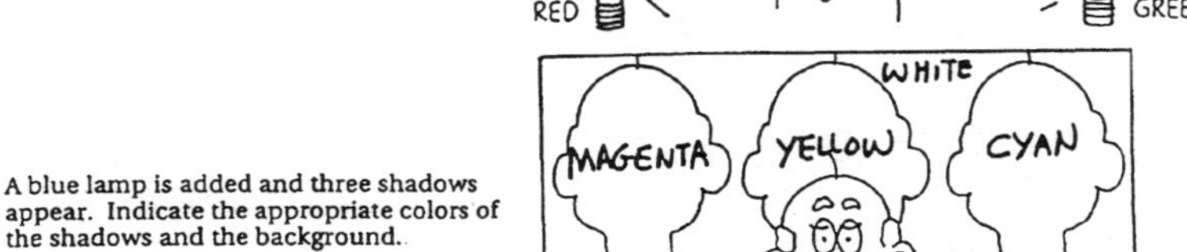

The lamps are placed closer together so the shadows overlap. Indicate the colors of all screen areas.

CONCEPTUAL Physics PRACTICE PAGE

If you have colored markers, have a go at these.

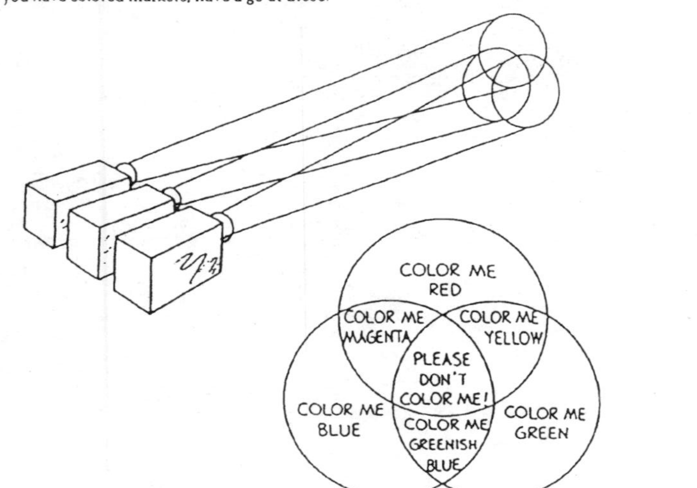

Name ______________________ Date ____________

CONCEPTUAL Physics PRACTICE PAGE

Chapter 28 Reflection and Refraction
Pool Room Optics

The law of reflection for optics is useful in playing pool. A ball bouncing off the bank of a pool table behaves like a photon reflecting off a mirror. As the sketch shows, angles become straight lines with the help of mirrors. The diagram to the right shows a top view of this, with a flattened "mirrored" region. Note that the angled path on the table appears as a straight line (dashed) in the mirrored region.

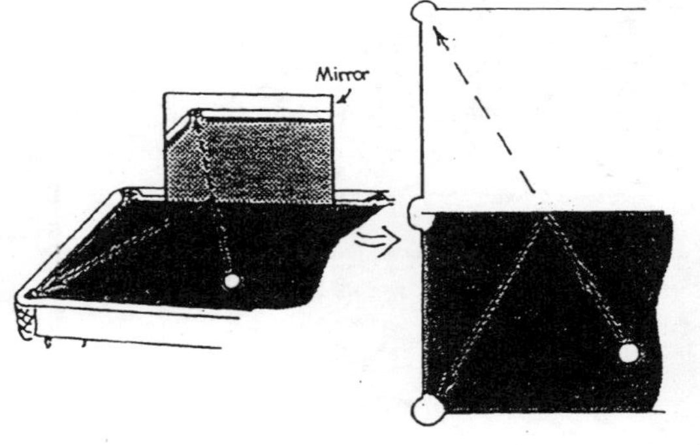

1. Consider a one-bank shot (one reflection) from the ball to the north bank and then into side pocket E.

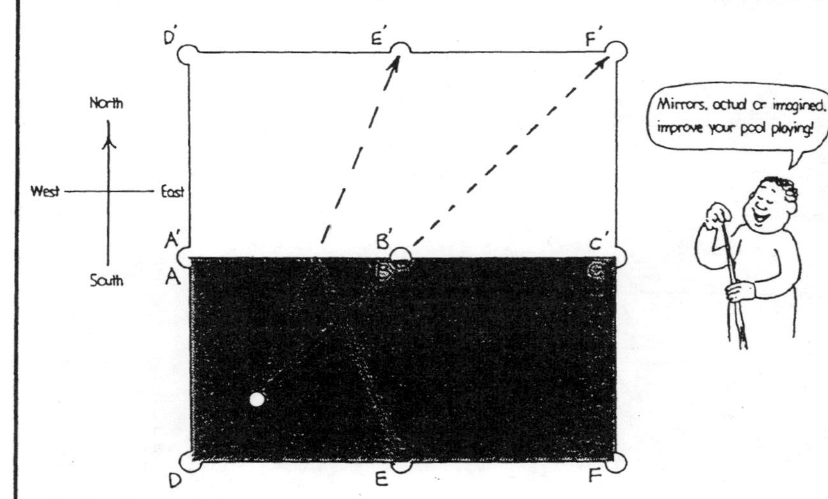

a. Use the mirror method to construct a straight line path to mirrored E'. Then construct the actual path to E.

b. Without using off-center strokes or other tricks, can a one-bank shot off the north bank put the ball in corner pocket F? NO; IT WOULD GO INTO B! Show why or why not using the diagram.

CONCEPTUAL Physics PRACTICE PAGE

2. Consider below a two-bank shot (two reflections) into corner pocket F. Here we use two mirrored regions. Note the straight line of sight to F", and how the north-bank impact point matches the intersection between B' and C'.
   a. Construct the similar path for a similar two-bank shot to get the ball in the side pocket E.

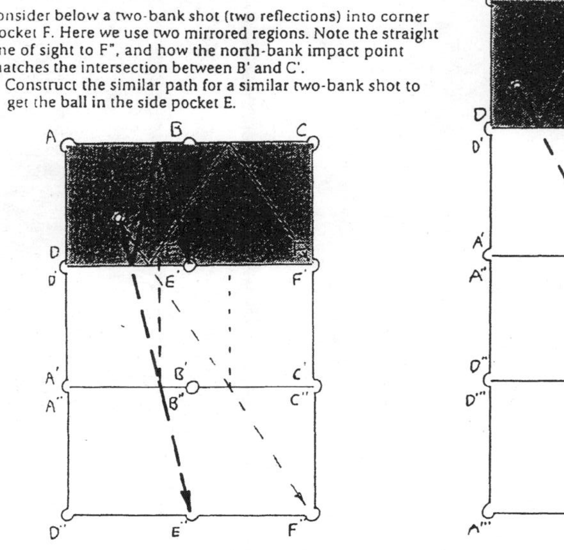

3. Consider above right a three-bank shot into corner pocket C, first bouncing against the south bank, then to the north, again to the south, and into pocket C.
   a. Construct the path. (First construct the single dashed line to C'''.)
   b. Construct the path to make a three-bank shot into side pocket B.

4. Let's try banking from adjacent banks of the table. Consider a two-bank shot to corner pocket F (first off the west bank, then to and off the north bank, then into F). Note how our two mirrored regions permit a straight-line path from the ball to F".

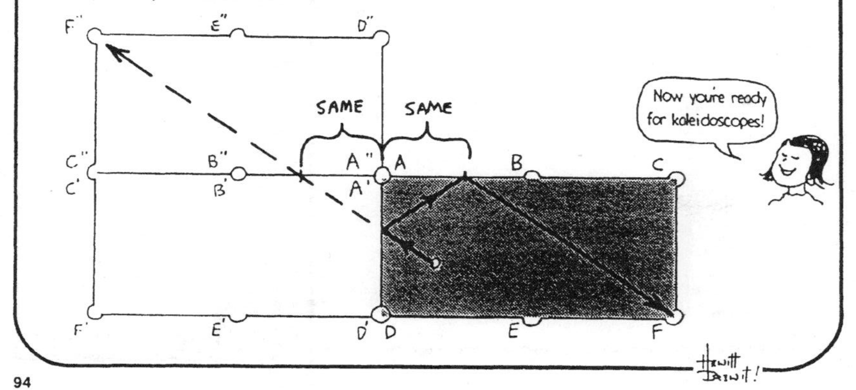

Name ______________________ Date ________

CONCEPTUAL Physics PRACTICE PAGE

Chapter 28 Reflection and Refraction

Reflection

Abe and Bev both look in a plane mirror directly in front of Abe (left, top view). Abe can see himself while Bev cannot see herself—but can Abe see Bev, and can Bev see Abe? To find the answer we construct their artificial locations "through" the mirror, the same distance behind as Abe and Bev are in front (right, top view). If straight-line connections intersect the mirror, as at point C, then each sees the other. The mouse, for example, cannot see or be seen by Abe and Bev.

Here we have eight students in front of a small plane mirror. Their positions are shown in the diagram below. Make appropriate straight-line constructions to answer the following:

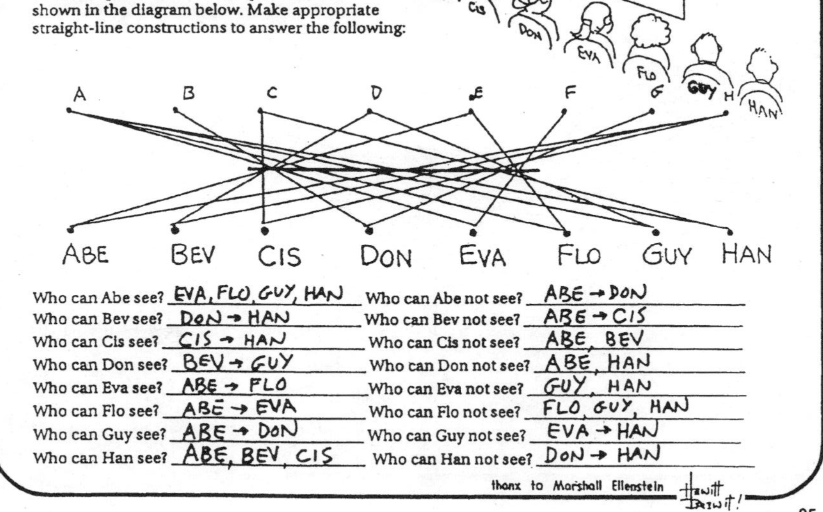

Who can Abe see?	EVA, FLO, GUY, HAN	Who can Abe not see?	ABE → DON
Who can Bev see?	DON → HAN	Who can Bev not see?	ABE → CIS
Who can Cis see?	CIS → HAN	Who can Cis not see?	ABE, BEV
Who can Don see?	BEV → GUY	Who can Don not see?	ABE, HAN
Who can Eva see?	ABE → FLO	Who can Eva not see?	GUY, HAN
Who can Flo see?	ABE → EVA	Who can Flo not see?	FLO, GUY, HAN
Who can Guy see?	ABE → DON	Who can Guy not see?	EVA → HAN
Who can Han see?	ABE, BEV, CIS	Who can Han not see?	DON → HAN

thanx to Marshall Ellenstein

CONCEPTUAL **Physics** PRACTICE PAGE

Six of our group are now arranged differently in front of the same mirror. Their positions are shown below. Make appropriate constructions for this more interesting arrangement, and answer the questions below.

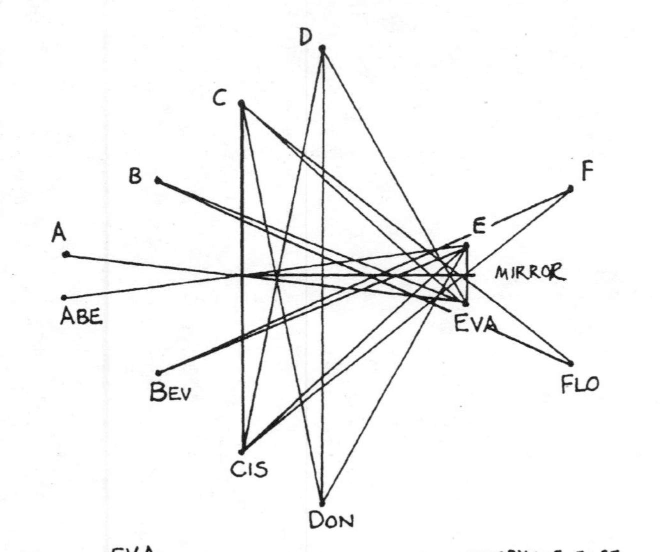

Who can Abe see? EVA	Who can Abe not see? EVERYONE ELSE
Who can Bev see? EVA, FLO	Who can Bev not see? EVERYONE ELSE
Who can Cis see? CIS, DON, EVA, FLO	Who can Cis not see? ABE, BEV
Who can Don see? CIS, DON, EVA	Who can Don not see? ABE, BEV, FLO
Who can Eva see? ABE, BEV, CIS, DON, EVA	Who can Eva not see? FLO
Who can Flo see? BEV, CIS	Who can Flo not see? ABE, DON, EVA, FLO

Harry Hotshot views himself in a full-length mirror (right). Construct straight lines from Harry's eyes to the image of his feet, and to the top of his head. Mark the mirror to indicate the minimum area Harry uses to see a full view of himself.

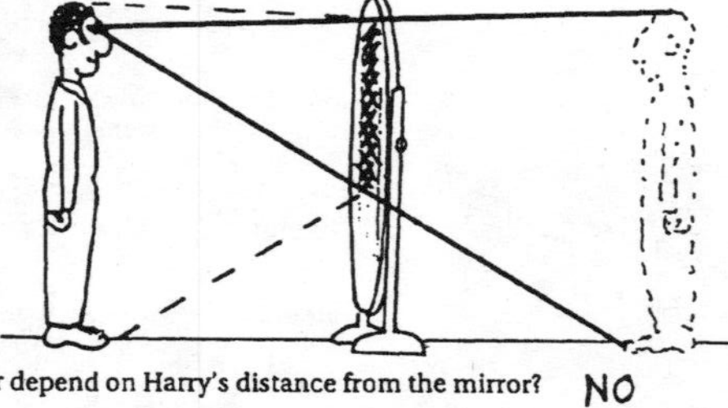

Does this region of the mirror depend on Harry's distance from the mirror? NO

CONCEPTUAL **Physics** PRACTICE PAGE

Chapter 28 Reflection and Refraction

Reflected Views

1. The ray diagram below shows the extension of one of the reflected rays from the plane mirror. Complete the diagram by (1) carefully drawing the three other reflected rays, and (2) extending them behind the mirror to locate the image of the flame. (Assume the candle and image are viewed by an observer on the left.)

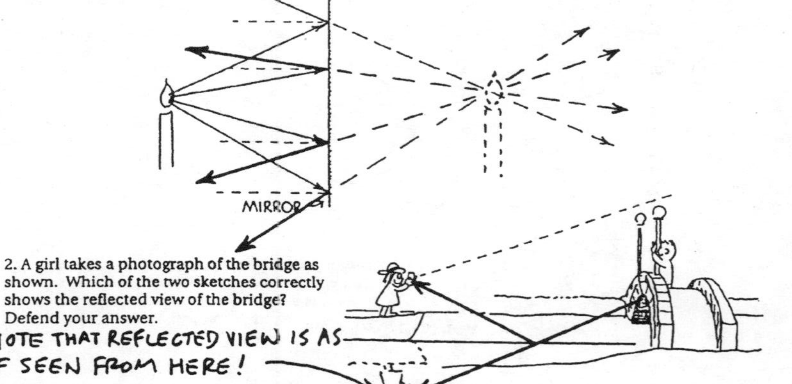

2. A girl takes a photograph of the bridge as shown. Which of the two sketches correctly shows the reflected view of the bridge? Defend your answer.

NOTE THAT REFLECTED VIEW IS AS IF SEEN FROM HERE!

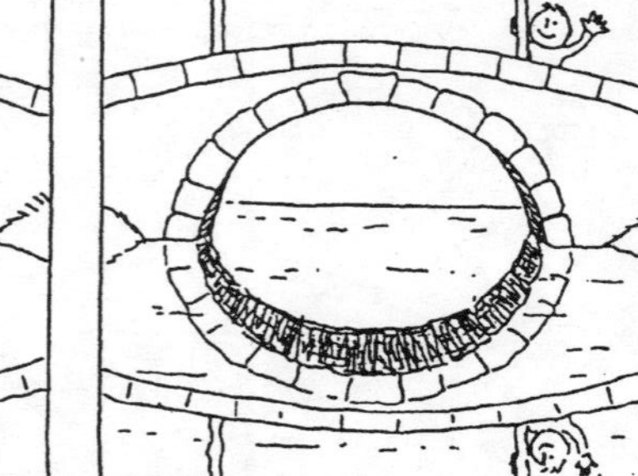

THE RIGHT-SIDE VIEW IS CORRECT, SHOWING THE UNDERSIDE OF BRIDGE, OR WHAT YOUR EYE WOULD SEE IF AS FAR BELOW THE REFLECTING SURFACE AS IT IS ABOVE. (PLACE A MIRROR ON THE FLOOR IN FRONT OF A TABLE. STUDENTS WILL SEE THAT THE REFLECTED VIEW OF THE TABLE SHOWS ITS BOTTOM!)

Name ______________________ Date __________

CONCEPTUAL **Physics** PRACTICE PAGE

Chapter 28 Reflection and Refraction

Refraction

1. A pair of toy cart wheels are rolled obliquely from a smooth surface onto two plots of grass — a rectangular plot as shown at the left, and a triangular plot as shown at the right. The ground is on a slight incline so that after slowing down in the grass, the wheels speed up again when emerging on the smooth surface. Finish each sketch and show some positions of the wheels inside the plots and on the other side. Clearly indicate their paths and directions of travel.

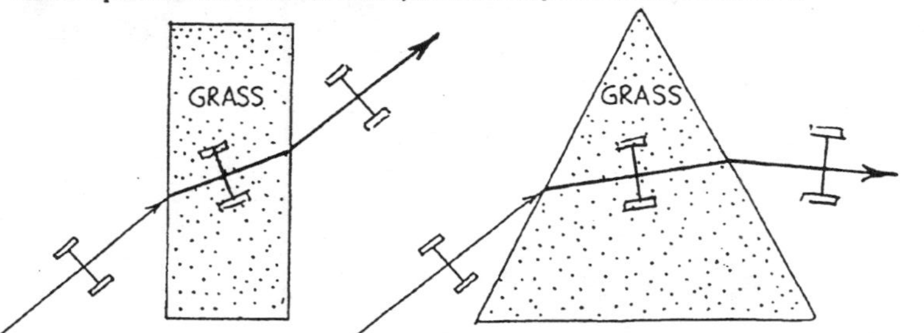

2. Red, green, and blue rays of light are incident upon a glass prism as shown. The average speed of red light in the glass is less than in air, so the red ray is refracted. When it emerges into the air it regains its original speed and travels in the direction shown. Green light takes longer to get through the glass. Because of its slower speed it is refracted as shown. Blue light travels even slower in glass. Complete the diagram by estimating the path of the blue ray.

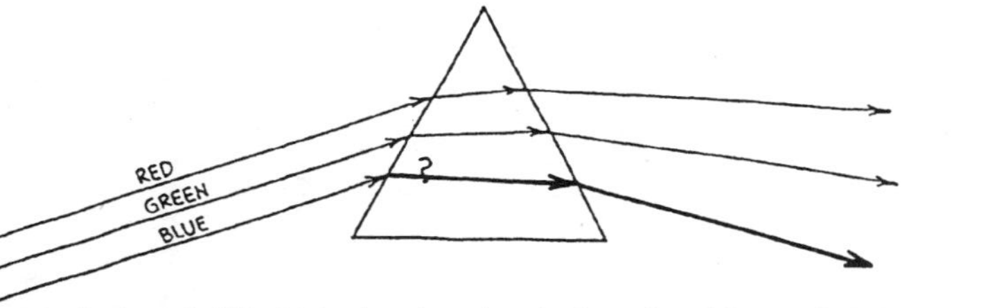

3. Suppose that instead of light slowing down in a prism, that it goes faster! Suppose that red light goes faster than green, and green light goes faster than blue. Show the relative directions of travel of these rays inside and beyond the prism.

LIGHT BENDS AWAY FROM NORMAL WHEN IT ENTERS PRISM

LIGHT BENDS TOWARD THE NORMAL WHEN EXITING

RED

GREEN

BLUE

CHALLENGING!

CONCEPTUAL **Physics** PRACTICE PAGE

4. Newton showed that when a second prism is placed in the dispersed light of a prism, the resulting beam is again white. Which arrangement of prisms will do this? ONLY IN b DO RAYS EMERGE PARALLEL

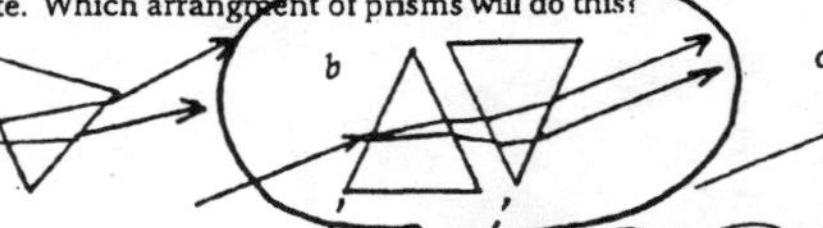

NOTE PARALLEL FACES!

5. The sketch shows that due to refraction, the man sees the fish closer to the water surface than it actually is.

a. Draw a ray beginning at the fish's eye to show the line of sight of the fish when it looks upward at 50° to the normal at the water surface. Draw the direction of the ray after it meets the surface of the water and continues in the air.

b. At the 50° angle, does the fish see the man, or does it see the reflected view of the starfish at the bottom of the pond? Explain.
FISH SEES REFLECTED VIEW OF STARFISH (50° > 48° CRITICAL ANGLE, SO THERE IS TOTAL INTERNAL REFLECTION)

c. To see the man, should the fish look higher or lower than the 50° path?
HIGHER, SO LINE OF SIGHT TO THE WATER IS LESS THAN 48° WITH NORMAL

d. If the fish's eye were barely above the water surface, it would see the world above in a 180° view, horizon to horizon. The fisheye view of the world above as seen beneath the water, however, is very different. Due to the 48° critical angle of water, the fish sees a normally 180° horizon-to-horizon view compressed within an angle of 96°.

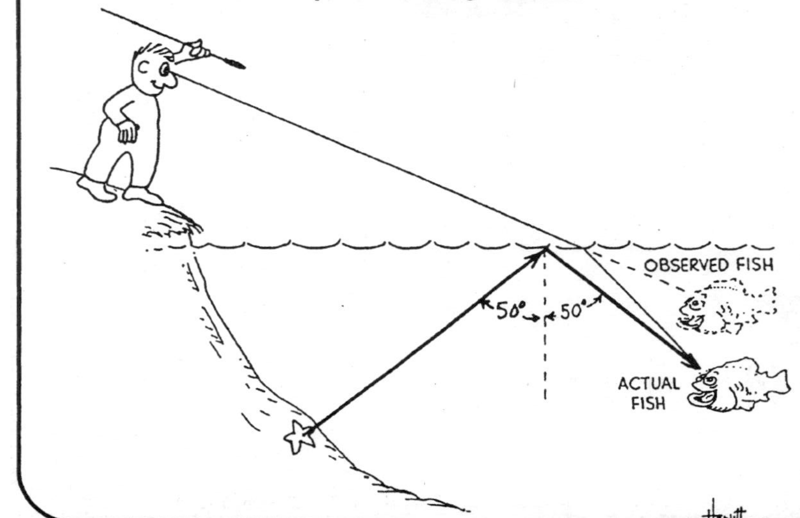

Name ______________________ Date __________

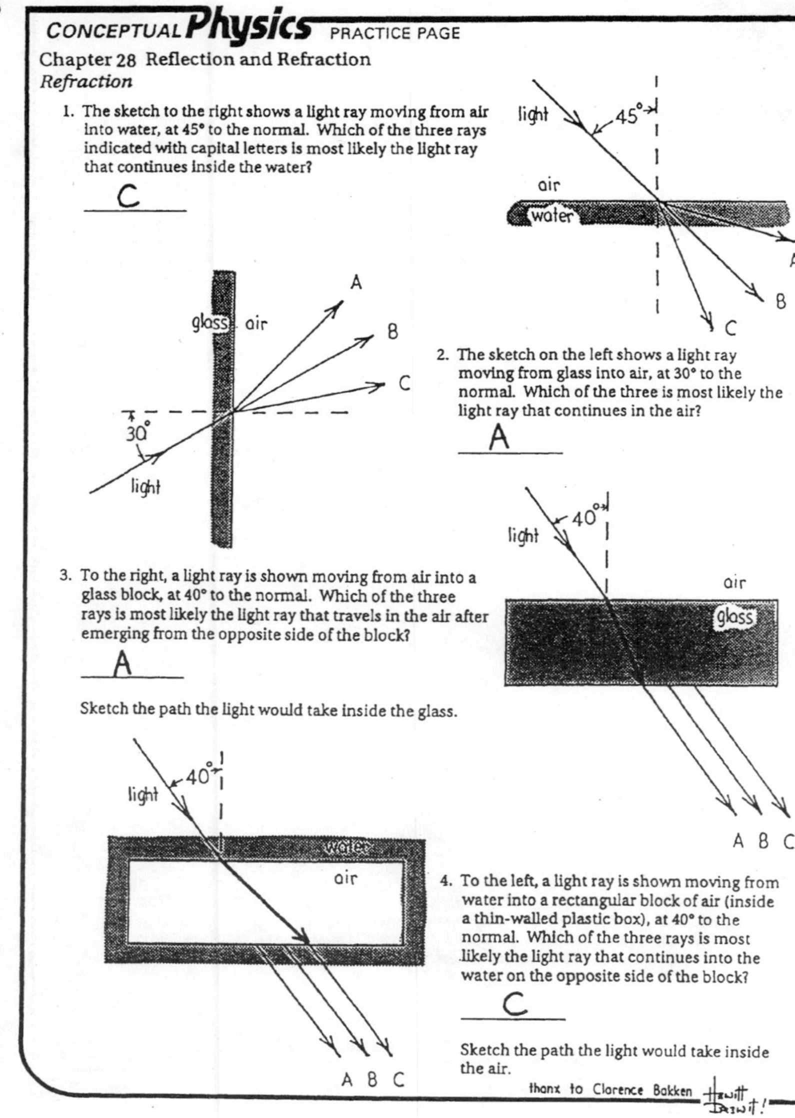

CONCEPTUAL **Physics** PRACTICE PAGE

Chapter 28 Reflection and Refraction

Refraction

1. The sketch to the right shows a light ray moving from air into water, at 45° to the normal. Which of the three rays indicated with capital letters is most likely the light ray that continues inside the water?

   C

2. The sketch on the left shows a light ray moving from glass into air, at 30° to the normal. Which of the three is most likely the light ray that continues in the air?

   A

3. To the right, a light ray is shown moving from air into a glass block, at 40° to the normal. Which of the three rays is most likely the light ray that travels in the air after emerging from the opposite side of the block?

   A

   Sketch the path the light would take inside the glass.

4. To the left, a light ray is shown moving from water into a rectangular block of air (inside a thin-walled plastic box), at 40° to the normal. Which of the three rays is most likely the light ray that continues into the water on the opposite side of the block?

   C

   Sketch the path the light would take inside the air.

thanx to Clarence Bakken

Hewitt Drew it!

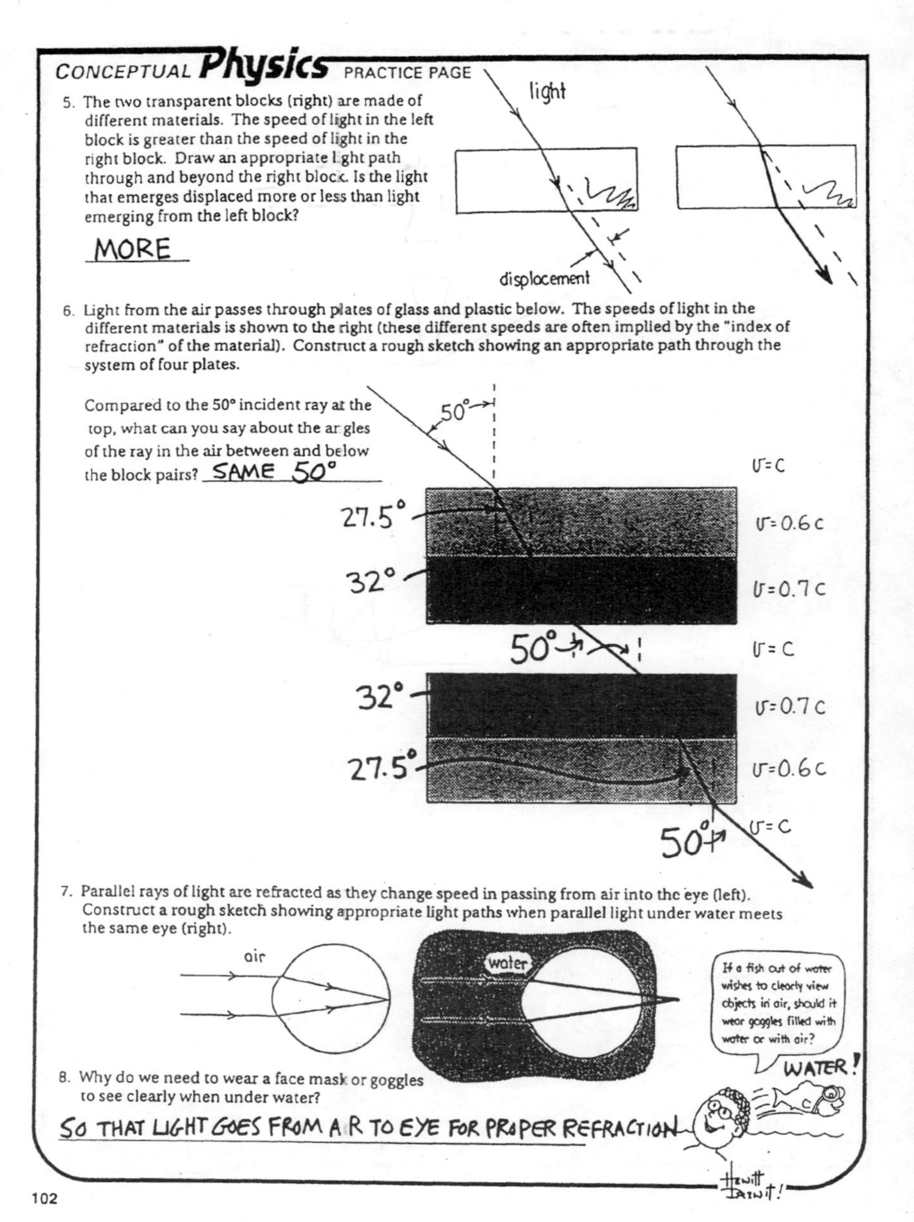

CONCEPTUAL **Physics** PRACTICE PAGE

5. The two transparent blocks (right) are made of different materials. The speed of light in the left block is greater than the speed of light in the right block. Draw an appropriate light path through and beyond the right block. Is the light that emerges displaced more or less than light emerging from the left block?

   MORE

6. Light from the air passes through plates of glass and plastic below. The speeds of light in the different materials is shown to the right (these different speeds are often implied by the "index of refraction" of the material). Construct a rough sketch showing an appropriate path through the system of four plates.

   Compared to the 50° incident ray at the top, what can you say about the angles of the ray in the air between and below the block pairs? SAME 50°

7. Parallel rays of light are refracted as they change speed in passing from air into the eye (left). Construct a rough sketch showing appropriate light paths when parallel light under water meets the same eye (right).

8. Why do we need to wear a face mask or goggles to see clearly when under water?

   SO THAT LIGHT GOES FROM AIR TO EYE FOR PROPER REFRACTION

Hewitt Drew it!

Name ____________________ Date __________

CONCEPTUAL **Physics** PRACTICE PAGE

Chapter 28 Reflection and Refraction

Lenses

Rays of light bend as shown when passing through the glass blocks.

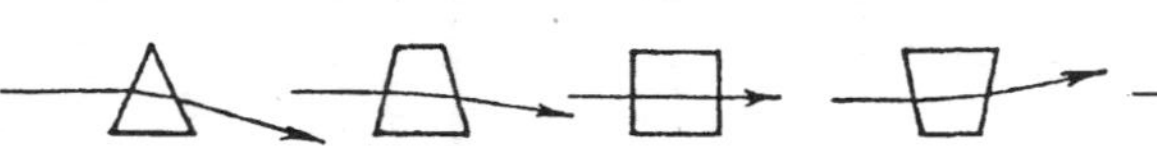

1. Show how light rays bend when they pass through the arrangement of glass blocks shown below.

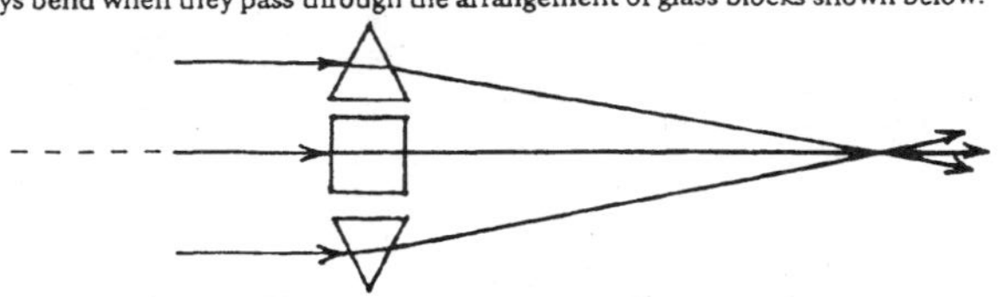

2. Show how light rays bend when they pass through the lens shown below. Is the lens a converging or a diverging lens? What is your evidence?
CONVERGING, AS EVIDENT IN THE CONVERGING RAYS

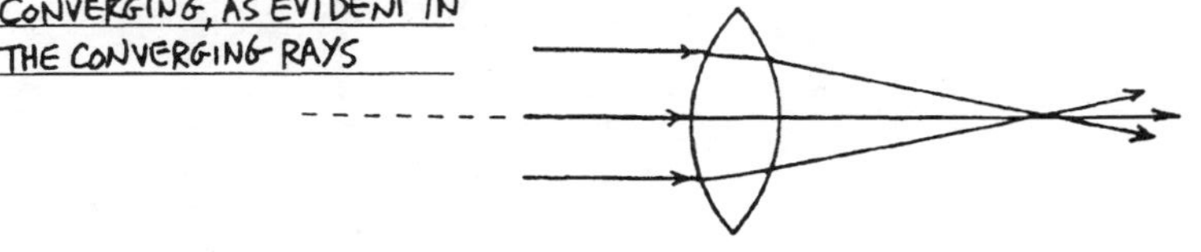

3. Show how light rays bend when they pass through the arrangement of glass blocks shown below.

4. Show how light rays bend when they pass through the lens shown below. Is the lens a converging or a diverging lens? What is your evidence?
DIVERGING LENS, AS EVIDENT IN THE DIVERGING RAYS

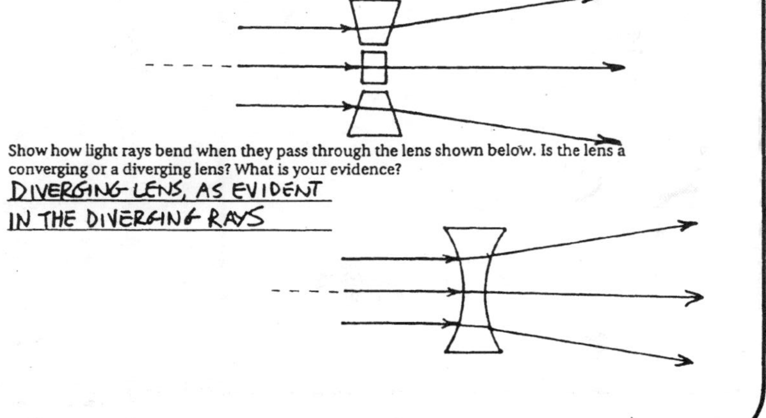

CONCEPTUAL **Physics** PRACTICE PAGE

5. Which type of lens is used to correct farsightedness? CONVERGING LENS

Nearsightedness? DIVERGING LENS

6. Construct rays to find the location and relative size of the arrow's image for each of the lenses below. Rays that pass through the middle of a lens continue undeviated. In a converging lens, rays from the tip of the arrow that are parallel to the optic axis extend through the far focal point after going through the lens. Rays that go through the near focal point go parallel to the axis after going through the lens. In a diverging lens, rays parallel to the axis diverge and appear to originate from the near focal point after passing though the lens. Have fun!

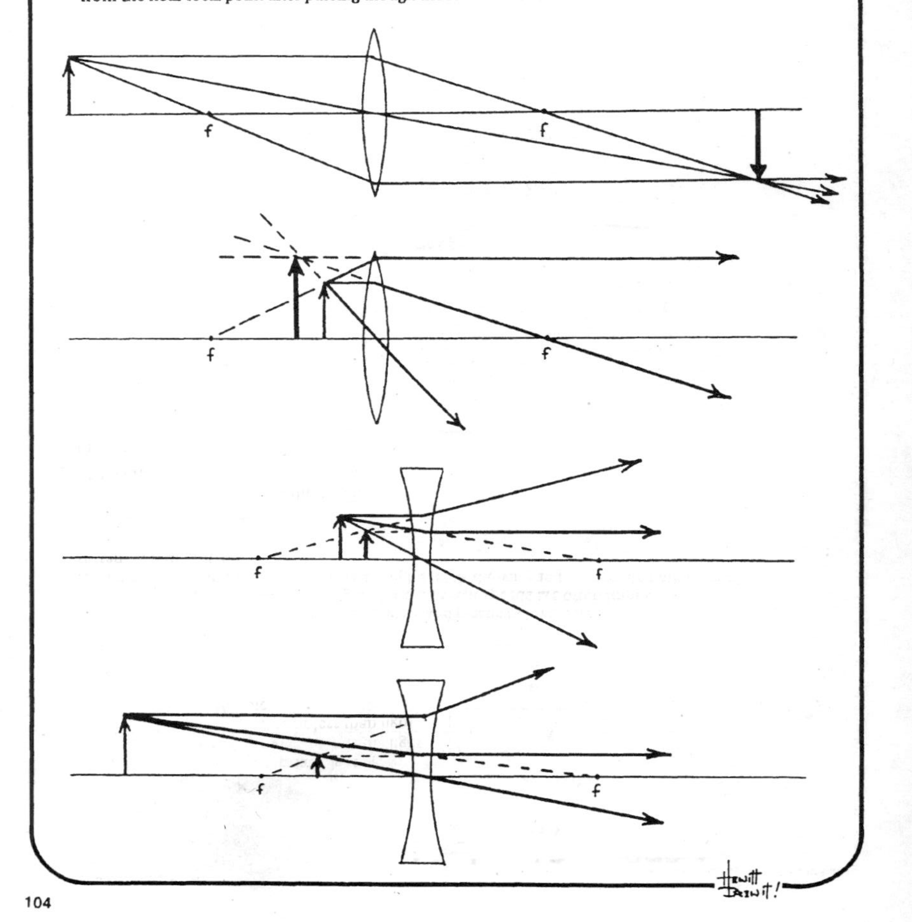

Name ______________________ Date ________

CONCEPTUAL **Physics** PRACTICE PAGE

Chapter 29 Light Waves
Diffraction and Interference

1. Shown below are concentric solid and dashed circles, each different in radius by 1 cm. Consider the circular pattern a top view of water waves, where the solid circles are crests and the dashed circles are troughs.

   a. Draw another set of the same concentric circles with a compass. Choose any part of the paper for your center (except the present central point). Let the circles run off the edge of the paper.

   b. Find where a dashed line crosses a solid line and draw a large dot at the intersection. Do this for ALL places where a solid and dashed line intersect.

   c. With a wide felt marker, connect the dots with smooth lines. These *nodal lines* lie in regions where the waves have cancelled — where the crest of one wave overlaps the trough of another (see Figures 29.15 and 29.16)

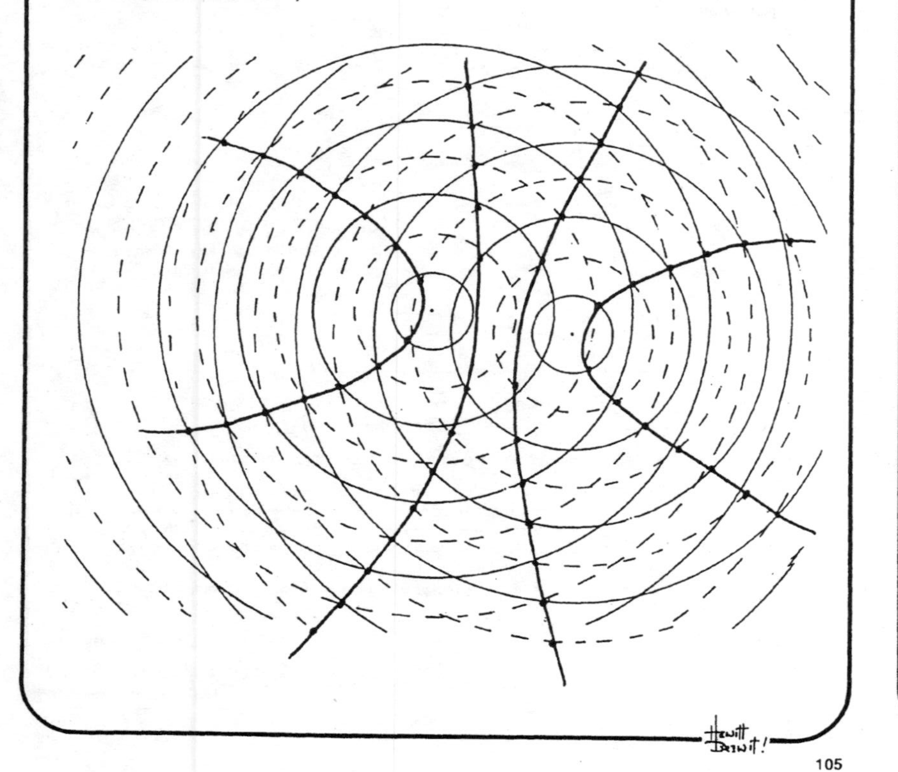

CONCEPTUAL **Physics** PRACTICE PAGE

2. Look at the construction of overlapping circles on your classmates' papers. Some will have more nodal lines than others, due to different starting points. How does the number of nodal lines in a pattern relate to the distance between centers of circles, (or sources of waves)?
   THE FARTHER APART THE CENTERS (OR WAVE SOURCES) THE MORE NODAL LINES. (NOTE IN FIGURE 29.15 IN YOUR TEXT MORE NODAL LINES IN THE RIGHT PATTERN COMPARED WITH CLOSER SOURCES IN THE CENTRAL PATTERN.)

3. Figure 29.19 from your text is repeated below. Carefully count the number of wavelengths (same as the number of wave crests) along the following paths between the slits and the screen.

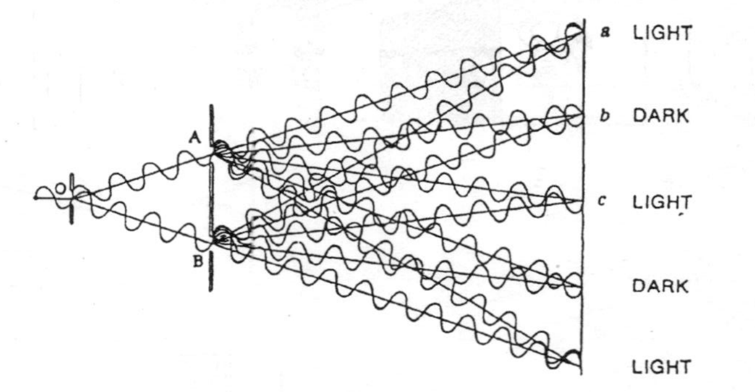

   a. Number of wavelengths between slit A and point a = 10.5
   b. Number of wavelengths between slit B and point a = 11.5
   c. Number of wavelengths between slit A and point b = 10.0
   d. Number of wavelengths between slit B and point b = 10.5
   e. Number of wavelengths between slit A and point c = 10.0
   f. Number of wave crests between slit B and point c = 10.0

When the number of wavelengths along each path is the same or differs by one or more whole wavelengths, interference is
(constructive) (destructive). [constructive circled]

and when the number of wavelengths differ by a half wavelength (or odd multiples of a half wavelength), interference is
(constructive) (destructive) [destructive circled]

It's nice how knowing some physics really changes the way we see things!

Name ____________________ Date ________

CONCEPTUAL Physics PRACTICE PAGE

Chapter 29 Light Waves
Polarization

The amplitude of a light wave has magnitude and direction, and can be represented by a vector. Polarized light vibrates in a single direction and is represented by a single vector. To the left the single vector represents vertically polarized light. The vibrations of non-polarized light are equal in all directions. There are as many vertical components as horizontal components. The pair of perpendicular vectors to the right represents non-polarized light.

1. In the sketch below non-polarized light from a flashlight strikes a pair of Polaroid filters.

NON-POLARIZED LIGHT VIBRATES IN ALL DIRECTIONS
HORIZONTAL AND VERTICAL COMPONENTS
VERTICAL COMPONENT PASSES THROUGH FIRST POLARIZER
...AND THE SECOND
VERTICAL COMPONENT DOES NOT PASS THROUGH THIS SECOND POLARIZER

a. Light is transmitted by a pair of Polaroids when their axes are

(aligned) (crossed at right angles) — answer circled: (aligned)

and light is blocked when their axes are

(aligned) (crossed at right angles) — answer circled: (crossed at right angles)

b. Transmitted light is polarized in a direction

(the same as) (different than) the polarization axis of the filter. — answer circled: (the same as)

2. Consider the transmission of light through a pair of Polaroids with polarization axes at 45° to each other. Although in practice the Polaroids are one atop the other, we show them spread out side by side below. From left to right: (*a*) Nonpolarized light is represented by its horizontal and vertical components. (*b*) These components strike filter A. (*c*) The vertical component is transmitted, and (*d*) falls upon filter B. This vertical component is not aligned with the polarization axis of filter B, but it has a component that is — component *t*; (*e*) which is transmitted.

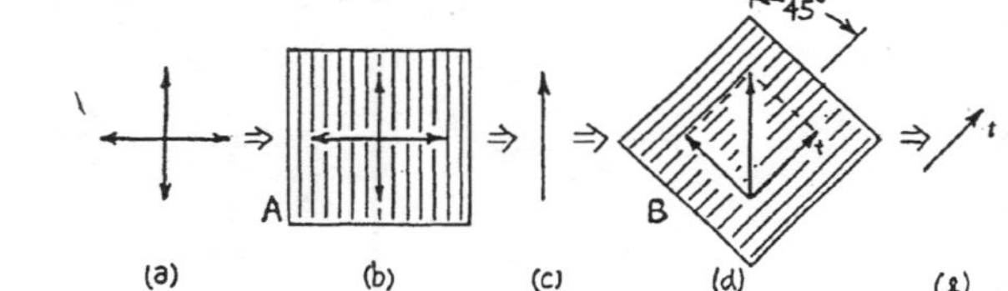

a. The amount of light that gets through Filter B, compared to the amount that gets through Filter A is

(more) (less) (the same) — answer circled: (less)

b. The component perpendicular to *t* that falls on Filter B is

(also transmitted) (absorbed) — answer circled: (absorbed)

CONCEPTUAL Physics PRACTICE PAGE

3. Below are a pair of Polaroids with polarization axes at 30° to each other. Carefully draw vectors and appropriate components (as in Question 2) to show the vector that emerges at *e*.

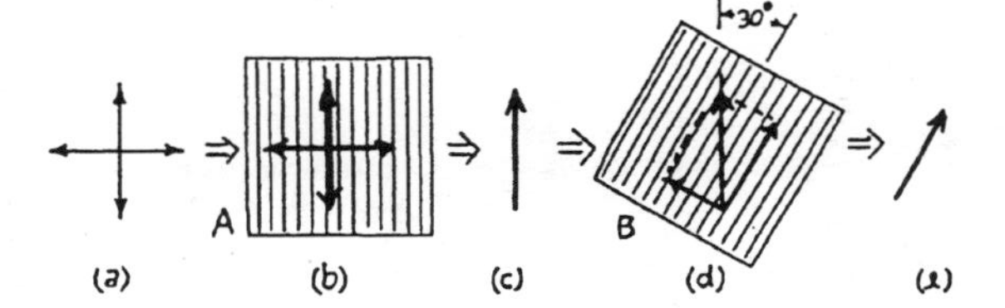

a. The amount of light that gets through the Polaroids at 30°, compared to the amount that gets though the 45° Polaroids is

(less) (more) (the same) — answer circled: (more)

4. Figure 29.35 in your textbook shows the smile of Ludmila Hewitt emerging through three Polaroids. Use vector diagrams to complete steps *b* through *g* below to show how light gets through the three-Polaroid system.

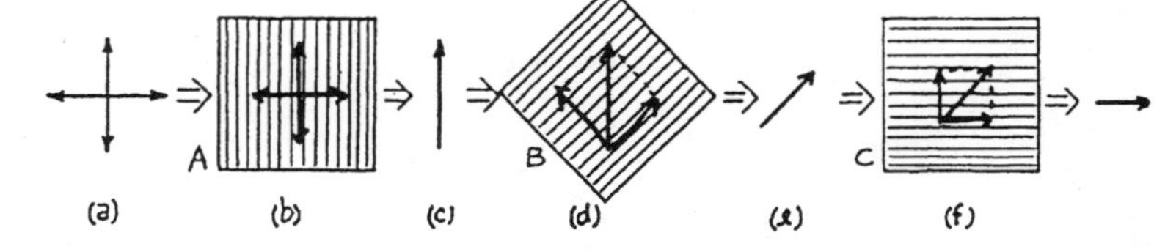

5. A novel use of polarization is shown below. How do the polarized side windows in these next-to-each-other houses provide privacy for the occupants? (Who can see what?)

OCCUPANTS CAN SEE OUTSIDE VIEWS NORMALLY, BUT IF SIDE WINDOWS ARE POLARIZED WITH AXES AT 90° TO EACH OTHER, THEN FROM INSIDE THEIR HOMES THEY CANNOT SEE THROUGH THE SIDE WINDOWS OF THEIR NEIGHBORS.

Name ______________________ Date __________

CONCEPTUAL Physics PRACTICE PAGE

Chapters 31 and 32 Light Quanta, and The Atom and the Quantum

Light Quanta

1. To say that light is quantized means that light is made up of
   (elemental units) (waves)

2. Compared to photons of low-frequency light, photons of higher-frequency light have more
   (energy) (speed) (quanta)

3. The photoelectric effect supports the
   (wave model of light) **(particle model of light)**

4. The photoelectric effect is evident when light shone on certain photosensitive materials ejects
   (photons) **(electrons)**

5. 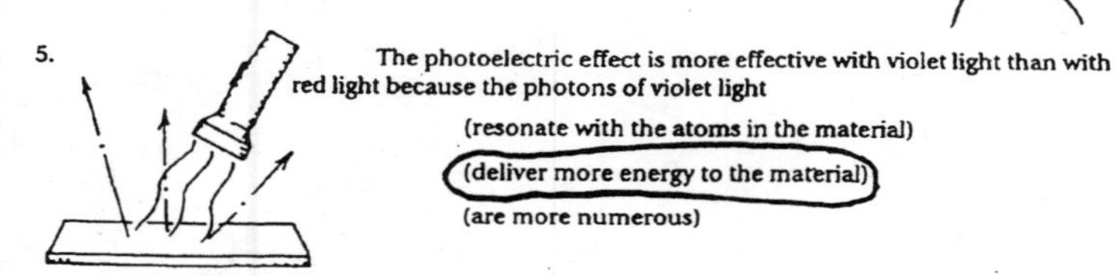

   The photoelectric effect is more effective with violet light than with red light because the photons of violet light
   (resonate with the atoms in the material)
   (deliver more energy to the material)
   (are more numerous)

6. According to De Broglie's wave model of matter, a beam of light and a beam of electrons
   (are fundamentally different) **(are similar)**

7. According to De Broglie, the greater the speed of an electron beam, the
   (greater is its wavelength) **(shorter is its wavelength)**

8. The discreteness of the energy levels of electrons about the atomic nucleus is best understood by considering the electron to be a
   (wave) (particle)

9. Heavier atoms are not appreciably larger in size than lighter atoms. The main reason for this is that the greater nuclear charge
   (pulls surrounding electrons into tighter orbits)
   (holds more electrons about the atomic nucleus) (produces a denser atomic structure)

10. Whereas in the everyday macroworld the study of motion is called *mechanics* in the microworld the study of quanta is called
    (Newtonian mechanics) **(quantum mechanics)**

A QUANTUM MECHANIC!

CONCEPTUAL Physics PRACTICE PAGE

Chapter 33 Atomic Nucleus and Radioactivity

1. *Complete the following statements.*

   a. A lone neutron spontaneously decays into a proton plus an ELECTRON.

   b. Alpha and beta rays are made of streams of particles, whereas gamma rays are streams of PHOTONS.

   c. An electrically charged atom is called an ION.

   d. Different ISOTOPES of an element are chemically identical but differ in the number of neutrons in the nucleus.

   e. Transuranic elements are those beyond atomic number 92.

   f. If the amount of a certain radioactive sample decreases by half in four weeks, in four more weeks the amount remaining should be ONE-FOURTH the original amount.

   g. Water from a natural hot spring is warmed by RADIOACTIVITY inside the earth.

2. The gas in the little girl's balloon is made up of former alpha and beta particles produced by radioactive decay.

   a. If the mixture is electrically neutral, how many more beta particles than alpha particles are in the balloon?
   THERE ARE TWICE AS MANY BETA PARTICLES AS ALPHA PARTICLES

   b. Why is your answer not "same"?
   AN ALPHA HAS DOUBLE CHARGE; THE CHARGE OF 2 BETAS = MAGNITUDE OF CHARGE OF 1 ALPHA PARTICLE.

   c. Why are the alpha and beta particles no longer harmful to the child?
   THEY HAVE LONG LOST THEIR HIGH KE, WHICH IS NOW REDUCED TO THE THERMAL ENERGY OF RANDOM MOLECULAR MOTION.

   d. What element does this mixture make?
   HELIUM

Name ______________________ Date ________

CONCEPTUAL **Physics** PRACTICE PAGE

Chapter 33 Atomic Nucleus and Radioactivity

Natural Transmutation

Fill in the decay-scheme diagram below, similar to that shown in Figure 33.13 in your textbook, but beginning with U-235 and ending up with an isotope of lead. Use the table at the left, and identify each element in the series with its chemical symbol.

Step	Particle emitted
1	Alpha
2	Beta
3	Alpha
4	Alpha
5	Beta
6	Alpha
7	Alpha
8	Alpha
9	Beta
10	Alpha
11	Beta
12	Stable

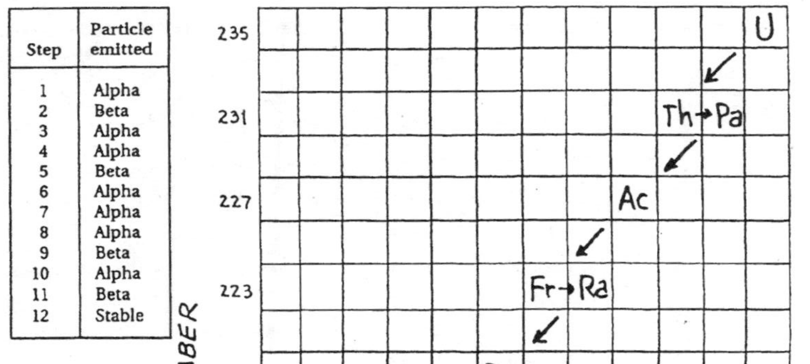

What isotope is the final product? $^{207}_{82}Pb$ (LEAD-207)

CONCEPTUAL **Physics** PRACTICE PAGE

Nuclear Reactions

Complete these nuclear reactions.

THORIUM LATE, I OVERTHLEPT !

1. $^{238}_{92}U \rightarrow {}^{234}_{90}Th + {}^{4}_{2}\underline{He}$
2. $^{234}_{90}Th \rightarrow {}^{234}_{91}Pa + {}^{0}_{-1}\underline{e}$
3. $^{234}_{91}Pa \rightarrow \underline{^{230}_{89}Ac} + {}^{4}_{2}He$
4. $^{220}_{86}Rn \rightarrow \underline{^{216}_{84}Po} + {}^{4}_{2}He$
5. $^{216}_{84}Po \rightarrow \underline{^{216}_{85}At} + {}^{0}_{-1}e$
6. $^{216}_{84}Po \rightarrow \underline{^{212}_{82}Pb} + {}^{4}_{2}He$
7. $^{210}_{83}Bi \rightarrow \underline{^{210}_{84}Po} + {}^{0}_{-1}e$
8. $^{1}_{0}n + {}^{10}_{5}B \rightarrow \underline{^{7}_{3}Li} + {}^{4}_{2}He$

NUCLEAR PHYSICS--- IT'S THE SAME TO ME WITH THE FIRST TWO LETTERS INTERCHANGED!

Name ______________________ Date ________

CONCEPTUAL Physics PRACTICE PAGE

Chapter 34 Nuclear Fission and Fusion

1. Complete the table for a chain reaction in which two neutrons from each step individually cause a new reaction.

EVENT	1	2	3	4	5	6	7
NO. OF REACTIONS	1	2	4	8	16	32	64

2. Complete the table for a chain reaction where three neutrons from each reaction cause a new reaction.

EVENT	1	2	3	4	5	6	7
NO. OF REACTIONS	1	3	9	27	81	243	729

3. Complete these beta reactions, which occur in a breeder reactor.

$$^{239}_{92}U \rightarrow {}^{239}_{93}Np + {}^{0}_{-1}e$$

$$^{239}_{93}Np \rightarrow {}^{239}_{94}Pu + {}^{0}_{-1}e$$

4. Complete the following fission reactions.

$$^{1}_{0}n + {}^{235}_{92}U \rightarrow {}^{143}_{54}Xe + {}^{90}_{38}Sr + 3\,({}^{1}_{0}n)$$

$$^{1}_{0}n + {}^{235}_{92}U \rightarrow {}^{152}_{60}Nd + {}^{80}_{32}Ge + 4\,({}^{1}_{0}n)$$

$$^{1}_{0}n + {}^{239}_{94}Pu \rightarrow {}^{141}_{54}Xe + {}^{97}_{40}Zr + 2\,({}^{1}_{0}n)$$

5. Complete the following fusion reactions.

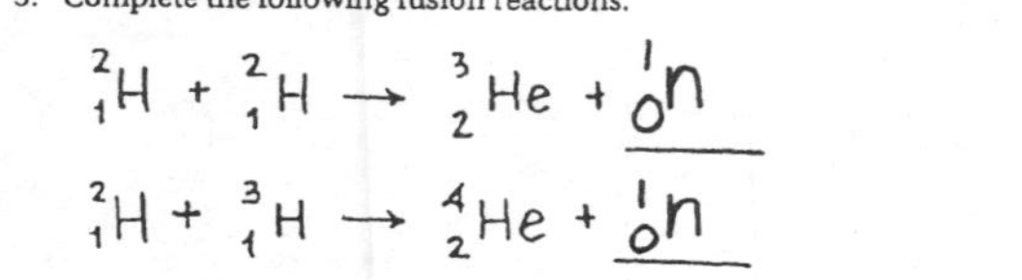

$$^{2}_{1}H + {}^{2}_{1}H \rightarrow {}^{3}_{2}He + {}^{1}_{0}n$$

$$^{2}_{1}H + {}^{3}_{1}H \rightarrow {}^{4}_{2}He + {}^{1}_{0}n$$

Name ______________________ Date ________

CONCEPTUAL Physics PRACTICE PAGE

Chapter 35 Special Theory of Relativity
Time Dilation

Chapter 35 in your textbook discusses *The Twin Trip*, in which a traveling twin takes a 2-hour journey while a stay-at-home brother records the passage of 2 1/2 hours. Quite remarkable! Times in both frames of reference are marked by flashes of light, sent each 6 minutes from the spaceship, and received on earth at 12-minute intervals for the ship going away, and 3-minute intervals for the ship returning. Read this section in the book carefully, and fill in the clock readings aboard the spaceship when each flash is emitted, and the clock reading on earth when each flash is received.

SHIP LEAVING EARTH

FLASH	TIME ON SHIP WHEN FLASH SENT	TIME ON EARTH WHEN FLASH SEEN
0	12:00	12:00
1	12:06	12:12
2	12:12	12:24
3	12:18	12:36
4	12:24	12:48
5	12:30	1:00
6	12:36	1:12
7	12:42	1:24
8	12:48	1:36
9	12:54	1:48
10	1:00	2:00

SHIP APPROACHING EARTH

FLASH	TIME ON SHIP WHEN FLASH SEEN	TIME ON EARTH WHEN FLASH SEEN
11	1:06	2:03
12	1:12	2:06
13	1:18	2:09
14	1:24	2:12
15	1:30	2:15
16	1:36	2:18
17	1:42	2:21
18	1:48	2:24
19	1:54	2:27
20	2:00	2:30

THIS CHECKS: FOR $v = 0.6c$

$$t = \frac{t_0}{\sqrt{1-(\frac{v}{c})^2}} = \frac{2\text{ HR}}{\sqrt{1-(\frac{0.6c}{c})^2}} = 2.5\text{ HR}$$

Answers to the Odd-Numbered Exercises and Problems from *Conceptual Physics—Ninth Edition*

Chapter 1: About Science
Answers to Odd-Numbered Exercises

1. a. This is a scientific hypothesis, for there is a test for wrongness. For example, you can extract chlorophyll from grass and note its color. b. This statement is without a means of proving it wrong and is not a scientific hypothesis. It is specul-ation. c. This is a scientific hypothesis. It coould be proved wrong, for example, by showing tides that do not correspond to the position of the moon.

3. To publicly change your mind about your ideas is a sign of strength rather than a sign of weakness. It takes more courage to change your ideas when confronted with counter evidence than to hold fast to your ideas. If a person's ideas and view of the world are no different after a lifetime of varied experience, then that person was either miraculously blessed with unusual wisdom at an early age, or learned nothing. The latter is more likely. Education is learning that which you don't yet know about. It would be arrogant to think you know it all in the later stages of your education, and stupid to think so at the beginning of your education.

5. The sun's radius is about 7×10^8 m. The distance between the Earth and moon is about 4×10^8 m. So the sun's radius is much larger, nearly twice the distance between the Earth and moon. The Earth and moon at their present distance from each other would easily fit inside the sun. The sun is *really* big — surprisingly big!

7. What is likely being misunderstood is the distinction between theory and hypothesis. In common usage, "theory" may mean a guess or hypothesis, something that is tentative or speculative. But in science a theory is a synthesis of a large body of validated information (e.g., cell theory or quantum theory). The value of a theory is its usefulness (not its "truth").

Problem Answer

By simple geometry, the ratios are equal. That is

$$\frac{\text{coin diameter}}{\text{coin distance}} = \frac{\text{moon diameter}}{\text{moon distance}}$$

Simple rearrangement gives

$$\text{moon diameter} = \frac{\text{coin diameter}}{\text{coin distance}} \times \text{moon distance} =$$

$$\frac{1}{110}\ 3.84 \times 10^5 \text{ km} = 3.5 \times 10^3 \text{ km}.$$

Chapter 2: Newton's First Law of Motion—Inertia
Answers to Odd-Numbered Exercises

1. Aristotle would say that the ball comes to rest because the ball seeks its natural state of rest. Galileo would likely have said it comes to rest because of some force acting on it—likely friction between the ball and table surface and with the air.

3. He discredited Aristotle's idea that the rate at which bodies fall is proportional to their weight.

5. Galileo came up with the concept of inertia before Newton was born.

7. Nothing keeps the probe moving. In the absence of a propelling force it would continue moving in a straight line.

9. You should disagree with your friend. In the absence of external forces, a body at rest tends to remain at rest; if moving, it tends to remain moving. Inertia is a *property* of matter to behave this way, not some kind of force.

11. The tendency of the ball is to remain at rest. From a point of view outside the wagon, the ball stays in place as the back of the wagon moves toward it. (Because of friction, the ball may roll along the cart surface—without friction the surface would slide beneath the ball.)

13. In a car at rest your head tends to stay at rest. When the car is rear ended, the car lurches forward and if the headrest isn't there, tends to leave your head behind. Hence a neck injury.

15. An object in motion tends to stay in motion, hence the discs tend to compress upon each other just as the hammer head is compressed onto the handle in Figure 2.4. This compression results in people being slightly shorter at the end of the day than in the morning. The discs tend to separate while sleeping in a prone position, so you regain your full height by morning. This is easily noticed if you find a point you can almost reach up to in the evening, and then find it is easily reached in the morning. Try it and see!

17. If there were no force acting on the ball it would continue in motion without slowing. But air drag does act, along with slight friction with the lane, and the ball slows. This doesn't violate the law of inertia because external forces indeed act.

19. If only a single non-zero force acts on an object, it will not be in mechanical equilibrium. There would have to be another or other forces to result in a zero net force for equilibrium.

21. If the puck moves in a straight line with unchanging speed, the forces of friction are negligible. Then the net force is practically zero, and the puck can be considered to be in dynamic equilibrium.

23. From the equilibrium rule, $\Sigma F = 0$, the upward forces are 400 N + tension in the right scale. This sum must equal the downward forces 250 N + 300 N + 300 N. Arithmetic shows the reading on the right scale is 450 N.

25. In the left figure, Harry is supported by two strands of rope that share his weight (like the little girl in the previous exercise). So each strand supports only 250 N, below the breaking point. Total force up supplied by ropes equals weight acting downward, giving a net force of zero and no acceleration. In the right figure, Harry is now supported by one strand, which for Harry's well-being requires that the tension be 500 N. Since this is above the breaking point of the rope, it breaks. The net force on Harry is then only his weight, giving him a downward acceleration of g. The sudden return to zero velocity changes his vacation plans.

27. By the parallelogram rule, the tension is less than 50 N.

29. When standing on a floor, the floor pushes upward against your feet with a force equal to that of your weight. This upward force (called the *normal force*) and your weight are oppositely directed, and since they both act on the same body, you, they cancel to produce a net force on you of zero—hence, you are not accelerated.

31. You can say that no net force acts on a body at rest, but there may be any number of forces that act — that produce a zero net force. When the net force is zero, the book is in static equilibrium.

33. Friction on the cart has to be 200 N, opposite to your 200-N pull.

35. A stone will fall vertically if released from rest. If the stone is dropped from the top of the mast of a moving ship, the horizontal motion is not changed when the stone is dropped — providing air drag on the stone is negligible and the ship's motion is steady and straight. From the frame of reference of the moving ship, the stone falls in a vertical straight line path, landing at the base of the mast.

37. We aren't swept off because we are traveling just as fast as the Earth, just as in a fast-moving vehicle you move along with the vehicle. Also, there is no atmosphere through which the Earth moves, which would do more than blow our hats off!

39. This is similar to Exercise 38. If the ball is shot while the train is moving at constant velocity (constant speed in a straight line), its horizontal motion before, during, and after being fired is the same as that of the train; so the ball falls back into the chimney as it would have if the train were at rest. If the train changes speed, the ball will miss, because the ball's horizontal speed will match the train speed as the ball is fired, but not when the ball lands. Similarly, on a circular track the ball will also miss the chimney because the ball will move along a tangent to the track while the train turns away from this tangent. So the ball returns to the chimney in the first case, and misses in the second and third cases because of the *change* in motion.

Chapter 3: Linear Motion

Answers to Odd-Numbered Exercises

1. The impact speed will be the relative speed, 2 km/h (100 km/h – 98 km/h = 2 km/h).

3. Your fine for speeding is based on your instantaneous speed; the speed registered on a speedometer or a radar gun.

5. Constant velocity means no acceleration, so the acceleration of light is zero.

7. Yes, again, velocity and acceleration need not be in the same direction. A ball tossed upward, for example, reverses its direction of travel at its highest point while its acceleration g, directed downward, remains constant (this idea will be explained further in Chapter 4). Note that if a ball had zero acceleration at a point where its speed is zero, its speed would *remain* zero. It would sit still at the top of its trajectory!

9. "The dragster rounded the curve at a constant *speed* of 100 km/h." Constant velocity means not only constant speed, but constant direction. A car rounding a curve changes its direction of motion.

11. You cannot say which car underwent the greater acceleration unless you know the times involved.

13. An object moving in a circular path at constant speed is a simple example of acceleration at constant speed, because it's velocity is changing direction. No example can be given for the second case, for constant velocity means zero acceleration. You can't have a non-zero acceleration while having a constant velocity.

There are no examples of things that accelerate while not accelerating.

15. The acceleration of an object is in a direction opposite to its velocity when velocity is decreasing, i.e., a ball rising, or a car braking to a stop.

17. The one in the middle. That ball gains speed more quickly at the beginning where the slope is steeper, so its average speed is greater even though it has less acceleration in the last part of its trip.

19. The greater change in speeds occurs for the slower object; Acceleration (30 km/h - 25 km/h = 5 km/h) than for the faster one (100 km/h - 96 km/h = 4 km/h). So for the same time, the slower one has the greater acceleration.

21. Free fall is defined as falling only under the influence of gravity, with no air resistance or other non-gravitational forces. So your friend should omit "free" and say something like, "Air resistance is more effective in slowing a falling feather than a falling coin."

23. Distance readings would indicate greater distances fallen in successive seconds. During each successive second the object falls faster and covers greater distance.

25. In the absence of air resistance, the acceleration will be g no matter how the ball is released. The acceleration of a ball and It's speed are entirely different.

27. Both will strike the ground below at the same speed. That's because the ball thrown upward will pass its starting point on the way down with the same speed it had when starting up. So its trip on down is the same as for a ball thrown down with that speed.

29. If air resistance is not a factor, its acceleration is the same 10 m/s^2 regardless of its initial velocity. Thrown downward, its velocity will be greater, but not its acceleration.

31. Counting to twenty means twice the time. In twice the time the ball will roll 4 times as far (distance moved is proportional to the square of the time).

33. If it were not for the slowing effect of the air, raindrops would strike the ground with the speed of high-speed bullets!

35. The ball on B finishes first, for its average speed along the lower part as well as the down and up slopes is greater than the average speed of the ball along track A.

37. How you respond may or may not agree with the author's response: There are few pure examples in physics, for most real situations involve a combination of effects. There is usually a "first order" effect that is basic to the situation, but then there are 2nd, 3rd, and even 4th or more order effects that interact also. If we begin our study of some concept by considering all effects together before we have studied their contributions separately, understanding is likely to be difficult. To have a better understanding of what is going on, we strip a situation of all but the first order effect, and then examine that. When we have a good grip on that, then we proceed to investigate the other effects for a fuller understanding. Consider Kepler, for example, who made the stunning discovery that planets move in elliptical paths. Now we know that they don't quite move in perfect ellipses because each planet affects the motion of every other one. But if Kepler had been stopped by these second-order effects, he would not have made his groundbreaking discovery. Similarly, if Galileo hadn't been able to free his thinking from real-world friction he may not have made his great discoveries in mechanics.

39. On the moon the acceleration due to gravity is considerably less, so jumping height would be considerably more (six times higher in the same time!).

Solutions to Chapter 2 Odd-Numbered Problems

1. From $v = d/t, t = d/v$. We convert 3 m to 3000 mm, and $t = \frac{3000 \text{ mm}}{1.5 \text{ mm/year}}$ = **2000 years**.

3. Since it starts going up at 30 m/s and loses 10 m/s each second, its time going up is 3 seconds. Its time returning is also 3 seconds, so it's in the air for a total of 6 seconds. Distance up (or down) is $1/2\ gt^2 = 5 \times 3^2 =$ **45 m**. Or from $d = vt$, where average velocity is (30 + 0)/2 = 15 m/s, and time is 3 seconds, we also get d = 15 m/s × 3 s = 45 m.

5. Using g = 10 m/s^2, we see that $v = gt$ = (10 m/s^2)(10 s) = 100 m/s;
$$v = \frac{(v_{\text{beginning}} + v_{\text{final}})}{2} = \frac{(0 + 100)}{2}$$
= **50 m/s**, downward.
We can get "how far" from either $d = vt$ = (50 m/s)(10 s) = **500 m**,
or equivalently, $d = 1/2\ gt^2 = 5\ (10)^2$ = 500 m. (Physics is nice...we get the same distance using either formula!)

7. Average speed = total distance traveled/time taken = 1200 km/total time. Time for first leg of trip = 600 km/200 km/h = 3 h. Time for last leg is 600 km/300 km/h = 2 h. So total time is 5 h. Then average speed = 1200 km/5 h = **240 km/h**. (Note you can't use the formula average speed = beginning speed + end speed divided by two—which applies for constant acceleration only.)

9. Drops would be in free fall and accelerate at g. Gain in speed = gt, so we need to find the time of fall.
From $d = 1/2\, gt^2$, $t = \sqrt{2d/g} = \sqrt{2000 \text{ m}/10 \text{ m/s}^2} =$ 14.1 s. So gain in speed = $(10 \text{ m/s}^2)(14.1 \text{ s}) =$ **141 m/s** (more than 300 miles per hour)!

Chapter 4: Newton's Second Law of Motion Exercises

1. The net force is zero because the Mercedes is traveling at constant velocity, which means with zero acceleration.

3. No, for there may be any number of applied forces acting on it. All you can conclude is that if it isn't accelerating no *net* force acts on it.

5. At constant velocity the net force is zero, so friction also equals 1 N.

7. The only force that acts on a tossed rock on the moon is the gravitational force between the rock and the moon, because there is no air and therefore no air drag on the moon.

9. Sliding down at constant velocity means acceleration is zero and the net force is zero. This can occur if friction equals the bear's weight, which is 4000 N. Friction = bear's weight = mg = $(400 \text{ kg})(10 \text{ m/s}^2) = 4000$ N.

11. Shake the boxes. The box that offers the greater resistance to acceleration is the more massive box, the one containing the sand.

13. A massive cleaver is more effective in chopping vegetables because its greater mass contributes to greater tendency to keep moving as the cleaver chops.

15. Ten kilograms weighs about 100 N on the Earth (weight = mg = $10 \text{ kg} \times 10 \text{ m/s}^2 = 100$ N, or 98 N if $g = 9.8 \text{ m/s}^2$ is used). On the moon the weight would be 1/6 of 100 N = 16.7 N (or 16.3 N if $g =$ 9.8 m/s^2 is used). The mass would be 10 kg everywhere.

17. Weight and mass are directly proportional, so in any locality, when your mass increases your weight also increases.

19. To see why the acceleration gains as a rocket burns its fuel, look at the equation $a = F/m$. As fuel is burned, the mass of the rocket becomes less. As m decreases, a increases! There is simply less mass to be accelerated as fuel is consumed.

21. Rate of gain in speed (acceleration), is the ratio force/mass (Newton's second law), which in free fall is just weight/mass. Since weight is proportional to mass, the ratio weight/mass is the same whatever the weight of a body. So all freely falling bodies undergo the same gain in speed — g (illustrated in Figure 4.14). Although weight doesn't affect speed in free fall, weight does affect falling speed when air resistance is present (non-free fall).

23. The forces acting horizontally are the driving force provided by friction between the tires and the road, and resistive forces—mainly air drag. These forces cancel and the car is in dynamic equilibrium with a net force of zero.

25. Note that 30 N pulls 3 blocks. To pull 2 blocks then requires a 20-N pull, which is the tension in the rope between the second and third block. Tension in the rope that pulls only the third block is therefore 10 N. (Note that the net force on the first block, 30 N – 20 N = 10 N, is the force needed to accelerate that block, having one-third of the total mass.)

27. The only upward force is that of the ground pushing up on you, in response to you pushing down on the ground.

29. At the top of your jump your acceleration is g. Let the equation for acceleration via Newton's second law guide your thinking: $a = F/m = mg/m = g$. If you said zero, you're implying the force of gravity ceases to act at the top of your jump— not so!

31. When you stop suddenly, your velocity changes rapidly, which means a large acceleration of stopping. By Newton's second law, this means the force that acts on you is also large. Experiencing a large force is what hurts you.

33. When you drive at constant velocity, the zero net force on the car is the resultant of the driving force that your engine supplies against the friction drag force. You continue to apply a driving force to offset the drag force that otherwise would slow the car.

35. When held at rest the upward support force equals the gravitational force on the apple and the net force is zero. When released, the upward support force is no longer there and the net force is the gravitational force, 1 N. (If the apple falls fast enough for air resistance to be important, then the net force will be less than 1 N, and eventually can reach zero if air resistance builds up to 1 N.)

37. Both forces have the same magnitude. This is easier to understand if you visualize the parachutist at rest in a strong updraft—static equilibrium. Whether equilibrium is static or dynamic, the net force is zero.

39. In each case the paper reaches terminal speed, which means air drag equals the weight of the paper. So air resistance will be the same on each! Of course the wadded paper falls faster for air resistance to equal the weight of the paper.

41. When anything falls at constant velocity, air drag and gravitational force are equal in magnitude. Rain drops are merely one example.

43. There are usually two terminal speeds, one before the parachute opens, which is faster, and one after, which is slower. The difference has mainly to do with the different areas presented to the air in falling. The large area presented by the open chute results in a slower terminal speed, slow enough for a safe landing.

45. The terminal speed attained by the falling cat is the same whether it falls from 50 stories or 20 stories. Once terminal speed is reached, falling extra distance does not affect the speed. (The low terminal velocities of small creatures enables them to fall without harm from heights that would kill larger creatures.)

47. Air resistance is not really negligible for so high a drop, so the heavier ball does strike the ground first. (This idea is shown in Figure 4.14.) But although a twice-as-heavy ball strikes first, it falls only a little faster, and not twice as fast, which is what followers of Aristotle believed. Galileo recognized that the small difference is due to friction and would not be present if there were no friction.

49. Air resistance decreases the speed of a moving object. Hence the ball has less than its initial speed when it returns to the level from which it was thrown. The effect is easy to see for a feather projected upward by a sling shot. No way will it return to its starting point with its initial speed!

Chapter 4 Problem Solutions

1. Acceleration = $F/m = 0.9\ mg/m = \mathbf{0.9g}$.

3. Weight of the pail is mg = 20 kg × 10 m/s^2 = 200 N.
So $a = F/m$ = (300 N – 200 N)/(50 kg) = **2 m/s^2**.

5. For the jumbo jet: $a = F/m$ = 4(30,000 N)/30,000 kg = **4 m/s^2**.

7. Net force (downward) = ma = (80 kg)(4 m/s^2) = 320 N. Gravity is pulling him downward with a force of (80 kg)(10 m/s^2) = 800 N, so the upward force of friction is 800 N – 320 N = **480 N**.

9. (a) a = (change of v)/t = (1 m/s)/(2 s) = **0.5 m/s^2**.
(b) $F = ma$ = (60 kg)(0.5 m/s^2) = **30 N**.

Chapter 5: Newton's Third Law of Motion Exercises

1. In accord with Newton's third law, Steve and Gretchen are touching each other. One may initiate the touch, but the physical interaction can't occur without contact between both Steve and Gretchen. Indeed, you cannot touch without being touched!

3. (a) Two force pairs act; Earth's pull on the apple (action), and the apple's pull on the Earth (reaction). Hand pushes apple upward (action), and apple pushes hand downward (reaction). (b) If air drag can be neglected, one force pair acts; Earth's pull on apple, and apple's pull on Earth. If air drag counts, then air pushes upward on apple (action) and apple pushes downward on air (reaction).

5. (a) Action; bat hits ball. Reaction; ball hits bat. (b) While in flight there are two interactions, one with the Earth's gravity and the other with the air. Action; Earth pulls down on ball (weight). Reaction; ball pulls up on Earth. And, action; air pushes ball, and reaction, ball pushes air.

7. The billions of force pairs are internal to the book, and exert no net force on the book. An external net force is necessary to accelerate the book.

9. When the barbell is accelerated upward, the force exerted by the athlete is greater than the weight of the barbell (the barbell, simultaneously, pushes with greater force against the athlete). When acceleration is downward, the force supplied by the athlete is less.

11. When you pull up on the handle bars, the handle bars in turn pull down on you. This downward force is transmitted to the pedals.

13. When the climber pulls the rope downward, the rope simultaneously pulls the climber upward—the direction desired by the climber.

15. The forces do not cancel because they act on different things—one acts on the horse, and the other acts on the wagon. It's true that the wagon pulls back on the horse, and this prevents the horse from running as fast as it could without the attached wagon. But the force acting on the wagon (the pull by the horse minus friction) divided by the mass of the wagon, produces the acceleration of the wagon. To accelerate, the horse must push against the ground with more force than it exerts on the wagon and the wagon exerts on it. So tell the horse to push backward on the ground.

17. As in the preceding exercise, the force on each cart will be the same. But since the masses are different, the accelerations will differ. The twice-as-massive cart will undergo only half the acceleration of the less massive cart and will gain only half the speed.

19. In accord with Newton's 3rd law, the force on each will be of the same magnitude. But the effect of the force (acceleration) will be different for each because of the different mass. The more massive truck undergoes less change in motion than the Civic.

21. The winning team pushes harder against the ground. The ground then pushes harder on them, producing a net force in their favor.

23. The girls win because they are able to benefit by more friction with the floor. The boys' feet slide!

25. The writer apparently didn't know that the reaction to exhaust gases does not depend on a medium for the gases. A gun, for example, will kick if fired in a vacuum. In fact, in a vacuum there is no air drag and a bullet or rocket operates even better.

27. Vector quantities are velocity and acceleration. All others are scalars.

29. A hammock stretched tightly has more tension in the supporting ropes than one that sags. The tightly stretched ropes are more likely to break.

31. The slanted streaks are composed of two components. One is the vertical velocity of the falling rain. The other is the horizontal velocity of the car. At 45° these components are equal, meaning the speed of falling drops equals the speed of the car.

33. The other interaction is between the stone and the ground on which it rests. The stone pushes down on the ground surface, say action, and the reaction is the ground pushing up on the stone. This upward force on the stone is called the *normal force*.

35. (a) As shown.

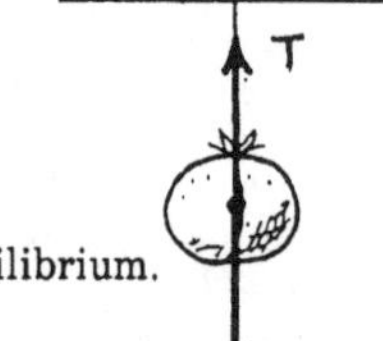

(b) Yes.
(c) Because the stone is in equilibrium.

37. (a) As shown.

(b) $A = F/m = mg/m = g$. .

39. (a) Weight and normal force only.
(b) As shown.

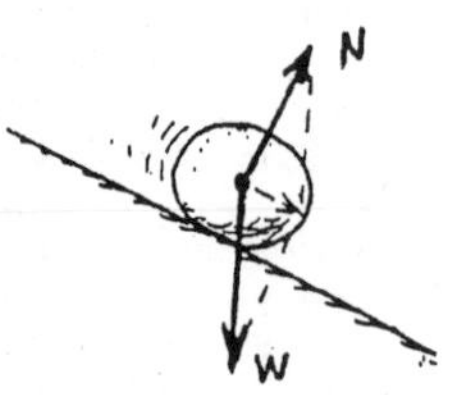

Chapter 5 Problem Solutions

1. $F = ma = m\Delta v/\Delta t$ = (0.003 kg)(25 m/s)/(0.05 s) = **1.5 N**, which is about 1/3 pound.

3. They hit your face with the resultant of the horizontal and vertical components: $R = \sqrt{(3.0\text{ m/s})^2 + (4.0\text{ m/s})^2}$ = **5 m/s.**

5. Ground velocity $V = \sqrt{(100\text{ km/h})^2 + (100\text{ km/h})^2}$ = **141 km/h, 45° northeast** (45° from the direction of the wind). The velocity relative to the ground makes the diagonal of a 45°-45°-90° triangle.

Chapter 6: Momentum Exercises

1. Supertankers are so massive, that even at modest speeds their motional inertia, or *momenta*, are enormous. This means enormous impulses are needed for changing motion. How can large impulses be produced with modest forces? By applying modest forces over long periods of time. Hence the force of the water resistance over the time it takes to coast 25 kilometers sufficiently reduces the momentum.

3. Air bags lengthen the time of impact thereby reducing the force of impact.

5. This illustrates the same point as the previous exercise. The time during which momentum decreases is lengthened, thereby decreasing the jolting force of the rope. Note that in all of these examples, bringing a person to a stop more gently does *not* reduce the impulse. It only reduces the force.

7. The blades impart a downward impulse to the air and produce a downward change in the momentum of the air. The air at the same time exerts an upward impulse on the blades, providing lift. (Newton's third law applies to impulses as well as forces.)

9. The impulse required to stop the heavy truck is considerably more than the impulse required to stop a skateboard moving with the same speed. The *force* required to stop either, however, depends on the time during which it is applied. Stopping the skateboard in a split second results in a certain force. Apply less than this amount of force on the moving truck and given enough time, the truck will come to a halt.

11. The large momentum of the spurting water is met by a recoil that makes the hose difficult to hold, just as a shotgun is difficult to hold when it fires birdshot.

13. Impulse is force × time. The forces are equal and opposite, by Newton's third law, and the times are the same, so the impulses are equal and opposite.

15. The momentum of the falling apple is transferred to the Earth. Interestingly enough, when the apple is released, the Earth and the apple move toward each other with equal and oppositely directed momenta. Because of the Earth's enormous mass, its motion is imperceptible. When the apple and Earth hit each other, their momenta are brought to a halt — zero, the same value as before.

17. The lighter gloves have less padding, and less ability to extend the time of impact, and therefore result in greater forces of impact for a given punch.

19. Without this slack, a locomotive might simply sit still and spin its wheels. The loose coupling enables a longer time for the entire train to gain momentum, requiring less force of the locomotive wheels against the track. In this way, the overall required impulse is broken into a series of smaller impulses. (This loose coupling can be very important for braking as well.)

21. In jumping, you impart the same momentum to both you and the canoe. This means you jump from a canoe that is moving away from the dock, reducing your speed relative to the dock, so you don't jump as far as you expected to.

23. To get to shore, the person may throw keys or coins or an item of clothing. The momentum of what is thrown will be accompanied by the thrower's oppositely-directed momentum. In this way, one can recoil towards shore. (One can also inhale facing the shore and exhale facing away from the shore.)

25. Regarding Exercise 23; If one throws clothing, the force that accelerates the clothes will be paired with an equal and opposite force on the thrower. This force can provide recoil toward shore. Regarding Exercise 24; According to Newton's third law, whatever forces you exert on the ball, first in one direction, then in the other, are balanced by equal forces that the ball exerts on you. Since the forces on the ball give it no final momentum, the forces it exerts on you also give no final momentum.

27. When two objects interact, the forces they exert on each other are equal and opposite and these forces act for the same time, so the impulses are equal and opposite. Therefore their changes of momenta are equal and opposite, and the total change of momentum of the two objects is zero.

29. Momentum is not conserved for the ball itself, because an impulse is exerted on it (gravitational force x time). So the ball gains momentum. It is in the *absence* of an external force that momentum doesn't change. If the whole Earth and the rolling ball are taken together as a system, then the gravitational interaction between the Earth and the ball are internal forces and no external impulse acts. Then the momentum of the ball is accompanied by an equal and opposite momentum of the Earth, which results in no change in momentum.

31. If the system is the stone only, its momentum certainly changes as it falls. If the system is enlarged to include the stone plus the Earth, then the downward momentum of the stone is cancelled by the equal but opposite momentum of the Earth "racing" up to meet the stone.

33. This exercise is similar to the previous one. If we consider Bronco to be the system, then a net force acts and momentum changes. In the system composed of Bronco alone, momentum is not conserved. If, however we consider the system to be Bronco and the world (including the air), then all the forces that act are internal forces and momentum is conserved. Momentum is conserved only in systems not subject to external forces.

35. The craft moves to the right. This is because there are two impulses that act on the craft: one is that of the wind against the sail, and the other is that of the fan recoiling from the wind it produces. These impulses are oppositely directed, but are they equal in magnitude? No, because of bouncing. The wind bounces from the sail and produces a greater impulse than if it merely stopped. This greater impulse on the sail produces a net impulse in the forward direction, toward the right. We can see this in terms of forces as well. Note in the sketch there are two force pairs to consider: (1) the fan-air force pair, and (2) the air-sail force pair. Because of bouncing, the air-sail pair is greater. Solid vectors show forces exerted on the craft; dashed vectors show forces exerted on the air. The net force on the craft is forward, to the right. The principle described here is applied in thrust reversers used to slow jet planes after they land. Also, you can see that after the fan is turned on, there is a net motion of air to the left, so the boat, to conserve momentum, will move to the right.

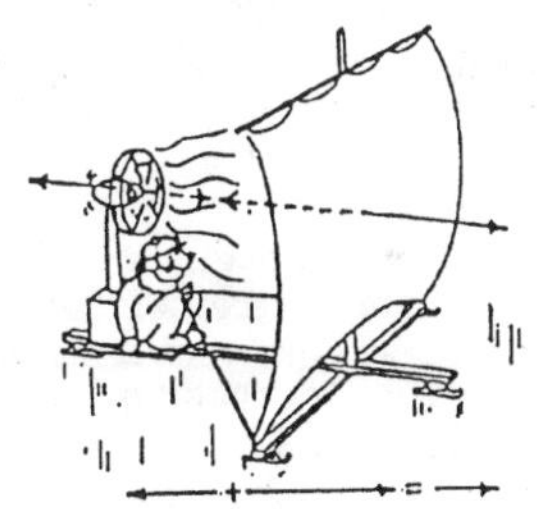

37. Removing the sail and turning the fan around is the best means of propelling the boat! Then maximum impulse is exerted on the craft. If the fan is not turned around, the boat is propelled backward, to the left. (Such propeller-driven boats are used where the water is very shallow, as in the Florida Everglades.)

39. By Newton's 3rd law, the force on the bug is equal in magnitude and opposite in direction to the force on the car windshield. The rest is logic: Since the time of impact is the same for both, the amount of impulse is the same for both, which means they both undergo the same change in momentum. The change in momentum of the bug is evident because of its large change in speed. The same change in momentum of the considerably more massive car is not evident, for the change in speed is correspondingly very small. Nevertheless, the magnitude of $m\Delta V$ for the bug is equal to $M\Delta v$ for the car!

41. Cars brought to a rapid halt experience a change in momentum, and a corresponding impulse. But greater momentum change occurs if the cars bounce, with correspondingly greater impulse and therefore greater damage. Less damage results if the cars stick upon impact than if they bounce apart.

43. We assume the equal strengths of the astronauts means that each throws with the same speed. Since the masses are equal, when the first throws the second, both the first and second move away from each other at equal speeds. Say the thrown astronaut moves to the right with velocity V, and the first recoils with velocity $-V$. When the third makes the catch, both she and the second move to the right at velocity $V/2$ (twice the mass moving at half the speed, like the freight cars in Figure 6.11). When the third makes her throw, she recoils at velocity V (the same speed she imparts to the thrown astronaut) which is added to the $V/2$ she acquired in the catch. So her velocity is $V + V/2 = 3V/2$, to the right—too fast to stay in the game. Why? Because the velocity of the second astronaut is $V/2 - V = -V/2$, to the left—too slow to catch up with the first astronaut who is still moving at $-V$. the game is over. Both the first and the third got to throw the second astronaut only once!

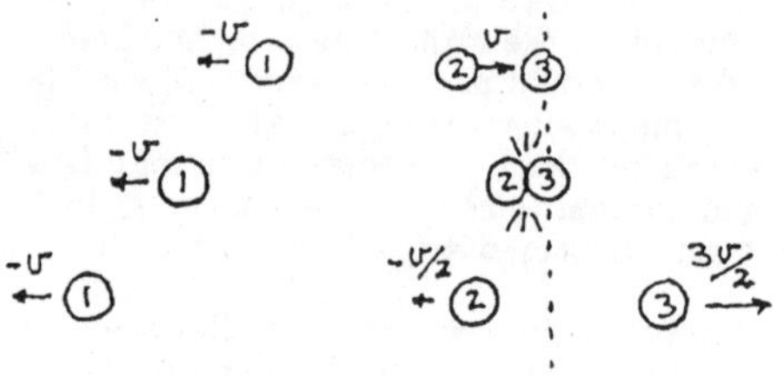

45. The impulse will be greater if the hand is made to bounce, because there is a greater change in the momentum of hand and arm, accompanied by a greater impulse. The force exerted on the bricks is equal and opposite to the force of the bricks on the hand. Fortunately, the hand is resilient and toughened by long practice.

47. Their masses are the same; half speed for the coupled particles means equal masses for the colliding and the target particles. This is like the freight cars of equal mass shown in Figure 6.11.

49. If a ball does not hit straight on, then the target ball flies off at an angle (to the left, say) and has a component of momentum sideways to the initial momentum of the moving ball. To offset this, the striking ball cannot be simply brought to rest, but must fly off in the other direction (say, the right). It will do this in such a way that its sideways component of momentum is equal and opposite to that of the target ball. This means the total sideways momentum is zero—what it was before collision. (See how the sideways components of momentum cancel to zero in Figure 6.16.)

Chapter 6 Problem Solutions

1. The bowling ball has a momentum of (10 kg)(6 m/s) = 60 kg·m/s, which has the magnitude of the impulse to stop it. That's **60 N·s**. (Note that units N·s = kg·m/s.)

3. From $Ft = \Delta mv$, $F = \dfrac{\Delta mv}{t}$
$= [(75\text{ kg})(25\text{ m/s})]/0.1\text{ s} =$ **18,750 N.**

5. Momentum of the caught ball is (0.15 kg)(40 m/s) = 6.0 kg·m/s. (a) The impulse to produce this change of momentum has the same magnitude, **6.0 N·s**. (b) From $Ft = \Delta mv$, $F = \Delta mv/t = [(0.15\text{ kg})(40\text{ m/s})]/0.03\text{ s} =$ **200 N.**

7. Momentum after collision is zero, which means the net momentum before collision must have been zero. So the 1-kg ball must be moving twice as fast as the 2-kg ball so that the magnitudes of their momenta are **equal**.

9. $\text{Momentum}_{\text{before}} = \text{momentum}_{\text{after}}$

$$(5\text{kg})(1\text{m/s}) + (1\text{kg})v = 0$$

$$5\text{m/s} + v = 0$$

$$v = -5\text{ m/s}$$

So if the little fish approaches the big fish at **5 m/s**, the momentum after lunch will be zero.

11. Momentum conservation can be applied in both cases. (a) For head-on motion the total momentum is zero, so the wreckage after collision is **motionless**. (b) As shown in Figure 5.14, the total momentum is directed to the northeast—the resultant of two perpendicular vectors, each of magnitude 20,000 kg m/s. It has magnitude 28,200 kg m/s. The speed of the wreckage is this momentum divided by the total mass, $v = (28{,}200\text{ kg m/s})/(2000\text{ kg}) =$ **14.1 m/s.**

Chapter 7: Energy
Exercises

1. It is easier to stop a lightly loaded truck than a heavier one moving at the same speed because it has less KE and will therefore require less work to stop. (An answer in terms of impulse and momentum is also acceptable.)

3. The PE of the drawn bow as calculated would be an overestimate, (in fact, about twice its actual value) because the force applied in drawing the bow begins at zero and increases to its maximum value when fully drawn. It's easy to see that less force and therefore less work is required to draw the bow halfway than to draw it the second half of the way to its fully-drawn position. So the work done is not *maximum force x distance drawn*, but *average force x distance drawn*. In this case where force varies almost directly with distance (and not as the square or some other complicated factor) the average force is simply equal to the initial

force + final force, divided by 2. So the PE is equal to the average force applied (which would be approximately half the force at its full-drawn position) multiplied by the distance through which the arrow is drawn.

5. The KE of the tossed ball relative to occupants in the airplane does not depend on the speed of the airplane. The KE of the ball relative to observers on the ground below, however, is a different matter. KE, like velocity, is relative. (This is similar to the Check Yourself Question in the text.)

7. If an object has KE, then it must have momentum—for it is moving. But it can have potential energy without being in motion, and therefore without having momentum. And every object has "energy of being"—stated in the celebrated equation $E = mc^2$. So whether an object moves or not, it has some form of energy. If it has KE, then with respect to the frame of reference in which its KE is measured, it also has momentum.

9. When the velocity is doubled, the momentum is doubled and the KE is increased by a factor of 4. Momentum is proportional to speed, KE to speed squared.

11. In the popular sense, conserving energy means not wasting energy. In the physics sense energy conservation refers to a law of nature that underlies natural processes. Although energy can be wasted (which really means transforming it from a more useful to a less useful form), it cannot be destroyed. Nor can it be created. Energy is transferred or transformed, without gain or loss. That's what a physicist means in saying energy is conserved.

13. The KE of a pendulum bob is maximum where it moves fastest, at the lowest point; PE is maximum at the uppermost points; When the pendulum bob swings by the point that marks half its maximum height, it has half its maximum KE, and its PE is half way between its minimum and maximum values. If we define PE = 0 at the bottom of the swing, the place where KE is half its maximum value is also the place where PE is half its maximum value, and KE = PE at this point. (In accordance with energy conservation: total energy = KE + PE).

15. The answers to both a and b are the same: When the direction of the force is perpendicular to the direction of motion, as is the force of gravity on both the bowling ball on the alley and the satellite in circular orbit, there is no force component in the direction of motion and no work is done by the force.

17. The string tension is everywhere perpendicular to the bob's direction of motion, which means there is no component of tension along the bob's path, and therefore no work done by the tension. The force of gravity, on the other hand, has a component along the direction of motion everywhere except at the bottom of the swing, and does work, which changes the bob's KE.

19. Whenever you're asked if work is done, you have to distinguish by what on what. Here we're considering work done by you on the package, not work done by the package on you. (a) Evidence that work is done on the package is the increase in energy (PE) of the package. (b) Your friend is discussing *two* work situations: work done by you on the package, and work the package does on you. Call the force you exert on the package *action*, then *reaction* is the force the package exerts on you. Direct your friend to Newton's third law in Chapter 5 and recall how action and reaction don't cancel when they act on different systems.

21. A Superball will bounce higher than its original height if thrown downward, but if simply dropped, no way. Such would violate the conservation of energy.

23. Kinetic energy is a maximum as soon as the ball leaves the hand. Potential energy is a maximum when the ball has reached its zenith.

25. You agree with your second classmate. The coaster could just as well encounter a low summit before or after a higher one, so long as the higher one is enough lower than the initial summit to compensate for energy dissipation by friction.

27. Except for the very center of the plane, the force of gravity acts at an angle to the plane, with a component of gravitational force along the plane — along the block's path. Hence the block goes somewhat against gravity when moving away from the central position, and moves somewhat with gravity when coming back. As the object slides farther out on the plane, it is effectively traveling "upward" against earth's gravity, and slows down. It finally comes to rest and then slides back and the process repeats itself. The block slides back and forth along the plane. From a flat-Earth point of view the situation is equivalent to that shown in the sketch.

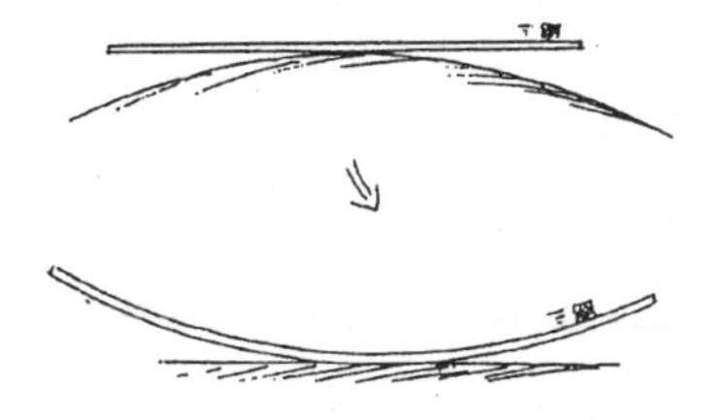

29. Yes, a car burns more gasoline when its lights are on. The overall consumption of gasoline does not depend on whether or not the engine is running. Lights and other devices run off the battery, which "run down" the battery. The energy used to recharge the battery ultimately comes from the gasoline.

31. Energy is dissipated into non-useful forms in an inefficient machine, and is "lost" only in the loose sense of the word. In the strict sense, it can be accounted for and is therefore not lost.

33. The work that the rock does on the ground is equal to its PE before being dropped, mgh = 100 joules. The force of impact, however, depends on the distance that the rock penetrates into the ground. If we do not know this distance we cannot calculate the force. (If we knew the time during which the impulse occurs we could calculate the force from the impulse-momentum relationship—but not knowing the distance or time of the rock's penetration into the ground, we cannot calculate the force.)

35. When air resistance is a factor, the ball will return with less speed (discussed in Exercise 48 in Chapter 4). It therefore will have less KE. You can see this directly from the fact that the ball loses mechanical energy to the air molecules it encounters, so when it returns to its starting point and to its original PE, it will have less KE. This does not contradict energy conservation, for energy is dissipated, not destroyed.

37. The quantities that are the same are (c) and (d). The changes of KE and PE are the same for both balls in the first *meter* of fall, because the work done by gravity on both is the same in 1 m (same force and same distance). In the first *second* of fall, the ball that is thrown moves farther, so it has greater changes in KE and PE.

39. Twice as far, because at twice the height it acquires twice the KE on its way down. This means twice the work done by the mud in bringing it to rest and therefore twice the distance if the force is the same. Note that we can bypass the KE stage and say work done by the mud = initial PE. (Practically speaking, however, the impact force would be greater so all else would not be equal, and the stone would penetrate less than twice as far.)

41. Momenta of birds in the flock can cancel because momentum is a vector quantity. But kinetic energy is a scalar that is always positive for a moving object (it is zero for a motionless object, but is never negative). The positive KEs of the birds in flight add up to a positive total.

43. The two skateboarders have equal momenta, but the lighter one has twice the KE and can do twice as much work on you. So choose the collision with the heavier, slower-moving kid and you'll endure less damage.

45. Exaggeration makes the fate of teacher Paul Robinson easier to assess: Paul would not be so calm if the cement block were replaced with the inertia of a small stone, for inertia plays a role in this demonstration. If the block were unbreakable, the energy that busts it up would instead be transferred to the beds of nails. So it is desirable to use a block that will break upon impact. If the bed consisted of a single nail, finding a successor to Paul would be very difficult, so it is important that the bed have plenty of nails!

47. An engine that is 100% efficient would not be warm to the touch, nor would its exhaust heat the air, nor would it make any noise, nor would it vibrate. This is because all these are transfers of energy, which cannot happen if all the energy given to the engine is transformed to useful work.

Chapter 7 Problem Solutions

1. The work done by 10 N over a distance o 5 m = 50 J. That by 20 N over 2 m = 40 J. So the 10-N force over 5 m does more work and could produce a greater change in KE.

3. $(Fd)_{input} = (Fd)_{output}$

$(100 \text{ N} \times 10 \text{ cm})_{input} = (? \times 1 \text{ cm})_{output}$

So we see that the output force is **1000 N** (or less if the efficiency is less than 100%).

5. By the work-energy theorem,

$$W = \Delta KE$$

Work done on the car is Fd, so

$$Fd = \Delta(1/2\ mv^2)$$

The only force F that does work to reduce the kinetic energy is the force of friction. This force acts through d, the distance of skidding. The mass of the car is m, and its initial speed is v. In this problem the final speed of the car will be zero, so the change in kinetic energy is simply the initial kinetic energy at speed v. You're looking for distance, so write the equation in a "d =" form. It becomes

$$d = \frac{\Delta(1/2\ mv^2)}{F} = \frac{1/2\ mv^2}{f} = \frac{1/2\ mv^2}{mg/2} = \frac{v^2}{g}$$

where F is half the car's weight, $mg/2$.
Note how the terms in the equation dictate subsequent steps and guide your thinking. The final expression tells you the stopping distance is proportional to speed squared, which is consistent with it being proportional to KE. It also tells you that if g were greater, the force of friction would be greater and skidding distance less—which is quite reasonable. Cancellation of mass tells you that the mass of the car doesn't matter. All cars skidding with the same initial speed, with friction equal to half their weights, will skid the same distance. And as for units, note that v^2/g has the unit $(m^2/s^2)/(m/s^2)$ = m, a distance, as it should be. How nice that much can be learned by a thoughtful examination of a simple equation.

7. From $p = mv$, you get $v = p/m$. Substitute this expression for v in KE = $(1/2)mv^2$ to get KE = $(1/2)\ m(p/m)^2 = p^2/2m$. (Alternatively, one

may work in the other direction, substituting $p = mv$ in $KE = p^2/2m$ to get $KE = (1/2)mv^2$.)

9. At 25% efficiency, only 1/4 of the 40 megajoules in one liter, or 10 MJ, will go into work. This work is
   $F \times d = 500 \text{ N} \times d = 10 \text{ MJ}$.
   Solve this for d and convert MJ to J, to get $d = 10 \text{ MJ}/500 \text{ N} = 10{,}000{,}000 \text{ J}/500 \text{ N} = 20{,}000 \text{ m} =$ **20 km**.
   So under these conditions, the car gets 20 kilometers per liter (which is 47 miles per gallon).

Chapter 8: Rotational Motion

Exercises

1. According to $v = r\omega$, if the RPM rate (ω) is doubled, the speed is doubled. Then if r is also doubled, the speed doubles again and the ladybug moves with four times its initial speed.

3. Large diameter tires mean you travel farther with each revolution of the tire. So you'll be moving faster than your speedometer indicates. (A speedometer actually measures the RPM of the wheels and displays this as mi/h or km/h. The conversion from RPM to the mi/h or km/h reading assumes the wheels are a certain size.) Oversize wheels give too low a reading, because they really travel farther per revolution than the speedometer indicates; and undersize wheels give too high a reading because the wheels do not go as far per revolution.

5. The amount of taper is related to the amount of curve the railroad tracks take. On a curve where the outermost track is say 10% longer than the inner track, the wide part of the wheel will also have to be at least 10% wider than the narrow part. If it's less than this, the outer wheel will rely on the rim to stay on the track, and scraping will occur as the train makes the curve. The "sharper" the curve, the more the taper needs to be on the wheels.

7. No, for the yo-yo continues to rotate in the same direction, in accord with the law of inertia for rotating systems. It is the continuation of this rotation that accounts for its climbing back up the string.

9. The meterstick against the wall will rotate to the floor, and its CM will follow a quarter-circle arc. On a smooth floor, with no wall to prevent sliding, the CM of a falling meterstick will be a vertical straight line. The lower end of the stick will slide as the stick falls.

11. Rotational inertia and torque are most predominantly illustrated here, and the conservation of angular momentum also plays a role. The long distance to the front wheels increases the rotational inertia of the vehicle relative to the back wheels and also increases the lever arm of the front wheels without appreciably adding to the vehicle's weight. As the back wheels are driven clockwise, the chassis tends to rotate counterclockwise (conservation of angular momentum) and thereby lift the front wheels off the ground. The greater rotational inertia and the increased clockwise torque of the more distant front wheels counter this effect.

13. Friction by the road on the tires produces a torque about the car's CM. When the car accelerates forward, the friction force points forward and rotates the car upward. When braking, the direction of friction is rearward, and the torque rotates the car in the **opposite** direction so the rear end rotates upward (and the nose downward).

15. If you roll them down an incline, the solid ball will roll faster. (The hollow ball has more rotational inertia compared with its weight.)

17. What your friend likely meant to say is that the rotation of a body cannot *change* when a net torque acts on it. Once rotating, a body will continue rotating when *no* torque acts on it. Again, emphasize *change*.

19. In the horizontal position the lever arm equals the length of the sprocket arm, but in the vertical position, the lever arm is zero because the line of action of forces passes right through the axis of rotation. (With cycling cleats, a cyclist pedals in a circle, which means they push their feet over the top of the stroke and pull around the bottom and even pull up on the recovery. This allows torque to be applied over a greater portion of the revolution.)

21. A rocking bus partially rotates about its center of mass, which is near its middle. The farther one sits from the center of mass, the greater is the up and down motion — as on a seesaw. Likewise for motion of a ship in choppy water or an airplane in turbulent air.

23. The wobbly motion of a star is an indication that it is revolving about a center of mass that is not at its geometric center, implying that there is some other mass nearby to pull the center of mass away from the star's center. This is the way in which astronomers have discovered that planets exist around stars other than our own.

25. Two buckets are easier because you may stand upright while carrying a bucket in each hand. With two buckets, the CG will be in the center of the support base provided by your feet, so there is no need to lean. (The same can be accomplished by carrying a single bucket on your head.)

27. The CG of a ball is not above a point of support when the ball is on an incline. The weight of the ball therefore acts at some distance from

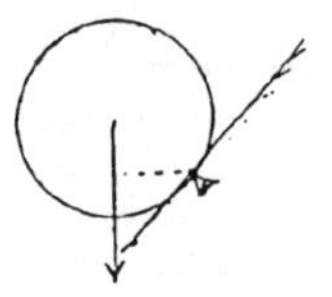

the point of support which behaves like a fulcrum. A torque is produced and the ball rotates. This is why a ball rolls down a hill.

29. The top brick would overhang 3/4 of a brick length as shown. This is best explained by considering the top brick and moving downward; i.e., the CG of the top brick is at its midpoint; the CG of the top two bricks is midway between their combined length. Inspection will show that this is 1/4 of a brick length, the overhang of the middle brick. (Interestingly, with a few more bricks, the overhang can be greater than a brick length, and with a limitless number of bricks, the overhang can be made as large as you like.)

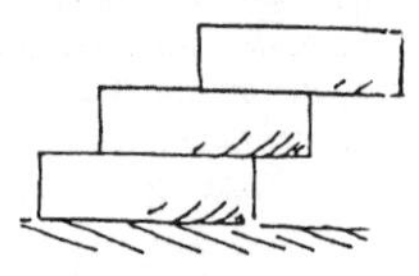

31. It is dangerous to pull open the upper drawers of a fully-loaded file cabinet that is not secured to the floor because the CG of the cabinet can easily be shifted beyond the support base of the cabinet. When this happens, the torque that is produced causes the cabinet to topple over.

33. The CG of Truck 1 is not above its support base; the CGs of Trucks 2 and 3 are above their support bases. Therefore, only Truck 1 will tip.

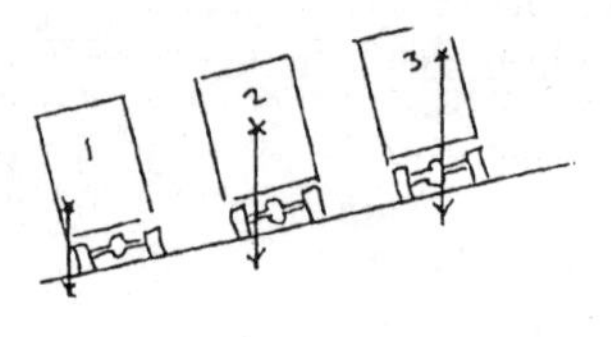

35. Acceleration g at the earth's surface effectively decreases as the spin of the Earth increases. This can be seen by the principle of exaggeration: If the Earth spins fast enough, like at 12.5 times its present rate, g at the equator would effectively be zero and things wouldn't fall at all. At a still greater rate, things would fly off instead of falling. (How much decrease depends on latitude. At the poles of the earth, for example, there would be no tangential speed and the value of g would not be affected as long as the pole-to-pole diameter of the Earth doesn't change.)

37. Newton's first and third laws provide a straightforward explanation. You tend to move in a straight line (Newton's first law) but are intercepted by the door. You press against the door because the door is pressing against you (Newton's third law). The push by the door provides the centripetal force that keeps you moving in a curved path. Without the door's push, you wouldn't turn with the car—you'd move along a straight line and be "thrown out." No need to invoke centrifugal force.

39.

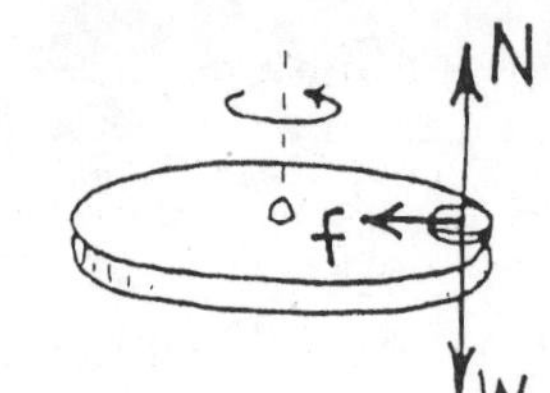

41. (a)

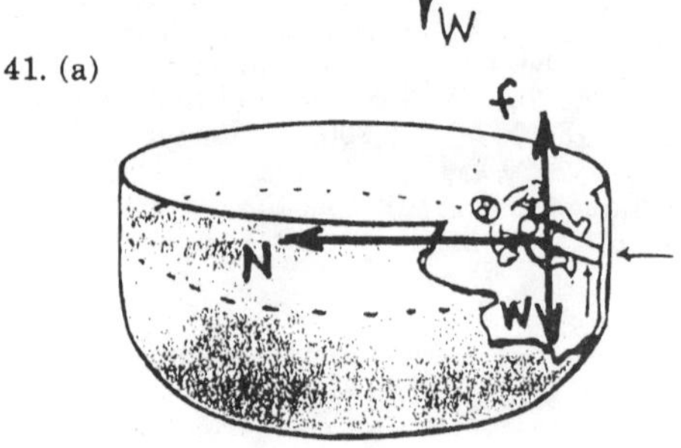

(b) The normal force provides the centripetal force. (In accord with Newton's third law, the motorcycle presses against the wall of the track and the track presses against the motorcycle, the normal force—which provides the centripetal force. This normal force increases with increasing speed. [Centripetal force is proportional to speed squared (footnote in the text) so, for example, twice the speed corresponds to four times the force.]

43. The rotational inertia of you and the rotating turntable is least when you are at the rotational axis. As you crawl outward, the rotational inertia of the system increases (like the masses held outward in Figure 8.42). In accord with the conservation of angular momentum, as you crawl toward the outer rim, you increase the rotational inertia of the spinning system and decrease the angular speed. You can also see that if you don't slip as you crawl out, you exert a friction force on the turntable opposite to its direction of rotation, thereby slowing it down.

45. In accord with the conservation of angular momentum, as the radial distance of mass increases, the angular speed decreases. The mass of material used to construct skyscrapers is lifted, slightly increasing the radial distance from the earth's spin axis. This would tend to slightly decrease the earth's rate of rotation, which in turn tends to make the days a bit longer. The opposite effect occurs for falling leaves, as their radial distance from the earth's axis decreases. As a practical matter, these effects are entirely negligible!

47. In accord with the conservation of angular momentum, if mass moves farther from the axis of rotation, rotational speed decreases. So the Earth would slow in its daily rotation.

49. Without the small rotor on its tail, the helicopter and the main rotor would rotate in opposite directions. The small rotor provides a torque to offset the rotational motion that the helicopter would otherwise have.

Chapter 8 Problem Solutions

1. Since the bicycle moves 2 m with each turn of the wheel, and the wheel turns once each second, the linear speed of the bicycle is **2 m/s.**

3. The center of mass of the two weights is where a fulcrum would balance both—where the torques about the fulcrum would balance to zero. Call the distance (lever arm) from the 1-kg weight to the fulcrum x. Then the distance (lever arm) from the fulcrum to the 3-kg weight is (100 – x). Equating torques:
   $1x = (100 - x)3$
   $x = 300 - 3x$
   $x = 75.$
   So the center of mass of the system is just below the **75-cm** mark. Then the three-times-as-massive weight is one-third as far from the fulcrum.

5. The mass of the stick is 1 kg (this is a "freebie"; in the textbook see the Check Yourself question and answer in the chapter)!

7. Centripetal force (and "weight" and "*g*" in the rotating habitat) is directly proportional to radial distance from the hub. At half the radial distance, the *g* force will be **half** that at his feet. The man will literally be "light-headed." (Gravitational variations of greater than 10% head-to-toe are uncomfortable for most people.)

9. The artist will rotate **3 times per second**. By the conservation of angular momentum, the artist will increase rotation rate by 3. That is
   $I\omega_{before} = I\omega_{after}$
   $I\omega_{before} = [(I/3)(3\omega)]_{after}$

Chapter 9: Gravity
Exercises

1. Nothing to be concerned about on this consumer label. It simply states the universal law of gravitation, which applies to *all* products. It looks like the manufacturer knows some physics and has a sense of humor.

3. In accord with the law of inertia, the moon would move in a straight line path instead of circling both the sun and Earth.

5. The force of gravity is the same on each because the masses are the same, as Newton's equation for gravitational force verifies. When dropped the crumpled paper falls faster only because it encounters less air drag than the sheet.

7. The force of gravity on moon rocks at the moon's surface is considerably stronger than the force of gravity of the distant Earth. Rocks dropped on the moon fall onto the moon surface. (the force of the moon's gravity is about 1/6 of the weight the rock would have on Earth; the force of the Earth's gravity at that distance is only about 1/3600 of the rock's Earth-weight.)

9. Astronauts are weightless because they lack a support force, but they are well in the grips of Earth gravity, which accounts for them circling the Earth rather than going off in a straight line in outer space.

11. In accord with Newton's 3rd law, the weight of the Earth in the gravitational field of the apple is 1 N; the same as the weight of the apple in the Earth's gravitational field.

13. Although the forces are equal, the accelerations are not. The much more massive Earth has much less acceleration than the moon. Actually Earth and moon *do* rotate around a common point, but it's not midway between them (which would require both Earth and moon to have the same mass). The point around which Earth and moon rotate (called the *barycenter*) is within the Earth about 4600 km from the Earth's center.

15. For the planet half as far from the sun, light would be four times as intense. For the planet ten times as far, light would be 1/100th as intense.

17. The gravitational force on a body, its weight, depends not only on mass but distance. On Jupiter, this is the distance between the body being weighed and Jupiter's center—the radius of Jupiter. If the radius of Jupiter were the same as that of the Earth, then a body would weigh 300 times as much because Jupiter is 300 times more massive than Earth. But Jupiter is also much bigger than the Earth, so the greater distance between its center and the CG of the body reduces the gravitational force. The radius is great enough to make the weight of a body only 3 times its Earth weight. How much greater is the radius of Jupiter? That will be Problem 2.

19. A person is weightless when the only force acting is gravity, and there is no support force. Hence the person in free fall is weightless. But more than gravity acts on the person falling at terminal velocity. In addition to gravity, the falling person is "supported" by air drag.

21. Gravitational force is indeed acting on a person who falls off a cliff, and on a person in a space shuttle. Both are falling under the influence of gravity.

23. First of all, it would be incorrect to say that the gravitational force of the distant sun on you is too small to be measured. It's small, but not immeasurably small. If, for example, the Earth's axis were supported such that the Earth could continue turning but not otherwise move, an 85-kg person would see a gain of 1/2 newton on a bathroom scale at midnight and a loss of 1/2 newton at noon. The key idea is *support*. There is no "sun support" because the Earth and all objects on the Earth—you, your bathroom scale, and everything else—are continually falling around the sun. Just as you wouldn't be pulled against

the seat of your car if it drives off a cliff, and just as a pencil is not pressed against the floor of an elevator in free fall, we are not pressed against or pulled from the Earth by our gravitational interaction with the sun. That interaction keeps us and the Earth circling the sun, but does not press us to the Earth's surface. Our interaction with the Earth does that.

25. The gravitational pull of the sun on the Earth is greater than the gravitational pull of the moon. The tides, however, are caused by the *differences* in gravitational forces by the moon on opposite sides of the Earth. The difference in gravitational forces by the moon on opposite sides of the Earth is greater than the corresponding difference in forces by the stronger pulling but much more distant sun.

27. No. Tides are caused by differences in gravitational pulls. If there are no differences in pulls, there are no tides.

29. Lowest tides occur along with highest tides—spring tides. So the spring tide cycle consists of higher-than-average high tides followed by lower-than-average low tides (best for digging clams!).

31. Because of its relatively small size, different parts of the Mediterranean Sea are essentially equidistant from the moon (or from the sun). As a result, one part is not pulled with any appreciably different force than any other part. This results in extremely tiny tides. The same argument applies, with even more force, to smaller bodies of water, such as lakes, ponds, and puddles. In a glass of water under a full moon you'll detect no tides because no part of the water surface is closer to the moon than any other part of the surface. Tides are caused by appreciable differences in pulls.

33. Yes, the Earth's tides would be due only to the sun. They'd occur twice per day (every 12 hours instead of every 12.5 hours) due to the Earth's daily rotation.

35. From the nearest body, the Earth.

37. In accord with the inverse-square law, twice as far from the Earth's center diminishes the value of g to 1/4 its value at the surface, or 2.45 m/s^2.

39. Your weight would be less in the mineshaft. One way to explain this is to consider the mass of the Earth above you which pulls upward on you. This effect reduces your weight, just as your weight is reduced if someone pulls upward on you while you're weighing yourself. Or more accurately, we see that you are effectively within a spherical shell in which the gravitational field contribution is zero; and that you are being pulled only by the spherical portion below you. You are lighter the deeper you go, and if the mine shaft were to theoretically continue to the Earth's center, your weight moves closer to zero.

41. More fuel is required for a rocket that leaves the Earth to go to the moon than the other way around. This is because a rocket must move against the greater gravitational field of the Earth most of the way. (If launched from the moon to the Earth, then it would be traveling with the Earth's field most of the way.)

43. $F \sim m_1 m_2/d^2$, where m_2 is the mass of the sun (which doesn't change when forming a black hole), m_1 is the mass of the orbiting Earth, and d is the distance between the center of mass of the Earth and the sun. None of these terms change, so the force F that holds the Earth in orbit does not change. (There may in fact be black holes in the galaxy around which stars or planets orbit.)

45. The misunderstanding here is not distinguishing between a theory and a hypothesis or conjecture. A theory, such as the theory of universal gravitation, is a synthesis of a large body of information that encompasses well-tested and verified hypothesis about nature. Any doubts about the theory have to do with its applications to yet untested situations, not with the theory itself. One of the features of scientific theories is that they undergo refinement with new knowledge. (Einstein's general theory of relativity has taught us that in fact there are limits to the validity of Newton's theory of universal gravitation.)

47. You weigh a tiny bit less in the lower part of a massive building because the mass of the building above pulls upward on you.

49. The total mass (or average density) of the universe. If it is less than a certain critical amount, expansion will continue. If it is greater than this critical amount, expansion will stop and give way to contraction. If it is exactly this amount, the expansion will coast to a stop. (Current theory says that in addition to mass, which is always attractive, there may be a form of energy spread through the universe that repels matter. It this were true, a universe with less than critical mass density could still expand without limit.)

Chapter 9 Problem Solutions

1. From $F = GmM/d^2$, five times d squared is 1/25 d, which means the force is **25 times** greater.

3. In accord with the inverse-square law, four times as far from the Earth's center diminishes the value of g to $g/4^2$, or $g/16$, or 0.6 m/s^2.

5. It is $g = GM/r^2 = (6.67 \times 10^{-11})(3.0 \times 10^{30})/(8.0 \times 10^3)^2 = 3.1 \times 10^{12}$ m/s^2, 300 billion times g on Earth.

7. (a) Mars: $F = G\frac{mM}{d^2} = 6.67 \times 10^{-11}\frac{(3\text{kg})(6.4 \times 10^{23})}{(5.6 \times 10^{10})^2}$
$= \mathbf{4.1 \times 10^{-8}\ N}$.

(b) Physician: $F = G\frac{mM}{d^2} = 6.67 \times 10^{-11}\frac{(3kg)(10^2)}{(0.5)^2}$

$= 8.0 \times 10^{-8}$ N.

(c) The gravitational force due to the physician is about twice that due to Mars.

9. Nearly 10,000 km thick, not much less than the diameter of the Earth itself! From the ratio 3.6×10^{22} N/x = 5×10^8 N/1 m^2, x = $(3.6 \times 10^{22})/(5 \times 10^8) = 7.2 \times 10^{13}$ m^2. This would be the cross-sectional area of the cable. From the area of a circle, $A = \pi D^2/4$, we find its diameter $D = \sqrt{4A/\pi} = 9.6 \times 10^6$ m = 9,600 km.

Chapter 10: Projectile and Satellite Motion

Exercises

1. The crate will not hit the Camaro, but will crash a distance beyond it determined by the height and speed of the plane.

3. When air resistance is negligible, the vertical component of motion for a projectile is identical to that of free fall.

5. Minimum speed occurs at the top, which is the same as the horizontal component of velocity anywhere along the path.

7. Kicking the ball at angles greater than 45° sacrifices some distance to gain extra time. A 45° kick doesn't go as far, but stays in the air longer, giving players on the kicker's team a chance to run down field and be close to the player on the other team who catches the ball.

9. The bullet falls beneath the projected line of the barrel. To compensate for the bullet's fall, the barrel is elevated. How much elevation depends on the velocity and distance to the target. Correspondingly, the gunsight is raised so the line of sight from the gunsight to the end of the barrel extends to the target. If a scope is used, it is tilted downward to accomplish the same line of sight

11. Any vertically projected object has zero speed at the top of its trajectory. But if it is fired at an angle, only its vertical component of velocity is zero and the velocity of the projectile at the top is equal to its horizontal component of velocity. This would be 100 m/s when the 141-m/s projectile is fired at 45°.

13. The hang time will be the same, in accord with the answer to the preceding exercise. Hang time is related to the vertical height attained in a jump, not on horizontal distance moved across a level floor.

15. The moon's tangential velocity is what keeps the moon coasting around the Earth rather than crashing into it. If its tangential velocity were reduced to zero, then it would fall straight into the Earth!

17. From Kepler's third law, $T^2 \sim R^3$, the period is greater when the distance is greater. So the periods of planets farther from the sun are longer than our year.

19. The initial vertical climb lets the rocket get through the denser, retarding part of the atmosphere most quickly, and is also the best direction at low initial speed, when a large part of the rocket's thrust is needed just to support the rocket's weight. But eventually the rocket must acquire enough tangential speed to remain in orbit without thrust, so it must tilt until finally its path is horizontal.

21. The moon has no atmosphere (because escape velocity at the moon's surface is less than the speeds of any atmospheric gases). A satellite 5 km above the Earth's surface is still in considerable atmosphere, as well as in range of some mountain peaks. Atmospheric drag is the factor that most determines orbiting altitude.
23. Consider "Newton's cannon" fired from a tall mountain on Jupiter. To match the wider curvature of much larger Jupiter, and to contend with Jupiter's greater gravitational pull, the cannonball would have to be fired significantly faster. (Orbital speed about Jupiter is about 5 times that for Earth.)

25. Upon slowing it spirals in toward the Earth and in so doing has a component of gravitational force in its direction of motion which causes it to gain speed. Or put another way, in circular orbit the perpendicular component of force does no work on the satellite and it maintains constant speed. But when it slows and spirals toward Earth there is a component of gravitational force that does work to increase the KE of the satellite.

27. At midnight you face away from the sun, and therefore cannot see the planets closest to the sun—Mercury and Venus (which lie inside the Earth's orbit).

29. Yes, a satellite needn't be above the surface of the orbiting body. It could orbit at any distance from the Earth's center of mass. Its orbital speed would be less because the effective mass of the Earth would be that of the mass below the tunnel radius. So interestingly, a satellite in circular orbit has its greatest speed near the surface of the Earth, and decreases with both decreasing and increasing distances.

31. In circular orbit there is no component of force along the direction of the satellite's motion so no work is done. In elliptical orbit, there is always a component of force along the direction of the satellite's motion (except at the apogee and perigee) so work is done on the satellite.

33. The period of any satellite at the same distance from Earth as the moon would be the same as the moon's, 28 days.

35. The plane of a satellite coasting in orbit intersects the Earth's center. If its orbit were tilted relative to the equator, it would be sometimes over the northern hemisphere, sometimes over the southern hemisphere. To stay over a fixed point off the equator, it would have to be following a circle whose center is not at the center of the Earth.

37. No, for an orbit in the plane of the Arctic Circle does not intersect the Earth's center. All Earth satellites orbit in a plane that intersects the center of the Earth. A satellite may pass over the Arctic Circle, but cannot remain above it indefinitely, as a satellite can over the equator.

39. When a capsule is projected rearward at 7 km/s with respect to the spaceship, which is itself moving forward at 7 km/s with respect to the Earth, the speed of the capsule with respect to the Earth will be zero. It will have no tangential speed for orbit. What will happen? It will simply drop vertically to Earth and crash.

41. This is similar to Exercises 26 and 27. The tangential velocity of the Earth about the sun is 30 km/s. If a rocket carrying the radioactive wastes were fired at 30 km/s from the Earth in the direction opposite to the Earth's orbital motion about the sun, the wastes would have no tangential velocity with respect to the sun. They would simply fall into the sun.

43. The half brought to rest will fall vertically to Earth. The other half, in accord with the conservation of linear momentum will have twice the initial velocity, overshoot the circular orbit, and enter an elliptical orbit whose apogee (highest point) is farther from the Earth's center.

45. The escape speeds from various planets refer to "ballistic speeds"—to the speeds attained *after* the application of an applied force at low altitude. If the force is sustained, then a space vehicle could escape the Earth at any speed, so long as the force is applied sufficiently long.

47. This is similar to the previous exercise. In this case, Pluto's maximum speed of impact on the sun, by virtue of only the sun's gravity, would be the same as the escape speed from the surface of the sun, which according to Table 10.1 in the text is 620 km/s.

49. The satellite experiences the greatest gravitational force at A, where it is closest to the Earth; and the greatest speed and the greatest velocity at A, and by the same token the greatest momentum and greatest kinetic energy at A, and the greatest gravitational potential energy at the farthest point C. It would have the same total energy (KE + PE) at all parts of its orbit, likewise the same angular momentum because it's conserved. It would have the greatest acceleration at A, where F/m is greatest.

Chapter 10 Problem Solutions

1. One second after being thrown, its horizontal component of velocity is 10 m/s, and its vertical component is also 10 m/s. By the Pythagorean theorem, $V = \sqrt{(10^2 + 10^2)}$ = **14.1 m/s**. (It is moving at a 45° angle.)

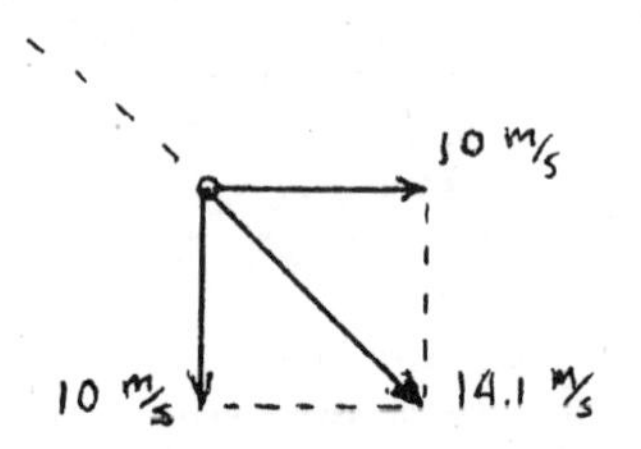

3. **100 m/s**. At the top of its trajectory, the vertical component of velocity is zero, leaving only the horizontal component. The horizontal component at the top or anywhere along the path is the same as the initial horizontal component, 100 m/s (the side of a square where the diagonal is 141).

5. John and Tracy's horizontal jumping velocity will be the horizontal distance traveled divided by the time of the jump. The horizontal distance will be a minimum of 20 m, but what will be the time? Aha, the same time it would take John and Tracy to fall straight down! From Table 3.3 we see such a fall would take 4 seconds. Or we can find the time from

$d = 5\,t^2$, where rearrangement gives $t = \sqrt{\frac{d}{5}}$

$= \sqrt{\frac{80}{5}} = 4$ s.

So to travel 20 m horizontally in this time means John and Tracy should jump horizontally with a velocity of 20 m/4 s = **5 m/s.** But this would put them at the edge of the pool, so they should jump a little faster. If we knew the length of the pool, we could calculate how much faster without hitting the far end of the pool. (John and Tracy would be better advised to take the elevator.)

7. Hang time depends only on the vertical component of initial velocity and the corresponding vertical distance attained. From $d = 5t^2$ a vertical 1.25 m drop corresponds to 0.5 s ($t = \sqrt{2d/g} = \sqrt{2(1.25)/10} = 0.5$ s). Double this (time up and time down) for a hang time of 1 s. Hang time is the **same** whatever the horizontal distance traveled.

9. $v = \sqrt{\frac{GM}{d}} = \sqrt{\frac{(6.67 \times 10^{-11})(6 \times 10^{24})}{3.8 \times 10^{8}}}$
= **1026 m/s.**

Part 2: Properties of Matter

Chapter 11: The Atomic Nature of Matter
Answers to Odd-Numbered Exercises

1. In a water molecule, H_2O, there are three atoms, one hydrogen and two oxygen.

3. The cat leaves a trail of molecules and atoms on the grass. These in turn leave the grass and mix with the air, where they enter the dog's nose, activating its sense of smell.

5. The speed at which the scent of a fragrance travels is much less than the speed of the individual molecules that make it up because of the many collisions among molecules. Although the molecular speed between collisions is great, the rate of migration in a particular direction through obstructing molecules is very much less.

7. The atoms that make up a newborn baby or anything else in this world originated in the explosions of ancient stars. (See Figure 11.1, my daughter Leslie.) The *molecules* that make up the baby, however, were formed from atoms ingested by the mother and transferred to her womb.

9. Water is not an element. It is a compound. It's molecules are made of the atoms of elements hydrogen and oxygen.

11. Brownian motion is apparent only for microscopic particles because of their small mass (meaning that they also have small size). Their low mass makes them respond more when they are bumped by the random collisions of surrounding atoms and molecules. Against a large particle, the random bumps exert nearly steady forces on each side that average to zero, but for a small particle there are moments where appreciably more hits occur on one side than the other, producing motion visible in a microscope.

13. No, it would not be the same. There are certain element pairs (cobalt - nickel, tellurium - iodine, thorium - protactinium, uranium - neptunium, plutonium - americium, 106 - 107) for which the order of increasing mass is opposite to the order of increasing atomic number. (Before atomic number was understood in 1913, this caused some confusion in arranging elements in the periodic table).

15. Nine.

17. The element is copper, atomic number 29. Any atom having 29 protons is by definition copper. (In Chapter 32 we'll learn that the total number of "nucleons," protons + neutrons, gives the "mass number" of the element, which is different for different isotopes. Here we have Cu-63.)

19. The source of oxygen and nitrogen is the air, which is needed for the burning (combustion) of gasoline.

21. The chemical properties of an element depend on the electrons in the atomic shells. But the number of electrons, in turn, is dictated by the number of protons in the nucleus. So in this indirect way, the number of protons in the atomic nucleus dictates the chemical properties of the element.

23. Carbon. (See the periodic table)

25. Radon.

27. Hydrogen and oxygen.

29. Oxygen.

31. The element below carbon in the periodic table, silicon, has similar properties and could conceivably be the basis of organic molecules elsewhere in the universe.

33. Electrical repulsion. Electrons speeding around within an atom create an electrified cloud that repels the similar clouds of other electrons, preventing the atoms from coalescing and keeping us from falling through our chairs. (For the record, quantum effects play a large role as well.)

35. (a) Heating gives the molecules more kinetic energy so they can shake loose from the bonds holding them in a solid, creating a liquid. (b) The solid must have stronger interatomic forces.

37. You really are a part of every person around you in the sense that you are composed of atoms not only from every person around you, but from every person who ever lived on Earth! The Richardson boy's statement that introduces Part 2 is indisputable. And the atoms that now compose you will make up the atomic pool that others will draw upon.

39. The amount of matter that a given amount of antimatter would annihilate is the same as the amount of antimatter, a pair of particles at a time. The whole world could not be annihilated by antimatter unless the mass of antimatter were at least equal to the mass of the world.

Chapter 11 Problem Solutions

1. There are 16 grams of oxygen in 18 grams of water. We can see from the formula for water, H_2O, there are twice as many hydrogen atoms (each of atomic mass 1) as oxygen atoms (each of atomic mass 16). So the molecular mass of H_2O is 18, with 16 parts oxygen by mass.

3. The atomic mass of element A is 3/2 the mass of element B. Why? Gas A has three times the mass of Gas B. If the equal number of molecules in A and B had equal numbers of atoms,

then the atoms in Gas A would simply be three times as massive. But there are twice as many atoms in A, so the mass of each atom must be half of three times as much — 3/2.

5. (a) 10^4 atoms (length 10^{-6} m divided by size 10^{-10} m). (b) 10^8 atoms ($10^4 \times 10^4$). (c) 10^{12} atoms ($10^4 \times 10^4 \times 10^4$). (d) $10,000 buys a good used car, for instance. $100 million buys a few jet aircraft and an airport on which to keep them, for instance. $1 trillion buys a medium-sized country, for instance. (Answers limited only by the imagination of the student.)

7. There are 10^{22} breaths of air in the world's atmosphere, which is the same number of atoms in a single breath. So for any one breath evenly mixed in the atmosphere, we sample one atom at any place or any time in the atmosphere.

Chapter 12: Liquids
Answers to Exercises

1. Physical properties have to do with the order, bonding, and structure of atoms that make up a material, and with the presence of other atoms and their interactions in the material. The silicon in glass is amorphous, whereas in semiconductors it is crystalline. Silicon in sand, from which glass is made, is bound to oxygen as silicon dioxide, while that in semiconductor devices is elemental and extremely pure. Hence their physical properties differ.

3. Iron is denser than cork, but not necessarily heavier. A common cork from a wine bottle, for example, is heavier than an iron thumbtack—but it wouldn't be heavier if the volumes of each were the same.

5. Its density increases.

7. Aluminum has more volume because it is less dense.

9. The top part of the spring supports the entire weight of the spring and stretches more than, say the middle, which only supports half the weight and stretches half as far. Parts of the spring toward the bottom support very little of the spring's weight and hardly stretch at all.

11. A twice-as-thick rope has four times the cross-section and is therefore four times as strong. The length of the rope does not contribute to its strength. (Remember the old adage, a chain is only as strong as its weakest link— the strength of the chain has to do with the thickness of the links, not the length of the chain.)

13. Concrete undergoes compression well, but not tension. So the steel rods should be in the part of the slab that is under tension, the top part.

15. The design to the left if better because the weight of water against the dam puts compression on the dam. Compression tends to jam the parts of the dam together, with added strength like the compression on an arch. The weight of water puts tension on the dam at the right, which tends to separate the parts of the dam.

17. A triangle is the most rigid of geometrical structures. Consider nailing four sticks together to form a rectangle, for example. It doesn't take much effort to distort the rectangle so that it collapses to form a parallelogram. But a triangle made by nailing three sticks together cannot collapse to form a tighter shape. When strength is important, triangles are used. That's why you see them in the construction of so many things.

19. Since each link in a chain is pulled by its neighboring links, tension in the hanging chain is exactly along the chain—parallel to the chain at every point. If the arch takes the same shape, then compression all along the arch will similarly be exactly along the arch—parallel to the arch at every point. There will be no internal forces tending to bend the arch. This shape is a catenary, and is the shape of modern-day arches such as the one that graces the city of St. Louis.

21. The candymaker needs less taffy for the larger apples because the surface area is less per kilogram. (This is easily noticed by comparing the peelings of the same number of kilograms of small and large apples.)

23. Kindling will heat to a higher temperature in a shorter time than large sticks and logs. Its greater surface area per mass results in most of its mass being very near the surface, which quickly heats from all sides to its ignition temperature. The heat given to a log, on the other hand, is not so concentrated as it conducts into the greater mass. Large sticks and logs are slower to reach the ignition temperature.

25. More heat is lost from the rambling house due to its greater surface area.

27. For a given volume, a sphere has less surface area than any other geometrical figure. A dome-shaped structure similarly has less surface area per volume than conventional block designs. Less surface exposed to the climate = less heat loss.

29. The wider, thinner burger has more surface area for the same volume. The greater the surface area, the greater will be the heat transfer from the stove to the meat.

31. Mittens have less surface than gloves. Anyone who has made mittens and gloves will tell you that much more material is required to make gloves. Hands in gloves will cool faster than hands in mittens. Fingers, toes, and ears have a disproportionately large surface area relative to other parts of the body and are therefore more prone to frostbite.

33. Small animals radiate more energy per body-weight, so the flow of blood is correspondingly greater, and the heartbeat faster.

35. The inner surface of the lungs is not smooth, but is sponge-like. As a result, there is an enormous surface exposed to the air that is breathed. This is nature's way of compensating for the proportional decrease in surface area for large bodies. In this way, the adequate amount of oxygen vital to life is taken in.

37. Large raindrops fall faster than smaller raindrops for the same reason that heavier parachutists fall faster than lighter parachutists. Both larger things have less surface area and therefore less air resistance relative to their weights.

39. Scaling plays a significant role in the design of the hummingbird and the eagle. The wings of a hummingbird are smaller than those of the eagle relative to the size of the bird, but are larger relative to the mass of the bird. The hummingbird's swift maneuvers are possible because the small rotational inertia of the short wings permits rapid flapping that would be impossible for wings as large as those of an eagle. If a hummingbird were scaled up to the size of an eagle, its wings would be much shorter than those of an eagle, so it couldn't soar. Its customary rate of flapping would be insufficient to provide lift for its disproportionately greater weight. Such a giant hummingbird couldn't fly, and unless its legs were disproportionately thicker, it would have great difficulty walking. The great difference in the design of hummingbirds and eagles is a natural consequence of the area to volume ratio of scaling. Interesting?

Chapter 12 Problem Solutions

1. Density = $\frac{\text{mass}}{\text{volume}} = \frac{5 \text{ kg}}{V}$. Now the volume of a cylinder is its round area x its height ($\pi r^2 h$).
So density = $\frac{5 \text{ kg}}{\pi r^2 h} = \frac{5000 \text{ g}}{(3.14)(3^2)(10)\text{cm}^3} =$ **17.7 g/cm^3**.

3. 45 N is three times 15 N, so the spring will stretch three times as far, **9 cm**. Or from Hooke's law; $F = kx$, $x = F/k$ = 45 N/(15 N/3 cm) = 9 cm. (The spring constant k = 15 N/3 cm.)

5. If the spring is cut in half, it will stretch as far as half the spring stretched before it was cut—half as much. This is because the tension in the uncut spring is the same everywhere, equal to the full load at the middle as well as the end. So the 10 N load will stretch it **2 cm**. (Cutting the spring in half doubles the spring constant k. Initially k = 10 N/4 cm = 2.5 N/cm; when cut in half, k = 10 N/2 cm = 5 N/cm.)

7. (a) **Eight** smaller cubes (see Figure 12.15). (b) Each face of the original cube has an area of 4 cm^2 and there are 6 faces, so the total area is **24 cm^2**. Each of the smaller cubes has an area of 6 cm^2 and there are eight of them, so their total surface area is **48 cm^2**, twice as great. (c) The surface-to-volume ratio for the original cube is (24 cm^2)/(8 cm^3) = **3 cm^{-1}**. For the set of smaller cubes, it is (48 cm^2)/(8 cm^3) = **6 cm^{-1}**, twice as great. (Notice that the surface-to-volume ratio has the unit inverse cm.)

9. The big cube will have the **same combined volume** of the eight little cubes, but **half their combined area**. The area of each side of the little cubes is 1 cm^2, and for its six sides the total area of each little cube is 6 cm^2. So all eight individual cubes have a total surface area 48 cm^2. The area of each side of the big cube, on the other hand, is 2^2 or 4 cm^2; for all six sides its total surface area is 24 cm^2, half as much as the separate small cubes.

Chapter 13: Liquids
Answers to Exercises

1. The scale measures force, not pressure, and is calibrated to read your weight. That's why your weight on the scale is the same whether you stand on one foot or both.

3. A sharp knife cuts better than a dull knife because it has a thinner cutting area which results in more cutting pressure for a given force.

5. A woman with spike heels exerts considerably more pressure on the ground than an elephant! Example: A 500-N woman with 1-cm^2 spike heels puts half her weight on each foot, distributed (let's say) half on her heel and half on her sole. So the pressure exerted by each heel will be (125 N/1 cm^2) = 125 N/cm^2. A 20,000-N elephant with 1000 cm^2 feet exerting 1/4 its weight on each foot produces (5000N/1000 cm^2) = 5N/cm^2; about 25 times less pressure. (So a woman with spike heels will make greater dents in a new linoleum floor than an elephant will.)

7. Your body gets more rest when lying than when sitting or standing because when lying, the heart does not have to pump blood to the heights that correspond to standing or sitting. Blood pressure is normally greater in the lower parts of your body simply because the blood is "deeper" there. Since your upper arms are at the same level as your heart, the blood pressure in your upper arms will be the same as the blood pressure in your heart.

9. (a) The reservoir is elevated so as to produce suitable water pressure in the faucets that it serves. (b) The hoops are closer together at the bottom because the water pressure is greater at the bottom. Closer to the top, the water pressure is not as great, so less reinforcement is needed there.

11. A one-kilogram block of aluminum is larger than a one-kilogram block of lead. The aluminum therefore displaces more water.

13. Your mass decreases, but not significantly. Your volume decreases as your lungs deflate, causing your density to increase significantly (nearly the same mass divided by less volume). Density most directly affects how high you float.

15. A typical plumbing design involves short sections of pipe bent at 45-degree angles between vertical sections two-stories long. The sewage therefore undergoes a succession of two-story falls which results in a moderate momentum upon reaching the basement level.

17. The use of a water-filled garden hose as an elevation indicator is a practical example of water seeking its own level. The water surface at one end of the hose will be at the same elevation above sea level as the water surface at the other end of the hose.

19. There is less blood pressure in your finger when held higher.

21. The diet drink is less dense than water, whereas the regular drink is denser than water. (Water with dissolved sugar is denser than pure water.)

23. Mountain ranges are very similar to icebergs: both float in a denser medium, and extend farther down into that medium than they extend above it. Mountains, like icebergs, are bigger than they appear to be. The concept of floating mountains is *isostacy*—Archimedes' principle for rocks.

25. The force needed will be the weight of 1 L of water, which is 9.8 N. If the weight of the carton is not negligible, then the force needed would be 9.8 N minus the carton's weight, for then the carton would be "helping" to push itself down.

27. The block of wood would float higher if the piece of iron is suspended below it rather than on top of it. By the law of flotation: The iron-and-wood unit displaces its combined weight and the same volume of water whether the iron is on top or the bottom. When the iron is on the top, more wood is in the water; when the iron is on the bottom, less wood is in the water. Or another explanation is that when the iron is below—submerged—buoyancy on it reduces its weight and less of the wood is pulled beneath the water line.

29. A sinking submarine will continue to sink to the bottom so long as the density of the submarine is greater than the density of the surrounding water. If nothing is done to change the density of the submarine, it will continue to sink because the density of water is practically constant. In practice, water is sucked into or blown out of a submarine's tanks to adjust its density to match the density of the surrounding water.

31. For the same reason as in the previous exercise, the water level will fall. (Try this one in your kitchen sink also. Note the water level at the side of the dishpan when a bowl floats in it. Tip the bowl so it fills and submerges, and you'll see the water level at the side of the dishpan fall.)

33. You are compressible, whereas a rock is not, so when you are submerged, the water pressure tends to squeeze in on you and reduce your volume. This increases your density. (Be careful when swimming—at shallow depths you may still be less dense than water and be buoyed to the surface without effort, but at greater depths you may be pressed to a density greater than water and you'll have to swim to the surface.).

35. The buoyant force does not change. The buoyant force on a floating object is always equal to that object's weight, no matter what the fluid.

37. No, there does not have to actually be 14.5 N of fluid in the skull to supply a buoyant force of 14.5 N on the brain. To say that the buoyant force is 14.5 N is to say that the brain is taking up the space that 14.5 N of fluid would occupy if fluid instead of the brain were there. The amount of fluid in excess of the fluid that immediately surrounds the brain does not contribute to the buoyancy on the brain. (A ship floats the same in the middle of the ocean as it would if it were floating in a small lock just barely larger than the ship itself. As long as there is enough water to press against the hull of the ship, it will float. It is not important that the amount of water in this tight-fitting lock weigh as much as the ship—think about that, and don't let a literal word explanation "a floating object displaces a weight of fluid equal to its own weight" and the idea it represents confuse you.)

39. When the ice cube melts the water level at the side of the glass is unchanged (neglecting temperature effects). To see this, suppose the ice cube to be a 5 gram cube; then while floating it will displace 5 grams of water. But when melted it becomes the same 5 grams of water. Hence the water level is unchanged. The same occurs when the ice cube with the air bubbles melts. Whether the ice cube is hollow or solid, it will displace as much water floating as it will melted. If the ice cube contains grains of heavy sand, however, upon melting, the water level at the edge of the glass will drop. This is similar to the case of the scrap iron of Exercise 30.

41. When the ball is submerged (but not touching the bottom of the container), it is supported partly by the buoyant force on the left and partly by the string connected to the right side. So the left pan must increase its upward force to provide the buoyant force in addition to whatever force it provided before, and the right pan's upward force decreases by the same amount, since it now supports a ball lighter by the amount of the buoyant force. To bring the scale back to balance, the additional weight

that must be put on the right side will equal twice the weight of water displaced by the submerged ball. Why twice? Half of the added weight makes up for the loss of upward force on the right, and the other half for the equal gain in upward force on the left. (If each side initially weighs 10 N and the left side gains 2 N to become 12 N, the right side loses 2 N to become 8 N. So an additional weight of 4 N, not 2 N, is required on the right side to restore balance.) Because the density of water is less than half the density of the iron ball, the restoring weight, equal to twice the buoyant force, will still be less than the weight of the ball.

43. Both you and the water would have half the weight density as on Earth , and you would float with the same proportion of your body above the water as on Earth . Water splashed upward with a certain initial speed would rise twice as high, since it would be experiencing only half the "gravity force". Waves on the water surface would move more slowly than on Earth (at about 70% as fast since $v_{wave} \sim \sqrt{g}$).

45. A Ping-Pong ball in water in a zero-g environment would experience no buoyant force. This is because buoyancy depends on a pressure difference on different sides of a submerged body. In this weightless state, no pressure difference would exist because no water pressure exists. (See the answer to Exercise 20, and Home Project 2.)

47. The strongman will be unsuccessful. He will have to push with 50 times the weight of the 10 kilograms. The hydraulic arrangement is arranged to his disadvantage. Ordinarily, the input force is applied against the smaller piston and the output force is exerted by the large piston—this arrangement is just the opposite.

49. When water is hot, the molecules are moving more rapidly and do not cling to one another as well as when they are slower moving, so the surface tension is less. The lesser surface tension of hot water allows it to pass more readily through small openings.

Problem Solutions

1. Pressure = weight density × depth = 9800 N/m^3 × 220 m = 2,160,000 N/m^2 = **2160 kPa**.

3. (a) The volume of the extra water displaced will weigh as much as the 400-kg horse. And the volume of extra water displaced will also equal the area of the barge times the extra depth. That is,
$V = Ah$, where A is the horizontal area of the barge; Then $h = \frac{V}{A}$.

Now A = 5m x 2m = 10 m^2; to find the volume V of barge pushed into the water by the horse's weight, which equals the volume of water displaced, we know that

density = $\frac{m}{V}$. Or from this, $V = \frac{m}{\text{density}}$
$= \frac{400\text{kg}}{1000\text{kg/m}^3} = 0.4\ \text{m}^3$.
So $h = \frac{V}{A} = \frac{0.4\ \text{m}^3}{10\ \text{m}^2}\ \frac{\text{m}^3}{\text{m}^2}$ = **0.04 m**, which is 4 cm deeper.
(b) If each horse will push the barge 4 cm deeper, the question becomes: How many 4-cm increments will make 15 cm? 15/4 = 3.75, so 3 horses can be carried without sinking. **4 horses** will sink the barge.

5. From Table 12.1 the density of gold is 19.3 g/cm^3. Your gold has a mass of 1000 grams, so $\frac{1000\ \text{g}}{\text{V}}$ = 19.3 g/cm^3. Solving for V,
$$V = \frac{1000\ \text{g}}{19.3\ \text{g/cm}^3} = \mathbf{51.8\ cm^3}.$$

7. 10% of ice extends above water. So 10% of the 9-cm thick ice would float above the water line; **0.9 cm**. So the ice pops up. Interestingly, when mountains erode they become lighter and similarly pop up! Hence it takes a long time for mountains to wear away.

9. The displaced water, with a volume 90 percent of the vacationer's volume, weighs the same as the vacationer (to provide a buoyant force equal to his weight). Therefore his density is 90 percent of the water's density. Vacationer's density = (0.90)(1,025 kg/m^3) = **923 kg/m^3**.

11. The pressure applied to the fluid in the reservoir must equal the pressure the piston exerts against the fluid in the cylinder. This is the weight of 2000 kg divided by 400 cm^2. That's (2000 kg x 9.8 N/kg)/400 cm^2 = (19,600 N)/(400 cm^2) = **49 N/cm^2**. (In standard units this is 490,000 N/m^2, or 490 kPa.)

Chapter 14: Gases and Plasmas
Answers to Exercises

1. Some of the molecules in the Earth 's atmosphere *do* go off into outer space—those like helium with speeds greater than escape speed. But the average speeds of most molecules in the atmosphere are well below escape speed, so the atmosphere is held to Earth by Earth gravity.

3. The weight of a truck is distributed over the part of the tires that make contact with the road. Weight/surface area = pressure, so the greater the surface area, or equivalently, the greater the number of tires, the greater the weight of the truck can be for a given pressure. What pressure? The pressure exerted by the tires on the road, which is determined by (but is somewhat greater than) the air pressure in its tires. Can you see how this relates to Home Project 1?

5. The density of air in a deep mine is greater than at the surface. The air filling up the mine

adds weight and pressure at the bottom of the mine, and according to Boyle's law, greater pressure in a gas means greater density.

7. The bubble's mass does not change. It's volume increases because its pressure decreases (Boyle's law), and its density decreases (same mass, more volume).]

9. To begin with, the two teams of horses used in the Magdeburg hemispheres demonstration were for showmanship and effect, for a single team and a strong tree would have provided the same force on the hemispheres. So if two teams of nine horses each could pull the hemispheres apart, a single team of nine horses could also, if a tree or some other strong object were used to hold the other end of the rope.

11. Airplane windows are small because the pressure difference between the inside and outside surfaces result in large net forces that are directly proportional to the window's surface area. (Larger windows would have to be proportionately thicker to withstand the greater net force—windows on underwater research vessels are similarly small.)

13. Because of the vacuum inside a TV tube, it implodes when it is broken. The pressure of the atmosphere simply pushes the parts of the broken tube inward.

15. A vacuum cleaner wouldn't work on the moon. A vacuum cleaner operates on Earth because the atmospheric pressure pushes dust into the machine's region of reduced pressure. On the moon there is no atmospheric pressure to push the dust anywhere.

17. If barometer liquid were half as dense as mercury, then to weigh as much, a column twice as high would be required. A barometer using such liquid would therefore have to be twice the height of a standard mercury barometer, or about 152 cm instead of 76 cm.

19. Mercury can be drawn a maximum of 76 cm with a siphon. This is because 76 vertical cm of mercury exert the same pressure as a column of air that extends to the top of the atmosphere. Or looked at another way; water can be lifted 10.3 m by atmospheric pressure. Mercury is 13.6 times denser than water, so it can only be lifted only 1/13.6 times as high as water.

21. Drinking through a straw is slightly more difficult atop a mountain. This is because the reduced atmospheric pressure is less effective in pushing soda up into the straw.

23. You agree with your friend, for the elephant displaces far more air than a small helium-filled balloon, or small anything. The *effects* of the buoyant forces, however, is a different story. The large buoyant force on the elephant is insignificant relative to its enormous weight. The tiny buoyant force acting on the balloon of tiny weight, however, is significant.

25. One's lungs, like an inflated balloon, are compressed when submerged in water, and the air within is compressed. Air will not of itself flow from a region of low pressure into a region of higher pressure. The diaphragm in one's body reduces lung pressure to permit breathing, but this limit is strained when nearly 1 m below the water surface. It is exceeded at more than 1 m.

27. The air tends to pitch toward the rear (law of inertia), becoming momentarily denser at the rear of the car, less dense in the front. Because the air is a gas obeying Boyle's law, its pressure is greater where its density is greater. Then the air has both a vertical and a horizontal "pressure gradient." The vertical gradient, arising from the weight of the atmosphere, buoys the balloon up. The horizontal gradient, arising from the acceleration, buoys the balloon forward. So the string of the balloon makes an angle. The pitch of the balloon will always be in the direction of the acceleration. Step on the brakes and the balloon pitches backwards. Round a corner and the balloon noticeably leans radially towards the center of the curve. Nice! (Another way to look at this involves the effect of two accelerations, g and the acceleration of the car. The string of the balloon will be parallel to the resultant of these two accelerations. Nice again!)

29. The buoyant force does not change, because the volume of the balloon does not change. The buoyant force is the weight of air displaced, and doesn't depend on what is doing the displacing.

31. The shape would be a catenary. It would be akin to Gateway Arch in St. Louis and the hanging chain discussed in Chapter 12.

33. The end supporting the punctured balloon tips upwards as it is lightened by the amount of air that escapes. There is also a loss of buoyant force on the punctured balloon, but that loss of upward force is less than the loss of downward force, since the density of air in the balloon before puncturing was greater than the density of surrounding air.

35. The rotating habitat is a centrifuge, and denser air is "thrown to" the outer wall. Just as on Earth, the maximum air density is at "ground level," and becomes less with increasing altitude (distance toward the center). Air density in the rotating habitat is least at the zero-g region, the hub.

37. The force of the atmosphere is on both sides of the window; the net force is zero, so windows don't normally break under the weight of the atmosphere. In a strong wind, however, pressure will be reduced on the windward side (Bernoulli's Principle) and the forces no longer cancel to zero. Many windows are blown *outward* in strong winds.

39. More air blows over the top of the beach ball than under it, causing the pressure above the ball to be less than the pressure under it (Bernoulli's principle). The pressure difference provides a support force to counteract the gravitational force.

41. (a) Speed increases (so that the same quantity of gas can move through the pipe in the same time). (b) Pressure decreases (Bernoulli's principle). (c) The spacing between the streamlines decreases, because the same number of streamlines fit in a smaller area.

43. An airplane flies upside down by tilting its fuselage so that there is an angle of attack of the wing with oncoming air. (It does the same when flying right side up, but then, because the wings are designed for right-side-up flight, the tilt of the fuselage may not need to be as great.)

45. Bernoulli's Principle. For the moving car the pressure will be less on the side of the car where the air is moving fastest— the side nearest the truck, resulting in the car's being pushed by the atmosphere towards the truck. Inside the convertible, atmospheric pressure is greater than outside, and the canvas roof top is pushed upwards towards the region of lesser pressure. Similarly for the train windows, where the interior air is at rest relative to the window and the air outside is in motion. Air pressure against the inner surface of the window is greater than the atmospheric pressure outside. When the difference in pressures is significant enough, the window is blown out.

47. The behavior of the flag is similar to that of the water waves in the previous exercise. See the sketch below. Where the flag is curving to the right (point A), the wind speed is greater on the right side, for the "trough" on the left is partially shielded. The pressure is therefore less on the right side. Where the flag is curving to the left (point B), the wind speed is greater to the left, and the pressure is less on the left. In both cases, the pressure difference on the two sides of the flag will tend to push the "wave" in the flag to greater amplitude. Points A and C will be pushed to the right; point B will be pushed to the left. With the wave crests amplified, the wind will push them along the flag (downward in this diagram), with the result that the free end flaps to and fro. (We should give some credit to your friend who stated a flag flaps because of Bernoulli's principle. To one who understands Bernoulli's principle, the answer is sufficient. To one who doesn't, more explanation should be given.)

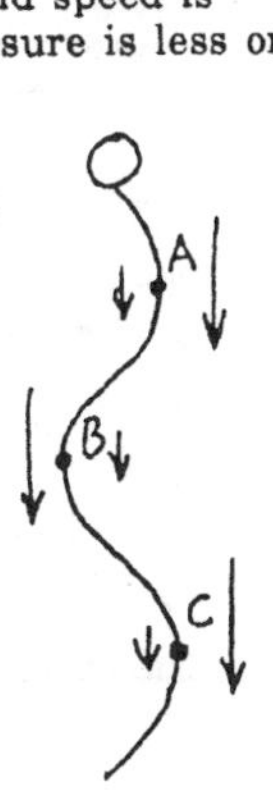

49. According to Bernoulli's principle, when a fluid gains speed in flowing through a narrow region, the pressure of the fluid is reduced. The gain in speed, the cause, produces reduced pressure, the effect. But one can argue that a reduced pressure in a fluid, the cause, will produce a flow in the direction of the reduced pressure, the effect. For example, if you decrease the air pressure in a pipe by a pump or by any means, neighboring air will rush into the region of reduced pressure. In this case the increase in air speed is the result, not the cause of, reduced pressure. Cause and effect are open to interpretation. Bernoulli's principle is a controversial topic with many physics types!

Chapter 13 Problem Solutions

1. According to Boyle's law, the pressure will increase to **three times** its original pressure.

3. To decrease the pressure ten-fold, back to its original value, in a fixed volume, 90% of the molecules must escape, leaving **one-tenth** of the original number.

5. If the atmosphere were composed of pure water vapor, the atmosphere would condense to a depth of 10.3 m. Since the atmosphere is composed of gases that have less density in the liquid state, their liquid depths would be more than 10.3 m, about **12 m**. (A nice reminder of how thin and fragile our atmosphere really is.)

7. (a) The weight of the displaced air must be the same as the weight supported, since the total force (gravity plus buoyancy) is zero. The displaced air weighs **20,000 N**. (b) Since weight = mg, the mass of the displaced air is $m = W/g$ = (20,000 N)/(10 m/s^2) = 2,000 kg. Since density is mass/volume, the volume of the displaced air is vol = mass/density = (2,000 kg)/(1.2 kg/m^3) = **1,700 m^3** (same answer to two figures if g = 9.8 m/s^2 is used).

9. Since the column of mercury has half the height it has at sea level (where it is 760 mm), air pressure is half its sea-level value. Air pressure does not decrease at a constant rate the way water pressure does, so it does not decrease as much in the second 5.6 km of altitude change as in the first 5.6 km. At 11.2 km, the mercury column will be shorter but not zero.

Part 3: Heat

Chapter 15: Temperature and Expansion

Answers to Exercises

1. Inanimate things such as tables, chairs, furniture, and so on, have the same temperature as the surrounding air (assuming they are in thermal equilibrium with the air—i.e., no sudden gush of different-temperature air or such). People and other mammals, however, generate their own heat and have body temperatures that are normally higher than air temperature.

3. Gas molecules move haphazardly and move at random speeds. They continually run into one another, sometimes giving kinetic energy to neighbors, sometimes receiving kinetic energy. In this continual interaction, it would be statistically impossible for any large number of molecules to have the same speed. Temperature has to do with average speeds.

5. The hot coffee has a higher temperature, but not a greater internal energy. Although the iceberg has less internal energy per mass, its enormously greater mass gives it a greater total energy than that in the small cup of coffee. (For a smaller volume of ice, the fewer number of more energetic molecules in the hot cup of coffee may constitute a greater total amount of internal energy—but not compared to an iceberg.)

7. Direct radiation from the sun can heat the thermometer to a higher temperature than the air around it when the thermometer absorbs radiation better than air. The thermometer is not "wrong." It is reading its *own* temperature, which is what every thermometer does.

9. When no more energy can be extracted from a material, it is at absolute zero, But there is no limit, in principle, to how much energy can be added to a material. (It is like kinetic energy, which has a minimum, zero, but no maximum.)

11. Other effects aside, the temperature should be slightly higher, because the PE of the water above has been transformed to KE below, which in turn is transformed to heat and internal energy when the falling water is stopped. (On his honeymoon, James Prescott Joule could not be long diverted from his preoccupation with heat, and he attempted to measure the temperature of the water above and below a waterfall in Chamonix. The temperature increase he expected, however, was offset by cooling due to evaporation as the water fell.)

13. Different substances have different thermal properties due to differences in the way energy is stored internally in the substances. When the same amount of heat produces different changes in temperatures in two substances of the same mass, we say they have different specific heat capacities. Each substance has its own characteristic specific heat capacity. Temperature measures the average kinetic energy of random motion, but not other kinds of energy.

15. Water has a high specific heat capacity, which is to say, it normally takes a long time to heat up, or cool down. The water in the watermelon resists changes in temperature, so once cooled it will stay cool longer than sandwiches or other non-watery substances under the same conditions. Be glad water has a high specific heat capacity the next time you're enjoying cool watermelon on a hot day!

17. The climate of Iceland, like that of Bermuda in the previous exercise, is moderated by the surrounding water.

19. As the ocean off the coast of San Francisco cools in the winter, the heat it loses warms the atmosphere it comes in contact with. This warmed air blows over the California coastline to produce a relatively warm climate. If the winds were easterly instead of westerly, the climate of San Francisco would be chilled by winter winds from dry and cold Nevada. The climate would be reversed also in Washington D.C., because air warmed by the cooling of the Atlantic Ocean would blow over Washington D.C. and produce a warmer climate in winter there.

21. Sand has a low specific heat, as evidenced by its relatively large temperature changes for small changes in internal energy. A substance with a high specific heat, on the other hand, must absorb or give off large amounts of internal energy for comparable temperature changes.

23. No, the different expansions are what bends the strip or coil. Without the different expansions a bimetallic strip would not bend when heated.

25. When doused, the outer part of the boulders cooled while the insides were still hot. This caused a difference in contraction, which fractured the boulders.

27. On a hot day the pendulum lengthens slightly, which increases its period. So the clock "slows." (Some clock pendula have countering devices to offset this.)

29. Cool the inner glass and heat the outer glass. If its done the other way around, the glasses will stick even tighter (if not break).

31. The photo was taken on a warm day. Note that the rocker is slightly tipped toward the end rather than toward the middle, which means the steel beam is slightly lengthened.

33. Every part of a metal ring expands when it is heated—not only the thickness, but the outer and inner circumference as well. Hence the ball that normally passes through the hole when the temperatures are equal will more easily pass through the expanded hole when the ring is heated. (Interestingly enough, the

hole will expand as much as a disk of the same metal undergoing the same increase in temperature. Blacksmiths mounted metal rims in wooden wagon wheels by first heating the rims. Upon cooling, the contraction resulted in a snug fit.)

35. The gap in the ring will become wider when the ring is heated. Try this: draw a couple of lines on a ring where you pretend a gap to be. When you heat the ring, the lines will be farther apart—the same amount as if a real gap were there. Every part of the ring expands proportionally when heated uniformly — thickness, length, gap and all.

37. It is not an exception. The air, when heated, *does* expand. Although the house may expand a little, it doesn't expand enough to make room for the larger volume of air. Some of the air leaks to the outside. (And, when you cool the house, air is drawn in from outside.)

39. In the construction of a light bulb, it is important that the metal leads and the glass have the same rate of heat expansion. If the metal leads expand more than glass, the glass may crack. If the metal expands less than glass upon being heated, air will leak in through the resulting gaps.

41. Temperature of 4°C.

43. The atoms and molecules of most substances are more closely packed in solids than in liquids. So most substances are denser in the solid phase than in the liquid phase. Such is the case for iron and aluminum and most all other metals. But water is different. In the solid state the structure is open-spaced and ice is less dense than water. Hence ice floats in water.

45. The curve for density versus temperature is:

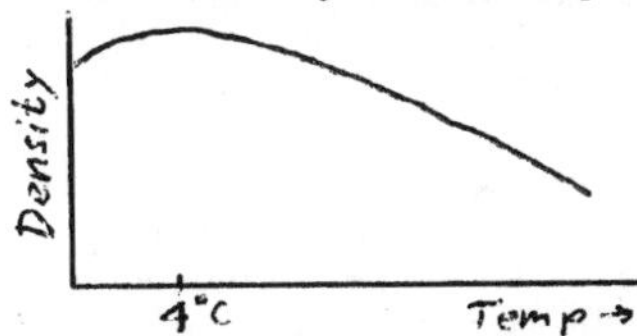

47. It is important to keep water in pipes from freezing because when the temperature drops below freezing, the water expands as it freezes whereas the pipe (if metal) will fracture if water in them freezes.

49. Ponds would be more likely to freeze if water had a lower specific heat. This is because the temperature would decrease more when water gives up energy; water would more readily be cooled to the freezing point.

Chapter 15 Problem Solutions

1. Heat gained by the cooler water = heat lost by the warmer water. Since the masses of water are the same, the final temperature is midway, 30°C. So you'll end up with 100 g of **30°C** water.

3. Raising the temperature of 10 gm of copper by one degree takes 10 x 0.092 = 0.92 calories, and raising it through 100 degrees takes 100 times as much, or 92 calories.

   By formula, $Q = mc\Delta T$ = (10 g)(0.092 cal/g°C)(100°C) = 92 cal.

   Heating 10 grams of water through the same temperature difference takes 1,000 calories, **more than ten times more** than for the copper—another reminder that water has a large specific heat capacity.

5. Heat gained by water = heat lost by nails

   $(cm\ \Delta T)_{water} = (cm\ \Delta T)_{nails}$

   $(1)(100)(T - 20) = (0.12)(100)(40 - T)$,

   giving T = **22.1°C.**

7. By formula: $\Delta L = L_o \alpha \Delta T$ =

   (1300 m)(11 x 10^{-6}/°C)(15°C) = **0.21 m.**

9. **Aluminum expands more** as evidenced by its greater coefficient of linear expansion. The ratio of the increases is equal to the ratios of coefficients of expansion, $24 \times 10^{-6}/11 \times 10^{-6} = 2.2$. So the same increase in temperature, the change in length of aluminum will be **2.2 times greater** than the change in length of steel.

Chapter 16: Heat Transfer

Answers to Exercises

1. No, the coat is not a source of heat, but merely keeps the thermal energy of the wearer from leaving rapidly.

3. When the temperatures of the blocks are the same as the temperature of your hand, then no heat transfer occurs. Heat will flow between your hand and something being touched only if there is a temperature difference between them.

5. Copper and aluminum are better conductors than stainless steel, and therefore more quickly transfer heat to the cookware's interior.

7. In touching the tongue to very cold metal, enough heat can be quickly conducted away from the tongue to bring the saliva to sub-zero temperature where it freezes, locking the tongue to the metal. In the case of relatively non-conducting wood, much less heat is conducted from the tongue and freezing does not take place fast enough for sudden sticking to occur.

9. Heat from the relatively warm ground is conducted by the gravestone to melt the snow in contact with the gravestone. Likewise for trees or any materials that are better conductors of heat than snow, and that extend into the ground.

11. Much of the energy of the flame is readily conducted through the paper to the water. The large amount of water relative to the paper absorbs the energy that would otherwise raise the temperature of the paper. The upper limit of 212°F for the water is well below the ignition temperature of the paper, 451°F (hence the title "451" of Ray Bradbury's science fiction novel about book burning).

13. Air is a poor conductor, whatever the temperature. So holding your hand in hot air for a short time is not harmful because very little heat is conducted by the air to your hand. But if you touch the hot conducting surface of the oven, heat readily conducts to you—ouch!

15. The conductivity of wood is relatively low whatever the temperature—even in the stage of red hot coals. You can safely walk barefoot across red hot wooden coals if you step quickly (like removing the wooden-handled pan with bare hands quickly from the hot oven in the previous exercise) because very little heat is conducted to your feet. Because of the poor conductivity of the coals, energy from within the coals does not readily replace the energy that transfers to your feet. This is evident in the diminished redness of the coal after your foot has left it. Stepping on red-hot *iron* coals, however, is a different story. Because of the excellent conductivity of iron, very damaging amounts of heat would transfer to your feet. More than simply ouch!

17. Agree, for your friend is correct.

19. More molecules are in the cooler room. The greater number of slower-moving molecules there produce air pressure at the door equal to the fewer number of faster-moving molecules in the warmer room.

21. At the same temperature, molecules of helium, nitrogen, and oxygen have the same average kinetic energy. But helium, because of its smaller mass, has greater average speed. So some helium atoms, high in the atmosphere, will be moving faster than escape speed from the Earth and will be lost to space. Through random collisions, every helium atom will eventually surpass escape speed.

23. Yes, gravity decreases upward movement of molecules, and increases downward movement. So the effect of gravity on molecular speed favors a downward migration of molecules. And yes, there is a greater "window" upward due to the smaller density of gas with altitude. The wider density window favors upward migration. When these two influences are equal, no convection occurs and the air is in thermal equilibrium.

25. Hydrogen molecules will be the faster moving when mixed with oxygen molecules. They will have the same temperature, which means they will have the same average kinetic energy. Recall that $KE = 1/2\, mv^2$. Since the mass of hydrogen is considerably less than oxygen, the velocity must correspondingly be greater.

27. Again, molecules of gas with less mass have higher average speeds. So molecules containing heavier U-238 are slower on the average. This favors the diffusion of the faster gas containing U-235 through a porous membrane (which is how U-235 was separated from U-238 by scientists in the 1940s!).

29. When we warm a volume of air, we add energy to it. When we expand a volume of air, we normally take energy out of it (because the expanding air does work on its surroundings). So the conditions are quite different and the results will be different. Expanding a volume of air actually lowers its temperature.

31. The mixture expands when it is ejected from the nozzle, and therefore cools. At the freezing temperature of 0°C, ice forms.

33. The heat that you received was from radiant energy.

35. If good absorbers were not also good emitters, then thermal equilibrium would not be possible. If a good absorber only absorbed, then its temperature would climb above that of poorer absorbers in the vicinity. And if poor absorbers were good emitters, their temperatures would fall below that of better absorbers.

37. The energy given off by rock at the Earth's surface transfers to the surroundings practically as fast as it is generated. Hence there isn't the buildup of energy that occurs in the Earth's interior.

39. Heat radiates into the clear night air and the temperature of the car goes down. Normally, heat is conducted to the car by the relatively warmer ground, but the rubber tires prevent the conduction of heat from the ground. So heat radiated away is not easily replaced and the car cools to temperatures below that of the surroundings. In this way frost can form on a below-freezing car in the above-freezing environment.

41. The white house is also a poor radiator, so when the house is warmed by fuel inside, its whiteness reduces the radiation it emits to the surrounding. This helps to keep it warmer inside.

43. When it is desirable to reduce the radiation that comes into a greenhouse, whitewash is applied to the glass to simply reflect much of the incoming radiation. Energy reflected is energy not absorbed.

45. If the upper atmosphere permitted the escape of more terrestrial radiation than it does presently, more energy would escape and the Earth's climate would be cooler.

47. Turn your heater off altogether and save fuel. When it is cold outside, your house is constantly losing heat. How much is lost depends on the insulation and the difference in inside and outside temperature (Newton's law of cooling). Keeping ΔT high consumes more fuel. To consume less fuel, keep ΔT low and turn your heater off altogether. Will more fuel be required to reheat the house when you return than would have been required to keep it warm while you were away? Not at all. When you return, you are replacing heat lost by the house at an average temperature below the normal setting, but if you had left the heater on, it would have supplied more heat, enough to make up for heat lost by the house at its normal, higher temperature setting. (Perhaps your instructor will demonstrate this with the analogy of leaking water buckets.)

49. Because warm air rises, there's a higher temperature at the ceiling than at the walls. With a greater difference in inside and outside temperatures, thicker insulation is needed to slow the transfer of heat.

Chapter 16 Problem Solutions

1. (a) The amount of heat absorbed by the water is $Q = cm\Delta T = (1.0 \text{ cal/g C°})(50.0 \text{ g})(50°\text{C} - 22°\text{C}) = 1400 \text{ cal}$. At 40% efficiency only 0.4 the energy from the peanut raises the water temperature, so the calorie content of the peanut is 1400/0.4 = **3500 cal**. (b) The food value of a peanut is 3500 cal/0.6 g = **5.8 kilocalories per gram**.

3. Work the hammer does on the nail is given by $F \times d$, and the temperature change of the nail can be found from using $Q = cm\ \Delta T$. First, we get everything into more convenient units for calculating: 5 grams = 0.005 kg; 6 cm = 0.06 m. Then $F \times d$ = 500 N x 0.06 m = 30 J, and 30 J = (0.005 kg)(450 J/kg°C)(ΔT) which we can solve to get ΔT = 30/(0.005 x 450) = **13.3°C**. (You will notice a similar effect when you remove a nail from a piece of wood. The nail that you pull out is noticeably warm.)

5. According to Newton's law of cooling, its rate of cooling is proportional to the temperature difference, so when the temperature difference is half as great, the rate of cooling will be half as great. After another eight hours, the coffee will lose 12.5 degrees, half as much as in the first eight hours, cooling from 50 °C to **37.5 °C**. (Newton's law of cooling leads to exponential behavior, in which the fractional change is the same in each equal increment of time.)

Chapter 17: Change of Phase Answers to Exercises

1. When a wet finger is held to the wind, evaporation is greater on the windy side, which feels cool. The cool side of your finger is windward.

3. When you blow over the top of a bowl of hot soup, you increase net evaporation and its cooling effect by removing the warm vapor that tends to condense and reduce net evaporation.

5. From our macroscopic point of view, it appears that nothing is happening in a covered glass of water, but at the atomic level there is chaotic motion as molecules are continually bumbling about. Molecules are leaving the water surface to the air above while vapor molecules in the air are leaving the air and plunging into the liquid. Evaporation and condensation are taking place continually, even though the net evaporation or net condensation is zero. Here we distinguish between the processes and the net effect of the processes.

7. In this hypothetical case evaporation would not cool the remaining liquid because the energy of exiting molecules would be no different than the energy of molecules left behind. Although internal energy of the liquid would decrease with evaporation, energy per molecule would not change. No temperature change of the liquid would occur. (The surrounding air, on the other hand, would be cooled in this hypothetical case. Molecules flying away from the liquid surface would be slowed by the attractive force of the liquid acting on them.)

9. If the perfume doesn't evaporate it will produce no odor. The odor of a substance is evidence for its evaporation.

11. A bottle wrapped in wet cloth will cool by the evaporation of liquid from the cloth. As evaporation progresses, the average temperature of the liquid left behind in the cloth can easily drop below the temperature of the cool water that wet it in the first place. So to cool a bottle of beer, soda, or whatever at a picnic, wet a piece of cloth in a bucket of cool water. Wrap the wet cloth around the bottle to be cooled. As evaporation progresses, the temperature of the water in the cloth drops, and cools the bottle to a temperature below that of the bucket of water.

13. Visibility of the windows is impaired if there is any condensation of water between the panes of glass. Hence the gas between the panes should contain no water vapor.

15. Aside from the connotation of kissing molecules and parking on a cool night, the warm air generated in the car's interior meets the cold glass and a lowering of molecular speed results in condensation of water on the inside of the windows.

17. Air swept upward expands in regions of less atmospheric pressure. The expansion is accompanied by cooling, which means molecules are moving at speeds low enough for coalescing upon collisions; hence the moisture that is the cloud.

19. Enormous thermal energy is released as molecular potential energy is transformed to molecular kinetic energy in condensation. (Freezing of the droplets to form ice adds even more thermal energy.)

21. As the bubbles rise, less pressure is exerted on them.

23. Decreased pressure lessens the squeezing of molecules, which favors their tendency to separate and form vapor.

25. Moisture in the cloth will convert to steam and burn you.

27. As in the answer to the previous exercise, high temperature and the resulting internal energy given to the food are responsible for cooking—if the water boils at a low temperature (presumably under reduced pressure), the food isn't hot enough to cook.

29. The air in the flask is very low in pressure, so that the heat from your hand will produce boiling at this reduced pressure. (Your instructor will want to be sure that the flask is strong enough to resist implosion before handing it to you!)

31. The lid on the pot traps heat which quickens boiling; the lid also slightly increases pressure on the boiling water which raises its boiling temperature. The hotter water correspondingly cooks food in a shorter time, although the effect is not significant unless the lid is held down as on a pressure cooker.

33. After a geyser has erupted, it must refill and then undergo the same heating cycle. If the rates of filling and heating don't change, then the time to boil to the eruption stage will be the same.

35. Yes, ice can be much colder than 0°C, which is the temperature at which ice will melt when it absorbs energy. The temperature of an ice-water mixture in equilibrium is 0°C. Iced tea, for example, is 0°C.

37. Regelation would not occur if ice crystals weren't open structured. The pressure of the wire on the open network of crystals caves them in and the wire follows. With the pressure immediately above the wire relieved, the molecules again settle to their low energy crystalline state. Interestingly, the energy score balances for these changes of state: The energy given up by the water that refreezes above the wire is conducted through the wire thickness to melt the ice being crushed beneath. The more conductive the wire, the faster regelation occurs. For an insulator like string, regelation won't occur at all. Try it and see!

39. The wood, because its greater specific heat capacity means it will release more energy in cooling.

41. This is an example that illustrates Figure 17.7. Water vapor in the warm air condenses on the relatively low-temperature surface of the can.

43. The dew point is higher on a moist summer day; less drop in temperature is required to produce condensation. (When the dew point and the temperature are equal, the air is saturated and fog is likely to form.)

45. Every gram of water that undergoes freezing releases 80 calories of thermal energy to the cellar. This continual release of energy by the freezing water keeps the temperature of the cellar from going below 0°C. Sugar and salts in the canned goods prevent them from freezing at 0°C. Only when all the water in the tub freezes will the temperature of the cellar go below 0°C and then freeze the canned goods. The farmer must, therefore, replace the tub before or just as soon as all the water in it has frozen.

47. As more freezing takes places and the polar icecaps grow, less water surface is exposed for evaporation. This results in less clouds and consequently, less precipitation. Without the pile-up of new snow, melting can more readily occur. The ice age withdraws.

49. Your friend has been bilked—better had he donated that sum of money to a worthy charity and walked across his own hot coals. Although a mind-over-matter manner of thinking may have its beneficial points in some cases, it is not demonstrated in firewalking. One can walk on hot coals of wood or charcoal without harm for two reasons that are straight physics. First, wood is a poor conductor, even when it is hot (which accounts for the wooden handles on cooking ware). Very little heat energy is transferred if contact is brief, so quick steps on red-hot coals will transfer safe amounts of heat to the feet. Secondly, just as you can touch a hot clothes iron with a wetted finger, vaporization of the moisture on your feet absorbs heat energy that would otherwise burn you. Nevertheless, this *is* a dangerous activity and many people have been harmed by stepping on unseen hot nails or coals that stick to the feet. Sometimes wet feet contribute to this sticking, and many firewalkers prefer to do this stunt with dry feet (Chapter 16). With either wet or dry feet, the practice is not advised. (As for the charlatan types who collect the money: If they really want to demonstrate mind over matter, let them walk barefoot on red hot coals of metal!)

Chapter 17 Problem Solutions

1. a. 1 kg 0°C ice to 0°C water requires 80 kilocalories.
b. 1 kg 0°C water to 100°C water requires 100 kilocalories.
c. 1 kg 100°C water to 100°C steam requires 540 kilocalories.
d. 1 kg 0°C ice to 100°C steam requires (80 + 100 + 540) = **720 kilocalories,**
or 720,000 calories.

3. First, find the number of calories that 10 g of 100°C steam will give in changing to 10 g of 0°C water.
10 g of steam changing to 10 g of boiling water at 100°C releases 5400 calories.
10 g of 100°C water cooling to 0°C releases 1000 calories.
So 6400 calories are available for melting ice.

$\frac{6400 \text{ cal}}{80 \text{ cal/g}}$ = **80 grams** of ice.

5. The quantity of heat lost by the iron is $Q = cm\Delta T$ = (0.11 cal/g°C)(50 g)(80°C) = 440 cal. The iron will lose a quantity of heat to the ice $Q = mL$. The mass of ice melted will therefore be $m = Q/L$ = (440 cal)/(80 cal/g) = **5.5 grams**. (The lower specific of heat of iron shows itself compared with the result of the previous problem.)

7. $0.5mgh = cm\Delta T$

$\Delta T = 0.5mgh/cm = 0.5gh/c$ = (0.5)(9.8 m/s^2)(100 m)/450 J/kg = **1.1°C**.

Again, note that the mass cancels, so the same temperature would hold for any mass ball, assuming half the heat generated goes into warming the ball. As in the previous problem, the units check because 1 J/kg = 1 m^2/s^2.

Chapter 18: Thermodynamics Answers to Exercises

1. In the case of the 500-degree oven it makes a lot of difference. 500 kelvins is 227°C, quite a bit different than 500°C. But in the case of the 50,000-degree star, the 273 increments either way makes practically no difference. Give or take 273, the star is still 50,000 K or 50,000°C when rounded off.

3. Its absolute temperature is 273 +10 = 283 K. Double this and you have 566 K. Expressed in Celsius; 566 - 273 = 293 °C.

5. You do work in compressing the air, which increases its internal energy. This is evidenced by an increase in temperature.

7. Gas pressure increases in the can when heated, and decreases when cooled. The pressure that a gas exerts depends on the average kinetic energy of its molecules, therefore on its temperature.

9. The high compression of fuel mixture in the cylinders of a diesel engine heats the mixture to the ignition point, so spark plugs are not needed.

11. Solar energy. The terms renewable and nonrenewable really refer to time scales for regeneration—tens of years for wood versus millions of years for coal and oil.

13. Solar energy.

15. The term pollution refers to an undesirable byproduct of some process. The desirability or undesirability of a particular by-product is relative, and depends on the circumstances. For example, using waste heat from a power plant to heat a swimming pool could be desirable whereas using the same heat to warm a trout stream could be undesirable.

17. It is advantageous to use steam as hot as possible in a steam-driven turbine because the efficiency is higher if there is a greater difference in temperature between the source and the sink (see Sadi Carnot's equation in the text).

19. Efficiency will increase, because back pressure is reduced. This is similar to reducing the back pressure on turbine blades in a steam generator. Efficiency increases (though not noticeably) on a cold day because of the greater ΔT in Carnot's equation.

21. As in the preceding exercise, inspection will show that decreasing T_{cold} will contribute to a greater increase in efficiency than by increasing T_{hot} by the same amount. For example, let T_{hot} be 600K and T_{cold} be 300K. Then efficiency = (600 - 300)/600 = 1/2. Now let T_{hot} be increased by 200K. Then efficiency = (800 - 300)/800 = 5/8. Compare this with T_{cold} decreased by 200K, in which case efficiency = (600 - 100)/600 = 5/6, which is clearly greater.

23. Even if the refrigerator were magically 100% efficient, the room wouldn't be cooled because the heat sink is also in the room. That's why the condensation coils are in a region outside the region to be cooled. What actually happens in the case of the refrigerator being operated with its door open in a closed room is that the room temperature increases. This is because the refrigerator motor warms the surrounding air. Net electric energy is coming into the room, heating it.

25. You *are* cooled by the fan, which blows air over you to increase the rate of evaporation from your skin, but you are a small part of the overall system, which warms.

27. Low efficiency for a fossil fuel or nuclear power plant means a greater amount of fuel is required for a given energy output. But an

OTEC power plant uses no fuel. Its low efficiency simply means a greater amount of ocean water must flow through it for a given energy output. Since ocean water about the plant is abundant, its low efficiency is not a problem.

29. Some of the electric energy that goes into lighting a lamp is transferred by conduction and convection to the air, some is radiated at invisible wavelengths ("heat radiation") and converted to internal energy when it is absorbed, and some appears as light. In an incandescent lamp, only about 5% goes into light, and in a fluorescent lamp, about 20% goes to light. But all of the energy that takes the form of light is converted to internal energy when the light is absorbed by materials upon which it is incident. So by the lst law, all the electrical energy is ultimately converted to internal energy. By the second law, organized electrical energy degenerates to the more disorganized form, internal energy.

31. Energy transfers from the bathing suit and its occupant to the surrounding air. No energy is lost, but is simply spread out as molecules evaporating from the suit become more disordered. The disorder associated with the escape of molecules more than offsets the order associated with the cooling of the bathing suit and its occupant—so the second law is "saved."

33. It is fundamental because it governs the general tendency throughout nature to move from order to disorder, yet it is inexact in the sense that it is based on probability, not certainty.

35. No, the freezing of water is not an exception to the entropy principle because work has been put into the refrigeration system to prompt this change of state. There is actually a greater net disorder when the environment surrounding the freezer is considered.

37. Such machines violate at least the second law of thermodynamics, and perhaps the first law as well. These laws are so richly supported by so many experiments over so long a time that the Patent Office wisely assumes that there is a flaw in the claimed invention.

39. As in the previous exercise, the smaller the number of random particles, the more the likelihood of them becoming "spontaneously" ordered. But the number of molecules in even the smallest room? Sleep comfortably!

Chapter 18 Problem Solutions

1. Heat added to system = change in system's internal energy + work done by the system, $Q = \Delta E + W$ so $W = Q - \Delta E = 0\text{J} - (-3000\text{ J})$ = **3000 J.**

3. Converting to kelvins; 25°C = 298 K; 4°C = 277 K. So Carnot efficiency = $\frac{T_h - T_c}{T_h}$ = $\frac{298 - 277}{298}$ = 0.07, or **7%**. This is very low, which means that large volumes of water (which there are) must be processed for sufficient power generation.

5. Adiabatic compression would heat the confined air by about 10°C for each kilometer decrease in elevation. The -35°C air would be heated 100C° and have a ground temperature of about (-35 + 100) = **65°C**. (This is 149°F, roasting hot!)

7. (a) For the room, W = **0.044 J**. (b) For the freezer, W = **0.69 J**. (c) For the helium refrigerator, W = **74 J**. The bigger the temperature "hill" relative to the lower temperature, the more work is needed to move the energy.

Part 3: Sound

Chapter 19: Vibrations and Waves Answers to Exercises

1. The period of a pendulum depends on the acceleration due to gravity. Just as in a stronger gravitational field a ball will fall faster, a pendulum will swing to and fro faster. (The exact relationship, $T = 2\pi\sqrt{l/g}$, is shown in the footnote in the chapter. So at mountain altitudes where the gravitational field of the Earth is slightly less, a pendulum will oscillate with a slightly longer period, and a clock will run just a bit slower and will "lose" time.

3. Assuming the center of gravity of the suitcase doesn't change when loaded with books, the pendulum rate of the empty case and loaded case will be the same. This is because the period of a pendulum is independent of mass. Since the length of the pendulum doesn't change, the frequency and hence the period is unchanged.

5. The frequency of a pendulum depends on the restoring force, which is gravity. Double the mass and you double the gravitational force that contributes to the torque acting on the pendulum. More mass means more torque, but also more inertia—so has no net effect. Similarly, mass doesn't affect free fall acceleration (see Figure 18.1).

7. The periods are equal. Interestingly, an edge-on view of a body moving in uniform circular motion is seen to vibrate in a straight line. How? Exactly in simple harmonic motion. So the up and down motion of pistons in a car engine are simple harmonic, and have the same period as the circularly rotating shaft that they drive.

9. To produce a transverse wave with a slinky, shake it to and fro in a direction that is perpendicular to the length of the slinky itself (as with the garden hose in the previous exercise). To produce a longitudinal wave, shake it to and fro along the direction of its length, so that a series of compressions and rarefactions is produced.

11. The fact that gas can be heard escaping from a gas tap before it is smelled indicates that the pulses of molecular collisions (the sound) travel more quickly than the molecules migrate. (There are three speeds to consider: (1) the average speed of the molecules themselves, as evidenced by temperature — quite fast, (2) the speed of the pulse produced as they collide — about 3/4 the speed of the molecules themselves, and (3) the very much slower speed of molecular migration.)

13. Violet light has the greater frequency.

15. Something that vibrates.

17. The frequency of vibration and the number of waves passing by each second are the same.

19. The energy of a water wave spreads along the increasing circumference of the wave until its magnitude diminishes to a value that cannot be distinguished from thermal motions in the water. The energy of the waves adds to the internal energy of the water.

21. The speed of light is 300,000 km/s, about a million times faster than sound. Because of this difference in speeds, lightning is seen a million times sooner than it is heard.

23. The nodes are at the fixed points, the two ends of the string. The wavelength is twice the length of the string (see Figure 19.14a).

25. The Doppler effect is a change in frequency as a result of the motion of source, receiver, or both. So if you move toward a stationary sound source, yes, you encounter wave crests more frequently and the frequency of the received sound is higher. Or if you move away from the source, the wave crests encounter you less frequently, and you hear sound of a lower frequency.

27. No, the effects of shortened waves and stretched waves would cancel one another.

29. Police use radar waves which are reflected from moving cars. From the shift in the returned frequencies, the speed of the reflectors (car bodies) is determined.

31. Oops, careful. The Doppler effect is about changes in *frequency*, not speed.

33. A boat that makes a bow wave is traveling faster than the waves of water it generates.

35. The fact that you hear an airplane in a direction that differs from where you see it simply means the airplane is moving, and not necessarily faster than sound (a sonic boom would be evidence of supersonic flight). If the speed of sound and the speed of light were the same, then you'd hear a plane where it appears in the sky. But because the two speeds are so different, the plane you see appears ahead of the plane you hear.

37. The speed of the sound source rather than the loudness of the sound is crucial to the production of a shock wave. At subsonic speeds, no overlapping of the waves will occur to produce a shock wave. Hence no sonic boom is produced.

39. Open ended.

Chapter 19 Problem Solutions

1. (a) f = 1/T = 1/0.10 s = **10 Hz**; (b) f = 1/5 = **0.2 Hz**; (c) f = 1/(1/60) s = **60 Hz**.

3. The skipper notes that 15 meters of wave pass each 5 seconds, or equivalently, that 3 meters pass each 1 second, so the speed of the wave must be
$$\text{Speed} = \frac{\text{distance}}{\text{time}} = \frac{15\text{ m}}{5\text{ s}} = 3\ \mathbf{m/s}.$$
Or in wave terminology:
Speed = frequency x wavelength = (1/5 Hz)(15 m) = **3 m/s.**

5. To say that the frequency of radio waves is 100 MHz and that they travel at 300,000 km/s, is to say that there are 100 million wavelengths packed into 300,000 kilometers of space. Or expressed in meters, 300 million m of space. Now 300 million m divided by 100 million waves gives a wavelength of 3 meters per wave. Or
$$\text{Wavelength} = \frac{\text{speed}}{\text{frequency}} = \frac{(300\text{ megameters/s})}{(100\text{ megahertz})} = 3\ \mathbf{m}.$$

7. (a) Period = 1/frequency = 1/(256 Hz) = 0.00391 s, or 3.91 ms. (b) Speed = wavelength x frequency, so wavelength = speed/frequency = (340 m/s)/(256 Hz) = **1.33 m.**

9. Below. Speed = frequency x wavelength, so frequency = speed/wavelength = (3×10^8 m/s)/(3.42 m) = 8.77×10^7 Hz = **87.7 MHz**, just below the FM band.

Chapter 20: Sound
Answers to Exercises

1. The same. The circles formed are relative to the water, and both will travel downstream together.

3. The shorter wavelengths are heard by bats (higher frequencies have shorter wavelengths).

5. The wavelength of sound from Source A is half the wavelength of sound from Source B.

7. Light travels about a million times faster than sound in air, so you see a distant event a million times sooner than you hear it.

9. When sound passes a particular point in air, the air is first compressed and then rarefied as

the sound passes. So its density is increased and then decreased as the wave passes.

11. Because snow is a good absorber of sound, it reflects little sound — which is responsible for the quietness.

13. The moon is described as a silent planet because it has no atmosphere to transmit sounds.

15. If the speed of sound were different for different frequencies, say, faster for higher frequencies, then the farther a listener is from the music source, the more jumbled the sound would be. In that case, higher-frequency notes would reach the ear of the listener first. The fact that this jumbling doesn't occur is evidence that sound of all frequencies travel at the same speed. (Be glad this is so, particularly if you sit far from the stage, or if you like outdoor concerts.)

17. Sound travels faster in warm air because the air molecules that compose warm air themselves travel faster and therefore don't take as long before they bump into each other. This lesser time for the molecules to bump against one another results in a faster speed of sound.

19. Refraction is the result of changing wave speeds, where part of a wave travels at a different speed than other parts. This occurs in nonuniform winds and nonuniform temperatures. Interestingly, if winds, temperatures, or other factors could not change the speed of sound, then refraction would not occur. (The fact that refraction does indeed occur is evidence for the changing speeds of sound.)

21. Sound is more easily heard when the wind traveling toward the listener at elevations above ground level travels faster than wind near the ground. Then the waves are bent downward as is the case of the refraction of sound shown in Figure 20.7.

23. An echo is weaker than the original sound because sound spreads and is therefore less intense with distance. If you are at the source, the echo will sound as if it originated on the other side of the wall from which it reflects (just as your image in a mirror appears to come from behind the glass). Also contributing to its weakness is the wall, which likely is not a perfect reflector.

25. The rule is correct: This is because the speed of sound in air (340 m/s) can be rounded off to 1/3 km/s. Then, from distance = speed x time, we have distance = (1/3)km/s x (number of seconds). Note that the time in seconds divided by 3 gives the same value.

27. Marchers at the end of a long parade will be out of step with marchers nearer the band because time is required for the sound of the band to reach the marchers at the end of a parade. They will step to the delayed beat they hear.

29. A harp produces relatively softer sounds than a piano because its sounding board is smaller and lighter.

31. The sound is louder when a struck tuning fork is held against a table because a greater surface is set into vibration. In keeping with the conservation of energy, this reduces the length of time the fork keeps vibrating. Loud sound over a short time spends the same energy as weak sound for a long time.

33. Certain dance steps set the floor into vibration that may resonate with the natural frequency of the floor. When this occurs, the floor heaves.

35. Long waves are most canceled, which makes the resulting sound so tinny. For example, when the speaker cones are, say, 4 centimeters apart, waves more than a meter long are nearly 180° out of phase, whereas 2-centimeter waves will be in phase. The higher frequencies are least canceled by this procedure. This must be tried to be appreciated.

37. Waves of the same frequency can interfere destructively or constructively, depending on their relative phase, but to *alternate* between constructive and destructive interference, two waves have to have different frequencies. Beats arise from such alternation between constructive and destructive interference.

39. The piano tuner should loosen the piano string. When 3 beats per second is first heard, the tuner knows he was 3 hertz off the correct frequency. But this could be either 3 hertz above or 3 hertz below. When he tightened the string and increased its frequency, a lower beat frequency would have told him he was on the right track. But the greater beat frequency told him he should have been loosening the string. When there is no beat frequency, the frequencies match.

Chapter 20 Problem Solutions

1. Wavelength = speed/frequency = $\frac{340 \text{ m/s}}{340 \text{ Hz}}$ = **1 m**.

   Similarly for a 34,000 hertz wave; wavelength = $\frac{340 \text{ m/s}}{34\,000 \text{ Hz}}$ = 0.01 m = **1 cm**.

3. The ocean floor is 4590 meters down. The 6-second time delay means that the sound reached the bottom in 3 seconds. Distance = speed x time = 1530 m/s x 3 s = **4590 m**.

5. The woman is about **340 meters** away. The clue is the single blow you hear after you see her stop hammering. That blow originated with the next-to-last blow you saw. The very first blow would have appeared as silent, and succeeding blows synchronous with successive strikes. In one second sound travels 340 meters in air.

7. (a) Constructively. (b) Constructively. (Even though each wave travels 1.5 wavelengths, they travel the *same* distance and are therefore in phase and interfere constructively.) (c) Destructively. The crest of one coincides with the trough of the other.

8. Wavelength = speed/frequency = (1,500 m/s)/(57 Hz) = **26 m**. Alternate method: For sounds of the same frequency in different media, wavelengths are proportional to wave speed. So (wavelength in water)/(wavelength in air) = (speed in water)/(speed in air) = (1,500 m/s)/(340 m/s) = 4.4. Multiply 6 m by 4.4 to get 26 m.

Chapter 21: Musical Sounds Answers to Exercises

1. The sound of commercials is concentrated at frequencies to which the ear is most sensitive. Whereas the overall sound meets regulations, our ears perceive the sound as distinctly louder.

3. A low pitch will be produced when a guitar string is (a) lengthened, (b) loosened so that tension is reduced, and (c) made more massive, usually by windings of wire around the string. That's why bass strings are thick— more inertia.

5. Pitch depends on frequency. It does not depend on loudness or quality.

7. If the wavelength of a vibrating string is reduced, such as by pressing it with your finger against the neck of a guitar, the frequency of the vibration increases. This is heard as an increased pitch.

9. The longer tines have greater rotational inertia, which means they'll be more resistant to vibrating, and will do so at lower frequency. Similarly, a long pendulum has greater rotational inertia and swings to and fro at a slower pace.

11. The sounding board's large surface is able to set more air vibrating than the string alone could do.

13. The fundamental for a string occurs when only two nodes exist; one at each end of the string, so that it vibrates in one segment. By touching the midpoint, a third node is imposed there and the string vibrates in two segments. The wavelength is diminished by one half, so the frequency increases by two. (This is because the speed of the wave along the string doesn't change; speed = frequency x wavelength)

15. Period = $1/f$ = 1/440 second (0.002 second, or 2 ms).

17. The amplitude in a sound wave corresponds to the overpressure of the compression or equivalently the underpressure of the rarefaction.

19. The pattern on the right has the greater amplitude and is therefore louder.

21. The range of human hearing is so wide that no single mechanical audio speaker can faithfully reproduce all the frequencies we can hear. So hi-fi speakers divide the range into two (and three) parts. A speaker with a relatively large surface has more inertia and is not as responsive to higher frequencies as a speaker with a smaller surface. So the larger speaker pushes the longer wavelengths, or lower frequencies, and the smaller speaker pushes the shorter wavelengths, or higher frequencies. (Ideally, the diameter of the speaker should be 1/2 the wavelength of a given sound — so a 12-inch speaker corresponds to about 550 hertz — and even larger speakers are best for bass notes).

23. The person with the more acute hearing is the one who can hear the faintest sounds—the one who can hear 5 dB.

25. Sound at 110 dB is a million times more intense than sound at 50 dB. The difference, 60 dB, corresponds to 10^6, a million.

27. Helium molecules and oxygen molecules at the same temperature have the same kinetic energies. Kinetic energy equals 1/2 mv^2. The smaller mass of helium is compensated for by a greater speed (Chapter 16).

29. The limited range of frequencies transmitted by a telephone can't match the full range in music Especially, it cuts off the higher-frequency overtones of music that contribute to its quality.

31. Frequency of second harmonic is twice the fundamental, or 440 Hz. The third is three times the harmonic, or 660 Hz.

33. The first harmonic is the fundamental, which is the same 440 Hz. The second harmonic is twice this, 880 Hz. The third harmonic is three times the first, 3 x 440 = 1320 Hz.

35. By controlling how hard he blows and how he holds his mouth, the bugler can stimulate different harmonics. The notes you hear from a bugle are actually harmonics; you don't hear the fundamental.

37. The CD will rotate slower when being read near the outer part of the disk. A single revolution there scans more pits than closer to the center of the disk. So when being read near the outside of the disk the disk spins at a lower RPM.

39. The likelihood is high that you subject yourself to louder sounds than your grandparents experienced — particularly via earphones.

Chapter 21 Problem Solutions

1. The decibel scale is based upon powers of 10. The ear responds to sound intensity in logarithmic fashion. Each time the intensity of sound is made 10 times larger, the intensity level in decibels increases by 10 units.
   So a sound of
   a. 10 dB is **ten** times more intense than the threshold of hearing
   b. 30 dB is **one thousand** times more intense than the threshold of hearing.
   c. 60 dB is **one million** times more intense than the threshold of hearing.

3. One octave above 1000 Hz is **2000 Hz**, and two octaves above 1000 Hz is **4000 Hz**. One octave below 1000 Hz is **500 Hz**, and two octaves below 100 Hz is **250 Hz**.

5. The wavelength of the fundamental is twice the length of the string, or 1.5 m. Then the speed of the wave is
   $v = f\lambda = 220 \text{ Hz} \times 1.5 \text{ m} = \mathbf{330 \text{ m/s}}$. (Note that this is for a transverse vibration of the string. A longitudinal sound-wave within the string could have a much greater speed.)

Part 5: Electricity and Magnetism

Chapter 22: Electrostatics
Answers to Exercises

1. There are no positives and negatives in gravitation — the interactions between masses are only attractive, whereas electrical interactions may be attractive as well as repulsive. The mass of one particle cannot "cancel" the mass of another, whereas the charge of one particle can cancel the effect of the opposite charge of another particle.

3. Clothes become charged when electrons from a garment of one material are rubbed onto another material. If the materials were good conductors, discharge between materials would soon occur. But the clothes are nonconducting and the charge remains long enough for oppositely charged garments to be electrically attracted and stick to one another.

5. Excess electrons rubbed from your hair leave it with a positive charge; excess electrons on the comb give it a negative charge.

7. More than two decades ago, before truck tires were made electrically conducting, chains or wires were commonly dragged along the road surface from the bodies of trucks. Their purpose was to discharge any charge that would otherwise build up because of friction with the air and the road. Electrically-conducting tires now in use prevent the buildup of static charge that could produce a spark — especially dangerous for trucks carrying flammable cargoes.

9. Cosmic rays produce ions in air, which offer a conducting path for the discharge of charged objects. Cosmic-ray particles streaming downward through the atmosphere are attenuated by radioactive decay and by absorption, so the radiation and the ionization are stronger at high altitude than at low altitude. Charged objects more quickly lose their charge at higher altitudes.

11. When an object acquires a positive charge, it loses electrons and its mass decreases. How much? By an amount equal to the mass of the electrons that have left. When an object acquires a negative charge, it gains electrons, and the mass of the electrons as well. (The masses involved are incredibly tiny compared to the masses of the objects. For a balloon rubbed against your hair, for example, the extra electrons on the balloon comprise less than a billionth of a billionth of a billionth the mass of the balloon.)

13. The crystal as a whole has a zero net charge, so any negative charge in one part is countered with as much positive charge in another part. So the net charge of the negative electrons has the same magnitude as the net charge of the ions. (This balancing of positive and negative charges within the crystal is almost, but not precisely, perfect because the crystal can gain or lose a few extra electrons.)

15. For the outer electrons, the attractive force of the nucleus is largely canceled by the repulsive force of the inner electrons, leaving a force on the outer electrons little different from the force on the single electron in a hydrogen atom. For the inner electrons, on the other hand, all of the electrons farther from the nucleus exert no net force (it is similar to the situation within the Earth, where only the Earth below, not the Earth above, exerts a gravitational force on a deeply buried piece of matter). So the inner electrons feel the full force of the nucleus, and a large amount of energy is required to remove them. Stripping all of the electrons from a heavy atom is especially difficult. Only in recent years have researchers at the University of California, Berkeley succeeded in removing all of the electrons from the atoms of heavy elements like uranium.

17. The law would be written no differently.

19. By the inverse-square law, the force reduces to one quarter when the particles are twice as far apart, and to one-ninth when three times as far apart.

21. (a) For charged pellets, the electric force is likely to be much greater than the gravitational force. (b) Both change by the same factor (to one-quarter of their original value) because both obey an inverse-square law.

23. The huge value of the constant k for electrical force indicates a relatively huge force between charges, compared with the small gravitational

force between masses and the small value of the gravitational constant G.

25. Planet Earth is negatively charged. If it were positive, the field would point outward.

27. The tree is likely to be hit because it provides a path of less resistance between the cloud overhead and the ground. The tree and the ground near it are then raised to a high potential relative to the ground farther away. If you stand with your legs far apart, one leg on a higher-potential part of the ground than the other, or if you lie down with a significant potential difference between your head and your feet, you may find yourself a conducting path. That, you want to avoid!

29. For both electricity and heat, the conduction is via electrons, which in a metal are loosely bound, easy flowing, and easy to get moving. (Many fewer electrons in metals take part in heat conduction than in electric conduction, however.)

31. An ion polarizes a nearby neutral atom, so that the part of the atom nearer to the ion acquires a charge opposite to the charge of the ion, and the part of the atom farther from the ion acquires a charge of the same sign as the ion. The side of the atom closer to the ion is then attracted more strongly to the ion than the farther side is repelled, making for a net attraction. (By Newton's third law, the ion, in turn, is attracted to the atom.)

33. The forces on the electron and proton will be equal in magnitude, but opposite in direction. Because of the greater mass of the proton, its acceleration will be less than that of the electron, and be in the direction of the electric field. How much less? Since the mass of the proton is nearly 2000 times that of the electron, its acceleration will be about 1/2000 that of the electron. The greater acceleration of the electron will be in the direction opposite to the electric field. The electron and proton accelerate in opposite directions.

35. By convention only, the direction of an electric field at any point is the direction of the force acting on a positive test charge placed at that point. A positive charge placed in the vicinity of a proton is pushed away from the proton, hence, the direction of the electric field vector is away from the proton.

37. Charge will be more concentrated on the corners. See Figure 22.21.

39. Yes, in both cases we have a ratio of energy per something. In the case of temperature, the ratio is energy/molecule. In the case of voltage it is energy/charge. Even with a small numerator, the ratio can be large if the denominator is small enough. Such is the case with the small energies involved to produce high-temperature sparklers and high-voltage metal balls.

41. Because charges of like sign mutually repel one another, when lightning hits a conductor such as an automobile, the charges will spread out over the outer conducting surface, and the electric field inside cancels to zero. Strictly speaking, it is only static charge that occupies only the outer surface of a conductor and produces a zero field within (see the answer to the next exercise), but the rule is obeyed approximately for currents that are not too large.

43. Increase the area of the plates and you'll increase energy storage. (You can also increase energy storage by bringing the plates closer together, but not touching. Or you can insert a non-conducting material, called a *dielectric*, between the plates.)

45. Agree with your friend. The hairs act like leaves in an electroscope. If your arms were as light, they'd stand out too.

Chapter 22 Problem Solutions

1. By the inverse-square law, twice as far is 1/4 the force; **5 N**.
   The solution involves relative distance only, so the magnitude of charges is irrelevant.

3. From Coulomb's law, $F = k\dfrac{q_1q_2}{d^2}$
   $= (9 \times 10^9)\dfrac{(1.0 \text{ x } 10^{-6})^2}{(0.03)^2}$ = **10 N**. This is the same as the weight of a 1-kg mass.

5. $F_{\text{(grav)}} = Gm_1m_2/d^2$
   $= (6.67 \times 10^{-11})\dfrac{(9.1 \text{ x } 10^{-31})(1.67 \text{ x } 10^{27})}{(1.0 \text{ x } 10^{-10})^2}$
   $= \mathbf{1.0 \times 10^{47}}$ **N**.
   $F_{\text{(elec)}} = kq_1q_2/d^2 = (9 \times 10^9)\dfrac{(1.6 \text{ x } 10^{-19})^2}{(1.0 \text{ x } 10^{-10})^2}$
   $= \mathbf{2.3 \times 10^{-8}}$ **N**.
   The electrical force between an electron and a proton is more than 1,000,000,000,000,000,000,000,000,000,000,000,000,000 times greater than the gravitational force between them!
   (Note that this ratio of forces is the same for any separation of the particles.)

7. Energy is charge x potential: PE = qV
   $= (2 \text{ C})(100 \times 10^6 \text{ V}) = \mathbf{2 \text{ x } 10^8}$ **J**.

9. a. From $E = \dfrac{F}{q}$ we see that $q = \dfrac{F}{E} = \dfrac{mg}{E}$
   $= \dfrac{(1.1 \text{ x } 10^{-14})(9.8)}{1.68 \text{ x } 10^5}$ $= \mathbf{6.4 \text{ x } 10^{-19}}$ **C**.
   b. Number of electrons $= \dfrac{6.4 \text{ x } 10^{-19}\text{C}}{1.6 \text{ x } 10^{-19}\text{C/electron}}$
   = **4 electrons.**

Chapter 23: Electric Currents
Answers to Exercises

1. You cannot say whether charge will flow if you only know the electric potential *energy* of each object. You must know that the electric potential energy PER CHARGE, or the electric potential (voltage), is different. This is analogous to the difference between heat and temperature. Just as heat does not necessarily flow from the object of greater internal energy to the object of lesser internal energy, but from the object of greater temperature to the object of lower temperature, charge flows from the object of higher electric potential to the object of lower electric potential. (If this is not clear, check to see that you know the difference between electric potential energy and electric potential.)

3. The cooling system of an automobile is a better analogy to an electric circuit because like an electric system it is a closed system, and it contains a pump, analogous to the battery or other voltage source in a circuit. The water hose does not re-circulate the water as the auto cooling system does.

5. The net charge in a wire, whether carrying current or not, is normally zero. The number of electrons is ordinarily offset by an equal number of protons in the atomic lattice. Thus current and charge are not the same thing: Many people think that saying a wire carries current is the same thing as saying a wire is charged. But a wire that is charged carries no current at all unless the charge moves in some uniform direction. And a wire that carries a current is typically not electrically charged and won't affect an electroscope. (If the current consists of a beam of electrons in a vacuum, then the beam would be charged. Current is not charge itself: current is the *flow* of charge.)

7. Only circuit number 5 is complete and will light the bulb. (Circuits 1 and 2 are "shortcircuits" and will quickly drain the cell of its energy. In circuit 3 both ends of the lamp filament are connected to the same terminal and are therefore at the same potential. Only one end of the lamp filament is connected to the cell in circuit 4.)

9. An electric device does not "use up" electricity, but rather *energy*. And strictly speaking, it doesn't "use up" energy, but transforms it from one form to another. It is common to say that energy is used up when it is transformed to less concentrated forms — when it is degraded. Electrical energy ultimately becomes heat energy. In this sense it is used up.

11. Your weak battery, once your engine is running, is being charged by your alternator. Assuming the alternator is working, you'll be able to restart your car after a few minutes of charging.

13. You have warmed it, and increased its resistance slightly. (Have you ever noticed that when bulbs burn out, it is usually a moment after they have been turned on? If the filament is weak, the initial pulse of higher current resulting from the lower resistance of the still-cool filament causes it to fail.)

15. A lie detector circuit relies on the likelihood that the resistivity of your body changes when you tell a lie. Nervousness promotes perspiration, which lowers the body's electrical resistance, and increases whatever current flows. If a person is able to lie with no emotional change and no change in perspiration, then such a lie detector will not be effective. (Better lying indicators focus on the eyes.)

17. Thick wires have less resistance and will more effectively carry currents without excessive heating.

19. (a) The resistance will be half, 5 ohms, when cut in half. (b) The resistance will be half again when the cross-sectional area is doubled, so it will be 2.5 ohms.

21. Current will be greater in the bulb connected to the 220-volt source. Twice the voltage would produce twice the current if the resistance of the filament remained the same. (In practice, the greater current produces a higher temperature and greater resistance in the lamp filament, so the current is greater than that produced by 110 volts, but appreciably less than twice as much for 220 volts. A bulb rated for 110 volts has a very short life when operated at 220 volts.)

23. In the first case the current passes through your chest; in the second case current passes only through your arm. You can cut off your arm and survive, but you cannot survive without your heart.

25. Auto headlights are wired in parallel. Then when one burns out, the other remains lit. If you've ever seen an automobile with one burned out headlight, you have evidence they're wired in parallel.

27. (a) volt, (b) ampere, (c) joule.

29. The equivalent resistance of resistors in parallel is less than the smaller resistance of the two. So connect a pair of resistors in parallel for less resistance.

31. The sign is a joke. High voltage may be dangerous, but high resistance is a property of all nonconductors.

33. If the parallel wires are closer than the wing span of birds, a bird could short circuit the wires by contact with its wings, be killed in the process, and possibly interrupt the delivery of power.

35. How quickly a lamp glows after an electrical switch is closed does not depend on the drift

velocity of the conduction electrons, but depends on the speed at which the electric field propagates through the circuit — about the speed of light.

37. A light bulb burns out when a break occurs in the filament or when the filament disintegrates or falls apart.

39. Most of the electric energy in a lamp filament is transformed to heat. For low currents in the bulb, the heat that is produced may be enough to feel but not enough to make the filament glow red or white hot.

41. As more bulbs are connected in series, more resistance is added to the single circuit path and the resulting current produced by the battery is diminished. This is evident in the dimmer light from the bulbs. On the other hand, when more bulbs are connected to the battery in parallel, the brightness of the bulbs is practically unchanged. This is because each bulb in effect is connected directly to the battery with no other bulbs in its electrical path to add to its resistance. Each bulb has its own current path.

43. What affects the other branches is the voltage impressed across them, and their own resistance — period. Opening or closing a branch doesn't alter either of these.

45. All are the same for identical resistors in parallel. If the resistors are not the same, the one of greater resistance will have less current through it and less power dissipation in it. Regardless of the resistances, the voltage across both will be identical.

47. Household appliances are not connected in series for at least two reasons. First, the voltage, current, and power for each appliance would vary with the introduction of other appliances. Second, if one device burns out, the current in the whole circuit ceases. Only if each appliance is connected in parallel to the voltage source can the voltage and current through each appliance be independent of the others.

49. More current flows in the 100-watt bulb. We see this from the relationship "power = current x voltage." More current for the same voltage means less resistance. So a 100-watt bulb has less resistance than a 60-watt bulb. Less resistance for the same length of the same material means a thicker filament. The filaments of high wattage bulbs are thicker than those of lower-wattage bulbs. (It is important to note that both watts and volts are printed on a light bulb. A bulb that is labeled 100 W, 120 V, is 100 W *only* if there are 120 volts across it. If there are only 110 volts across it, and the resistance remains unchanged, then the power output would be only 84 watts!)

Chapter 23 Problem Solutions

1. From "Power = current x voltage," 60 watts = current × 120 volts, current = $\frac{60\text{W}}{120\text{V}}$ = **0.5 A.**

3. From power = current × voltage, current = $\frac{\text{power}}{\text{voltage}} = \frac{1200\text{W}}{120\text{V}}$ = **10 A.**
From the formula derived above, resistance = $\frac{\text{voltage}}{\text{current}} = \frac{120\text{V}}{10\text{A}}$ = **12 Ω.**

5. **$2.52**. First, 100 watts = 0.1 kilowatt. Second, there are 168 hours in one week (7 days × 24 hours/day = 168 hours). So 168 hours × 0.1 kilowatt = 16.8 kilowatt-hours, which at 15 cents per kWh comes to $2.52.

7. The iron's power is $P = IV$ = (110 V)(9 A) = 990 W = 990 J/s. The heat energy generated in 1 minute is E = power × time = (990 J/s)(60 s) = **59,400 J**.

9. It was designed for use in a 120-V circuit. With an applied voltage of 120 V, the current in the bulb is $I = V/R$ = (120 V)/(95 W) = 1.26 A. The power dissipated by the bulb is then $P = IV$ = (1.26 A)(120 V) = 151 W, close to the rated value. If this bulb is connected to 220 V, it would carry twice as much current and would dissipate four times as much power (twice the current x twice the voltage), more than 600 W. It would likely burn out. (This problem can also be solved by first carrying out some algebraic manipulation. Since current = voltage/resistance, we can write the formula for power as $P = IV = (V/R)V = V^2/R$. Solving for V gives $V = \sqrt{PR}$. Substituting for the power and the resistance gives $V = \sqrt{(150)(95)}$ = 119 V.)

Chapter 24: Magnetism
Answers to Exercises

1. All magnetism originates in moving electric charges. For an electron there is magnetism associated with its spin about its own axis, with its motion about the nucleus, and with its motion as part of an electric current.

3. Attraction will occur because the magnet induces opposite polarity in a nearby piece of iron. North will induce south, and south will induce north. This is similar to charge induction, where a balloon will stick to a wall whether the balloon is negative or positive.

5. The poles of the magnet attract each other and will cause the magnet to bend, even enough for the poles to touch if the material is flexible enough.

7. An electric field surrounds a stationary electric charge. An electric field and a magnetic field surround a moving electric charge. (And a gravitational field also surrounds both).

9. A magnet will induce the magnetic domains of a nail or paper clip into alignment. Opposite poles in the magnet and the iron object are then closest to each other and attraction results (this is similar to a charged comb attracting bits of electrically neutral paper — Figure 22.13). A wooden pencil, on the other hand, does not have magnetic domains that will interact with a magnet.

11. Domains in the paper clip are induced into alignment in a manner similar to the electrical charge polarization in an insulator when a charged object is brought nearby. Either pole of a magnet will induce alignment of domains in the paper clip: attraction results because the pole of the aligned domains closest to the magnet's pole is always the opposite pole, resulting in attraction.

13. The needle is not pulled toward the north side of the bowl because the south pole of the magnet is equally attracted southward. The net force on the needle is zero. (The net torque, on the other hand, will be zero only when the needle is aligned with the Earth's magnetic field.)

15. The dip needle will point most nearly vertically near the Earth's magnetic poles, where the field points toward or away from the poles (which are buried deep beneath the surface). It will point most nearly horizontally near the Equator (see Figure 24.19).

17. The north and south poles of a magnet are so named because they are "north-seeking" and "south-seeking." Suppose the magnetic poles of the Earth were marked by wooden poles. If we had to paint either an N or an S on the wooden pole near the north geographical pole, we should paint an S, which is the magnetic pole that attracts the N pole of a compass. Similarly, we should paint an N on the wooden pole erected in the southern hemisphere, for that is the pole that attracts the S pole of a magnetic compass.

19. Yes, for the compass aligns with the Earth's magnetic field, which extends from the magnetic pole in the southern hemisphere to the magnetic pole in the northern hemisphere.

21. Moving electrons are deflected from their paths by a magnetic field. A magnet held in front of a TV picture deflects the electron beam from its intended path and distorts the picture.

23. Yes, it does. Since the magnet exerts a force on the wire, the wire, according to Newton's third law, must exert a force on the magnet.

25. Beating on the nail shakes up the domains, allowing them to realign themselves with the Earth's magnetic field. The result is a net alignment of domains along the length of the nail. (Note that if you hit an already magnetized piece of iron that is not aligned with the Earth's field, the result can be to weaken, not strengthen, the magnet.)

27. An electron must move across lines of magnetic field in order to feel a magnetic force. So an electron at rest in a stationary magnetic field will feel no force to set it in motion. In an electric field, however, an electron will be accelerated whether or not it is already moving. (A combination of magnetic and electric fields is used in particle accelerators such as cyclotrons. The electric field accelerates the charged particle in its direction, and the magnetic field accelerates it perpendicular to its direction, making it follow a nearly circular path.)

29. When we write *work = force × distance*, we really mean the component of force in the direction of motion multiplied by the distance moved (Chapter 7). Since the magnetic force that acts on a beam of electrons is always perpendicular to the beam, there is no component of magnetic force along the instantaneous direction of motion. Therefore a magnetic field can do no work on a charged particle. (Indirectly, however, a *time-varying magnetic field* can induce an electric field that *can* do work on a charged particle.)

31. Associated with every moving charged particle, electrons, protons, or whatever, is a magnetic field. Since a magnetic field is not unique to moving electrons, there is a magnetic field about moving protons as well. However, it differs in direction. The field lines about the proton beam circle in one way, the field lines about an electron beam in the opposite way. (Physicists use a "right-hand rule." If the right thumb points in the direction of motion of a positive particle, the curved fingers of that hand show the direction of the magnetic field. For negative particles, the left hand can be used.)

33. The Van Allen radiation belts are filled with swarms of high-energy charged particles that can damage living tissue. Astronauts, therefore, make an effort to keep below these belts.

35. Cosmic ray intensity at the Earth's surface would be greater when the Earth's magnetic field passed through a zero phase. Fossil evidence suggests the periods of no protective magnetic field may have been as effective in changing life forms as X rays have been in the famous heredity studies of fruit flies.

37. A habitat in space could be shielded from cosmic radiation if a magnetic field were set up about the habitat, just as the magnetic field of the Earth shields us from much of the cosmic radiation that would otherwise strike the Earth. (As to the idea of a blanket, some have proposed that a thick layer of slag from mining operations on planets or asteroids could be placed around the habitat.)

39. Magnetic levitation will reduce surface friction to near zero. Then only air friction will remain. It can be made relatively small by aerodynamic design, but there is no way to elimi-

nate it (short of sending vehicles through evacuated tunnels). Air friction gets rapidly larger as speed increases.

Chapter 25: Electromagnetic Induction Answers to Exercises

1. Magnetic induction will not occur in nylon, since it has no magnetic domains. That's why electric guitars use steel strings.

3. The magnetic field of the iron core adds to the magnetic field of the coil, as stated in the answer to the previous exercise. Greater magnetic field means greater torque on the armature.

5. A cyclist will coast farther if the lamp is disconnected from the generator. The energy that goes into lighting the lamp is taken from the bike's kinetic energy, so the bike slows down. The work saved by not lighting the lamp will be the extra "force x distance" that lets the bike coast farther.

7. As in the previous answer, eddy currents induced in the metal change the magnetic field, which in turn changes the ac current in the coils and sets off an alarm.

9. Copper wires were not insulated in Henry's time. A coil of non-insulated wires touching one another would comprise a short circuit. Silk was used to insulate the wires so current would flow along the wires in the coil rather than across the loops touching one another.

11. There is no fundamental difference between an electric motor and electric generator. When mechanical energy is put into the device and electricity is produced, we call it a generator. When electrical energy is put in and it spins and does mechanical work, we call it a motor. (While there are usually some practical differences in the designs of motors and generators, some devices are designed to operate either as motors or generators, depending only on whether the input is mechanical or electrical.)

13. In accord with Faraday's law of induction, the greater the rate of change of magnetic field in a coil or armature, the greater the induced voltage. So voltage output increases when the generator spins faster.

15. In accord with electromagnetic induction, if the magnetic field alternates in the hole of the ring, an alternating voltage will be induced in the ring. Because the ring is metal, its relatively low resistance will result in a correspondingly high alternating current. This current is evident in the heating of the ring.

17. If the light bulb is connected to a wire loop that intercepts changing magnetic field lines from an electromagnet, voltage will be induced which can illuminate the bulb. Change is the key, so the electromagnet should be powered with ac.

19. Induction occurs only for a *change* in the intercepted magnetic field. The galvanometer will display a pulse when the switch in the first circuit is closed and current in the coil increases from zero. When the current in the first coil is steady, no current is induced in the secondary and the galvanometer reads zero. The galvanometer needle will swing in the opposite direction when the switch is opened and current falls to zero.

21. A transformer requires alternating voltage because the magnetic field in the primary winding must change if it is to induce voltage in the secondary. No change, no induction.

23. A transformer is analogous to a mechanical lever in that work is transferred from one part to another. What is multiplied in a mechanical lever is *force*, and in an electrical lever, *voltage*. In both cases, energy and power are conserved, so what is not multiplied is energy, a conservation of energy no no!

25. High efficiency requires that the maximum number of magnetic field lines produced in the primary are intercepted by the secondary. The core guides the lines from the primary through the secondary. Otherwise some of the magnetic field generated by the primary would go into heating metal parts of the transformer instead of powering the secondary circuit.

27. Oops! This is a dc circuit. Unless there is a changing current in the primary, no induction takes place. No voltage and no current are induced in the meter.

29. No, no, no, a thousand times no! No device can step up energy. This principle is at the heart of physics. Energy can not be created or destroyed.

31. The moving magnet will induce a current in the loop. This current produces a field that tends to repel the magnet as it approaches and attract it as it leaves, slowing it in its flight. From an energy point of view, the energy that the coil transfers to the resistor is equal to the loss of kinetic energy of the magnet.

33. A voltage difference is induced across the wings of a moving airplane. This produces a momentary current and charge builds up on the wing tips to create a voltage difference that counteracts the induced voltage difference. So charge is pulled equally in both directions and doesn't move.

35. Waving it changes the "flux" of the Earth's magnetic field in the coil, which induces voltage and hence current. You can think of the flux as the number of field lines that thread through the coil. This depends on the orientation of the coil, even in a constant field.

37. The incident radio wave causes conduction electrons in the antenna to oscillate. This oscillating charge (an oscillating current) provides the signal that feeds the radio.

39. Agree with your friend, for light is electromagnetic radiation having a frequency that matches the frequency to which our eyes are sensitive.

Chapter 25 Problem Solutions

1. If power losses can be ignored, in accord with energy conservation, the power provided by the secondary is also **100W.**

3. From the transformer relationship,

$$\frac{\text{primary voltage}}{\text{primary turns}} = \frac{\text{secondary voltage}}{\text{secondary turns}},$$

$$\frac{120\text{V}}{240\text{ turns}} = \frac{6\text{V}}{\text{x turns}}$$

Solve for x: x = (6 V)(240 turns)/(120 V) = **12 turns**

5. (a) Since power is voltage x current, the current supplied to the users is

$$\text{current} = \frac{\text{power}}{\text{voltage}} = \frac{100000\text{ W}}{12000\text{ V}} = \mathbf{8.3\text{ A.}}$$

(b) Voltage in each wire = current x resistance of the wire = (8.3 A)(10 Ω) = **83 V.**

(c) In each line, power = current × voltage = (8.3 A)(83 V) = 689 W. The total power wasted as heat is twice this, **1.38 kW.**
This is a small and tolerable loss. If the transmission voltage were ten times less, the losses to heat in the wires would be 100 times more! Then more energy would go into heat in the wires than into useful applications for the customers. That would not be tolerable. That's why high-voltage transmission is so important.

Part 4: Light

Chapter 26: Properties of Light Answers to Exercises

1. Your friend is correct. Also in a profound voice, your friend could say that sound is the only thing we hear!

3. The fundamental source of electromagnetic radiation is oscillating electric charges, which emit oscillating electric and magnetic fields.

5. Ultraviolet has shorter waves than infrared. Correspondingly, ultraviolet also has the higher frequencies.

7. What waves in a light wave are the electric and magnetic fields. Their oscillation frequency is the frequency of the wave.

9. The shorter wavelength corresponds to a higher frequency, so the frequency of the blue-green light from the argon laser has higher frequency than the red light from the helium-neon laser.

11. Radio waves most certainly travel at the speed of every other electromagnetic wave — the speed of light.

13. Radio waves and light are both electromagnetic, transverse, move at the speed of light, and are created and absorbed by oscillating charge. They differ in their frequency and wavelength and in the type of oscillating charge that creates and absorbs them.

15. The average speed of light will be less where it interacts with absorbing and re-emitting particles of matter, such as in the atmosphere. The greater the number of interactions along the light's path, the less the average speed.

17. The person walking across the room and pausing to greet others is analogous to the transmission-of-light model in that there is a pause with each interaction. However, the same person that begins the walk ends the walk, whereas in light transmission there is a "death-birth" sequence of events as light is absorbed and "new light" is emitted in its place. The light to first strike the glass is not the same light that finally emerges. (Another analogy is a relay race, where the runner to begin the race is not the runner to cross the finish line.)

19. The greater number of interactions per distance tends to slow the light and result is a smaller average speed.

21. Clouds are transparent to ultraviolet light, which is why clouds offer no protection from sunburn. Glass, however, is opaque to ultraviolet light, and will therefore shield you from sunburn.

23. Any shadow cast by a faraway object such as a high-flying plane is filled in mainly by light tapering in from the sun, which is not a point source. This tapering is responsible for the umbra and penumbra of solar eclipses (Figure 26.12). If the plane is low to the ground, however, the tapering of light around the airplane may be insufficient to fill in the shadow, part of which can be seen. This idea is shown in Figure 26.10.

25. A lunar eclipse occurs when the Earth, sun, and moon all fall on a straight line, with the Earth between the sun and the moon. During perfect alignment the Earth's shadow falls on the moon. Not-quite-perfect alignment gives Earth observers a full view of the moon.

Moonlight is brightest and the moon is always fullest when the alignment is closest to perfect — on the night of a lunar eclipse. At the time of a half moon, however, lines from Earth to moon and from Earth to sun are at right angles to each other. This is as non-aligned as the Earth, moon, and sun can be, with the moon nowhere near the Earth's shadow — so no eclipse is possible. Similarly for the non-aligned times of a crescent moon.

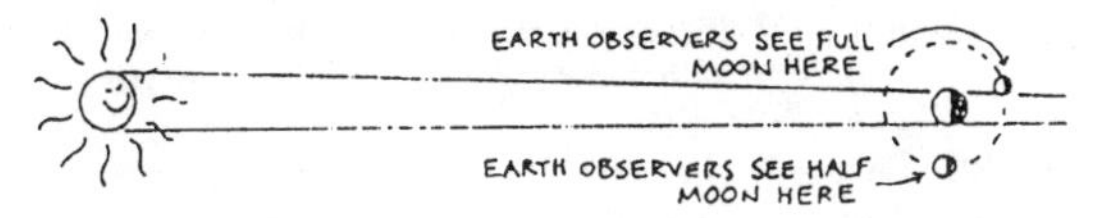

27. Rods, not cones, will respond to weak light, so you want to focus low-intensity light on a part of the retina that is composed of rods. That would be off to the side of the fovea. If you're looking at a weak star, look a bit off to the side of where you expect to see it. Then its image will fall on a part of your eye where rods may pick it up.

29. The sky is black on the moon because there is no atmosphere there. Something can have color; nothing has no color. There is nothing to reflect or emit light in the space surrounding the moon.

31. The blind spot is located on the side of the fovea away from your nose.

33. We cannot infer that people with large pupils are generally happier than people with small pupils. The size of a person's pupils has to do with the sensitivity of the retina to light intensity. Your pupils tend to become smaller with age as well. It is the *change* in pupil size that suggests one's psychological disposition.

35. Light from the flash spreads via the inverse-square law to the ground below, and what little returns to the airplane spreads further. The passenger will find that the flash makes no difference at all. Taking pictures at great distances, whether from an airplane or the football stands, with the flash intentionally turned on is rather foolish.

37. In accord with the inverse-square law, brightness is about 1/25 that from Earth — actually less since the distance is more than five times as far.

39. You see your hand in the past! How much? To find out, simply divide the distance between your hands and your eyes by the speed of light. (At 30 cm, this is about a billionth of a second.)

Chapter 26 Problem Solutions

1. In seconds, this time is 16.5 min × 60 s = 990 s.

$$\text{Speed} = \frac{\text{distance}}{\text{time}} = \frac{300{,}000 \text{ km}}{990 \text{ s}} = \mathbf{303{,}030 \text{ km/s}}.$$

This calculated value is very close to the currently accepted value of 300,000 km/s.

3. From $v = \frac{d}{t}$,

$$t = \frac{d}{v} = \frac{d}{c} = \mathbf{500 \text{ s}}$$

(which equals 8.3 min).

The time to cross the diameter of the Earth's orbit is twice this, or 1000 s, about 24% less than the 1,320 s measured by Roemer.

5. As in the previous problem, $t = \frac{d}{v}$

$$= \frac{4.2 \times 10^{16} \text{ m}}{3 \times 10^{8} \text{ m/s}} = \mathbf{1.4 \times 10^{8} \text{ s}}.$$

Converting to years by dimensional analysis,

$$1.4 \times 10^{8} \text{ s} \times \frac{1 \text{ h}}{3600 \text{ s}} \times \frac{1 \text{ day}}{24 \text{ h}} \times \frac{1 \text{ yr}}{365 \text{ day}}$$

$$= \mathbf{4.4 \text{ yr}}.$$

7. From $c = f\lambda$, $\lambda = \frac{c}{f} = \frac{3 \times 10^{8} \text{m/s}}{6 \times 10^{14} \text{Hz}}$

$= \mathbf{5 \times 10^{-7} \text{ m}}$, or 500 nanometers. This is 5000 times larger than the size of an atom, which is 0.1 nanometer. (The nanometer is a common unit of length in atomic and optical physics.)

9. (a) Frequency = speed/wavelength $= (3 \times 10^{8} \text{ m/s})/(0.03 \text{ m}) = 1.0 \times 10^{10}$ Hz = 10 GHz. (b) Distance = speed × time, so time = distance/speed = $(10{,}000 \text{ m})/(3 \times 10^{8} \text{ m/s})$ $= \mathbf{3.3 \times 10^{-5} \text{ s}}$. (Note the importance of consistent SI units to get the right numerical answers.)

Chapter 27: Color
Answers to the Exercises

1. The customer is being reasonable in requesting to see the colors in the daylight. Under fluorescent lighting, with its predominant higher frequencies, the bluer colors rather than the redder colors will be accented. Colors will appear quite different in sunlight.

3. Either a white or green garment will reflect incident green light and be cooler. The complementary color, magenta, will absorb green light and be the best garment color to wear when the absorption of energy is desired.

5. The interior coating absorbs rather than reflects light, and therefore appears black. A black interior in an optical instrument will absorb any stray light rather than reflecting it and passing it around the interior of the instrument to interfere with the optical image.

7. Tennis balls are yellow green in color so that they match the color to which we are most sensitive.

9. Red cloth appears red in sunlight, and red by the illumination of the red light from a neon tube. But because the red cloth absorbs cyan light, it appears black when illuminated by cyan light.

11. The color that will emerge from a lamp coated to absorb yellow is blue, the complementary color. (White - yellow = blue.)

13. The overlapping blue and yellow beams will produce white light. When the two panes of glass are overlapped and placed in front of a single flashlight, however, little or no light will be transmitted.

15. Red and green produce yellow; red and blue produce magenta; red, blue, and green produce white.

17. Blue.

19. Deep in water, red is no longer present in light. So blood looks black. But there is plenty of red in a camera flash, so the blood looks red when so illuminated.

21. The green shirt in the photo is seen as magenta in the negative, and the red shirt appears cyan — the complementary colors. When white light shines through the negative, green is transmitted where magenta is absorbed. Likewise, red is transmitted where cyan is absorbed.

23. Blue illumination produces black. A yellow banana reflects yellow and the adjacent colors, orange and green, so when illuminated with any of these colors it reflects that color and appears that color. A banana does not reflect blue, which is too far from yellow in the spectrum, so when illuminated with blue it appears black.

25. You see the complimentary colors due to retina fatigue. The blue will appear yellow, the red cyan, and the white black. You should try this and see!

27. At higher altitudes, there are fewer molecules above you and therefore less scattering of sunlight. This results in a darker sky. The extreme, no molecules at all, results in a black sky, as on the moon.

29. Ultraviolet light is reflected by the sand, so although you are not in direct light, you are in indirect light, including ultraviolet. Also, just as visible light is scattered by particles that make up the atmosphere, ultraviolet radiation is scattered even more. So you can get a sunburn in the shade — by both reflection and scattering. (The author was quite sunburned while sitting in the shade of a mangrove tree at a sandy beach brainstorming exercises for this book! I learned this one the hard way.)

31. Light travels faster through the upper atmosphere where the density is less and there are fewer interactions with molecules in the air.

33. The statement is true. A more positive tone would omit the word "just," for the sunset is not *just* the leftover colors, but *is* those colors that weren't scattered in other directions.

35. Through the volcanic emissions, the moon appears cyan, the complementary color of red.

37. Rain clouds are composed of relatively big particles that absorb much of the incident light. If the rain clouds were composed only of absorbing particles, then the cloud would appear black. But its mixture of particles includes tiny high-frequency scattering particles, so the cloud is not completely absorbing, and is simply dark instead of black.

39. If we assume that Jupiter has an atmosphere which is similar to that of the Earth in terms of transparency, then the sun would appear to be a deep reddish orange, just as it would when sunlight grazes 1000 kilometers of the Earth's atmosphere for a sunset from an elevated position. Interestingly enough, there is a thick cloud cover in Jupiter's atmosphere that blocks all sunlight from reaching its "surface." And it doesn't even have a solid surface! Your grandchildren may visit one of Jupiter's moons, but will not "land" on Jupiter itself — not intentionally, anyway. (Incidentally, there are only 4 1/3 planets with "solid" surfaces: Mercury, Venus, Mars, Pluto, and 1/3 of Earth!)

Chapter 28 Reflection and Refraction Answers to Exercises

1. Fermat's principle for refraction is of least time, but for reflection it could be of least distance as well. This is because light does not go from one medium to another for reflection, so no change in speed occurs and least-time paths and least-distance paths are equivalent. But for refraction, light goes from a medium where it has a certain speed to another medium where its speed is different. When this happens the least-distance straight-line paths take a longer time to travel than the non-straight-line least-time paths. See, for example, the difference in the least-distance and least-time paths in Figure 28.13 on page 492.

3. Cowboy Joe should simply aim at the mirrored image of his assailant, for the ricocheting bullet will follow the same changes in direction when its momentum changes (angle of incidence = angle of rebound) that light follows when reflecting from a plane surface.

5. Such lettering is seen in proper form in the rear view mirrors of cars ahead.

7. Two surfaces of the mirror reflect light. The front surface reflects about 4% of incident light, and the silvered surface reflects most of the rest. When the mirror is tilted in the "daytime" position, the driver sees light reflecting from the silvered surface. In the "nighttime" position, with the mirror tilted upward, light reflecting from the silvered surface is directed above the driver's view and the driver sees light reflected from the front surface of the mirror. The 4% of rearview light is adequate for night driving.

9. A window both transmits and reflects light. Window glass typically transmits about 92% of incident light, and the two surfaces reflect about 8%. Percentage is one thing, total amount is another. The person outside in the daylight who looks at the window of a room that is dark inside sees 8% of the outside light reflected back and 92% of the inside light transmitted out. But 8% of the bright outside light might be more intense than 92% of the dim inside light, making it difficult or impossible for the outside person to see in. The person inside the dark room, on the other hand, receiving 92% of the bright outside light and 8% of the dim inside light, reflected, easily sees out. (You can see how the reverse argument would be applied to a lighted room at night. Then the person inside may not be able to see out the window while the person outside easily sees in.)

11. Whereas diffuse reflection from a rough road allows a motorist to see the road illuminated by headlights on a dry night, on a rainy night the road is covered with water and acts like a plane mirror. Very little of the illumination from the headlights returns to the driver, and is instead reflected ahead (causing glare for oncoming motorists!).

13. The half-height mirror works at any distance, as shown in the sketch above. This is because if you move closer, your image moves closer as well. If you move farther away, your image does the same. Many people must actually try this before they believe it. The confusion arises because people know that they can see whole distant buildings or even mountain ranges in a hand-held pocket mirror. Even then, the distance the object is from the mirror is the same as the distance of the virtual image on the other side of the mirror. You can see all of a distant person in your mirror, but the distant person cannot see all of herself in your mirror.

15. The wiped area will be half as tall as your face.

17. First of all, the reflected view of a scene is different than an inverted view of the scene, for the reflected view is seen from lower down. Just as a view of a bridge may not show its underside where the reflection does, so it is with the bird. The view reflected in water is the inverted view you would see if your eye were positioned as far beneath the water level as your eye is above it (and there were no refraction). Then your line of sight would intersect the water surface where reflection occurs. Put a mirror on the floor between you and a distant table. If you are standing, your view of the table is of the top. But the reflected view shows the table's bottom. Clearly, the two views are not simply inversions of each other. Take notice of this whenever you look at reflections (and of paintings of reflections — it's surprising how many artists are not aware of this).

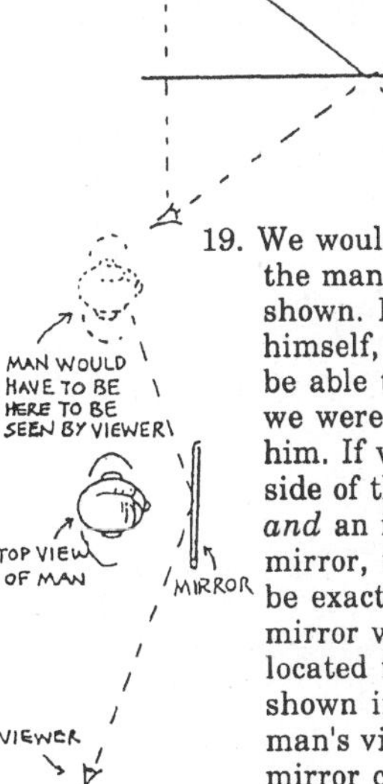

19. We would not see an image of the man in the mirror as shown. If he is viewing himself, then we wouldn't also be able to see his image unless we were in back (or in front) of him. If we are to stand to the side of the man and see him *and* an image of him in the mirror, then the mirror cannot be exactly in front of him. The mirror would have to be located to the man's right, as shown in the sketch. The man's view would miss the mirror completely. Such arrangements are made when staging an actor who is supposed to be viewing himself in a mirror. Actually, however, the actor pretends to be looking at himself. If he really were, his image in the mirror wouldn't be shared by the audience. That's Hollywood!

21. Red light travels faster through glass and will exit first.

23. The speeds in both glass and soybean oil are the same, so there is no refraction between the glass and oil.

25. You would throw the spear below the apparent position of the fish, because the effect of refraction is to make the fish appear closer to the surface than it really is. But in zapping a fish with a laser, make no corrections and simply aim directly at the fish. This is because the light from the fish you see has been refracted in getting to you, and the laser light will refract along the same path in getting to the fish. A slight correction may be necessary, depending on the colors of the laser beam and the fish — see the next exercise.

27. A fish sees the sky (as well as some reflection from the bottom) when it looks upward at 45°, for the critical angle is 48° for water. If it looks at and beyond 48° it sees only a reflection of the bottom.

29. In sending a laser beam to a space station, make no corrections and simply aim at the station you see. This is like zapping the fish in Exercise 25. The path of refraction is the same in either direction.

31. We cannot see a rainbow "off to the side," for a rainbow is not a tangible thing "out there." Colors are refracted in infinite directions and fill the sky. The only colors we see that aren't washed out by others are those that are along the conical angles between 40° and 42° to the sun-antisun axis. To understand this, consider a paper-cone cup with a hole cut at the bottom. You can view the circular rim of the cone as an ellipse when you look at it from a near side view. But if you view the rim only with your eye at the apex of the cone, through the hole, you can see it only as a circle. That's the way we view a rainbow. Our eye is at the apex of a cone, the axis of which is the sun-antisun axis, and the "rim" of which is the bow. From every vantage point, the bow forms part (or all) of a circle.

33. When the sun is high in the sky and people on the airplane are looking down toward a cloud opposite to the direction of the sun, they may see a rainbow that makes a complete circle. The shadow of the airplane will appear in the center of the circular bow. This is because the airplane is directly between the sun and the drops or rain cloud producing the bow.

35. A projecting lens with chromatic aberration casts a rainbow-colored fringe around a spot of white light. The reason these colors don't appear inside the spot is because they overlap to form white. Only at the edges, which act as a circular prism, do they not overlap.

37. The average intensity of sunlight at the bottom is the same whether the water is moving or is still. Light that misses one part of the bottom of the pool reaches another part. Every dark region is balanced by a bright region — "conservation of light."

39. Normal sight depends on the amount of refraction that occurs for light traveling from air to the eye. The speed change ensures normal vision. But if the speed change is from water to eye, then light will be refracted less and an unclear image will result. A swimmer uses goggles to make sure that the light travels from air to eye, even if underwater.

41. The diamond sparkles less because there are smaller angles of refraction between the water and the diamond. Light is already slowed when it meets the diamond so the amount of further slowing, and refraction, is reduced.

43. If light had the same average speed in glass lenses that it has in air, no refraction of light would occur in lenses, and no magnification would occur. Magnification depends on refraction, which in turn depends on speed changes.

45. A pinhole camera with two holes simply produces two images. If the holes are close together, the images overlap. Multiple holes produce multiple images. Overlapping can be prevented by placing a converging lens at the holes. If you make the holes into one big hole covered with one big lens, you'll have a conventional camera!

47. For very distant objects, effectively at "infinity," light comes to focus at the focal plane of the lens. So your film is one focal length in back of the lens for very distant shots. For shorter distances, the film is farther from the lens.

49. Moon maps are upside-down views of the moon to coincide with the upside-down image that moon watchers see in a telescope.

Chapter 28 Problem Solutions

1 When a mirror is rotated, its normal rotates also. Since the angle that the incident ray makes with the normal is the same angle that the reflected ray makes, the total deviation is twice. In the sample diagram, if the mirror is rotated by 10°, then the normal is rotated by 10° also, which results in a 20° total deviation of the reflected ray. This is one reason that mirrors are used to detect delicate movements in instruments such as galvanometers. The more important reason is the amplification of displacement by having the beam arrive at a scale some distance away.

3. Set your focus for 4 m, for your image will be as far in back of the mirror as you are in front.

5. If 96% is transmitted through the first face, and 96% of 96% is transmitted through the second face, 92% is transmitted through both faces of the glass.

7. Use ratios: (1440 min)/(360 deg) = (unknown time)/(0.53 deg). So the unknown time is 0.53 × 1440/360 = 2.1 minutes. So the sun moves a solar diameter across the sky every 2.1 minutes. At sunset, time is somewhat extended, depending on the the extent of refraction. Then the disk of the setting sun disappears over the horizon in a little longer than 2.1 minutes.

Chapter 29: Light Waves
Answers to Exercises

1. The Earth intercepts such a tiny fraction of the expanding spherical wave from the sun that it can be approximated as a plane wave (just as a small portion of the spherical surface of the Earth can be approximated as flat). The spherical waves from a nearby lamp have noticeable curvature. (See Figures 28.3 and 28.4).

3. The wavelengths of AM radio waves are hundreds of meters, much larger than the size of buildings, so they are easily diffracted around buildings. FM wavelengths are a few meters, borderline for diffraction around buildings. Light, with wavelengths a tiny fraction of a centimeter, show no appreciable diffraction around buildings.

5. The signals of lower channel numbers are broadcast at lower frequencies and longer wavelengths, which are more diffracted into regions of poor reception than are higher-frequency signals.

7. The alternation of sound from loud to soft is evidence of interference. Where the sound is loud, the waves from each loudspeaker are interfering constructively; where it is soft, destructive interference from the speakers is taking place.

9. Both interference fringes of light and the varying intensities of sound are the result of the superposition of waves that interfere constructively and destructively.

11. Blue light will produce narrower-spaced fringes.

13. You'll photograph what you see through the lens — a spectrum of colors on either side of the streetlights. We'll see in the following chapter that the colors diffracted correlate with the illuminating gas in the streetlights.

15. Young's interference experiment produces a clearer fringe pattern with slits than with pinholes because the pattern is of parallel straight-line-shaped fringes rather than the fringes of overlapping circles. Circles overlap in relatively smaller segments than the broader overlap of parallel straight lines. Also, the slits allow more light to get through; the pattern with pinholes is dimmer.

17. For complete cancellation, the amplitudes of each part of the wave must be identical. If the amplitudes are not the same, then partial cancellation results.

19. Interference colors result from double reflections from the upper and lower surfaces of the thin transparent coating on the butterfly wings. Some other butterfly wings produce colors by diffraction, where ridges in the surface act as diffraction gratings.

21. Interference of light from the upper and lower surfaces of the soap or detergent film is taking place.

23. Light from a pair of stars will not produce an interference pattern because the waves of light from the two separate sources are incoherent; when combined they smudge. Interference occurs when light from a single source divides and recombines.

25. Blue, the complementary color. The blue is white minus the yellow light that is seen above. (Note this exercise goes back to information in Chapter 27.)

27. The problem is serious, for depending on the orientation of the polarization axes of the displ and the glasses, no display may be seen.

29. If the sheet is aligned with the polarization of the light, all the light gets through. If it is aligned perpendicular to the polarization of the light, none gets through. At any other angle, some of the light gets through because the polarized light can be "resolved" (like a vector) into components parallel and perpendicular to the alignment of the sheet.

31. You can determine the polarization axis for a single sheet of Polaroid by viewing the glare from a flat surface, as in Figure 29.33. The glare is most intense when the polarization axis is parallel to the flat surface.

33. The axis of the filter should be vertical, not allowing the passage of the glare, which is parallel to the plane of the floor — horizontal.

35. You can determine that the sky is partially polarized by rotating a single sheet of Polaroid in front of your eye while viewing the sky. You'll notice the sky darken when the axis of the Polaroid is perpendicular to the polarization axis of the skylight.

37. When you look at a nearby object, your two eyes look in slightly different directions to see it, and your brain translates this difference into an estimated distance. This is parallax, a principal means of judging distance for nearby objects. If you stand close to a painting and look with two eyes, parallax tells you that you are looking at a surface, all parts of it equally distant from your eyes. But if you look with one eye, there is no parallax to help you judge distance and you must rely instead on distance cues provided by the painter, such as relative size, brightness, or gradations of color. Looking with one eye, you are seeing the scene the way the painter wanted you to see it, without your brain telling you how far away the surface is. (This difference between one-eyed and two-eyed viewing is evident only when you are close to the painting. If you look at it from a distance, parallax is no longer important. Then, even using two eyes, you rely on the painter's cues to judge distance.)

39. Magnification is accomplished by making the hologram with short wavelength light and viewing it with a longer wavelength light. This is similar to the wider spacings between fringes when long wavelength light illuminates thin slits.

Chapter 30: Light Emission

Answers to Exercises

1. In accord with $E = hf$, a gamma ray photon has a higher energy because it has a higher frequency.

3. Higher-frequency higher-energy blue light corresponds to a greater change of energy in the atom.

5. Doubling the wavelength of light halves its frequency. Light of half frequency has half the energy per photon. Think in terms of the equation $c = f\lambda$. Since the speed of light c is constant, λ is inversely proportional to f.

7. A slit can be made very thin — thinner than the diameter of a circle and still be visible. If the thin slit in a spectroscope were replaced with a round hole, the "lines" would appear as round spots. This would be disadvantageous because the wider circles might overlap. If the diameter of the hole were made as small as the width of the slit, insufficient light would get through.

9. When a spectrum of the sun is compared with the spectrum of the element iron, the iron lines overlap and perfectly match certain Fraunhofer lines. This is evidence for the presence of iron in the sun.

11. No spectral-line patterns appear in starlight that don't appear in the spectra of elements here on Earth. Since the spectra of light from distant stars matches the spectra of elements here on Earth, we conclude that we and the observable universe are made of the same stuff. Spectral lines are the "atomic fingerprints" that reveal the presence of the same atoms with the same properties elsewhere in the universe.

13. The stars are incandescent sources, where peak radiation frequency is proportional to stellar temperature. But light from gas discharge tubes is not a function of gas temperatures, but depends on the states of excitation in the gas. These states are not dependent on the temperature of the gas, and can occur whether the gas is hot or cool.

15. Atomic excitation occurs in solids, liquids, and gases. Because atoms in a solid are close packed, radiation from them (and liquids) is smeared into a broad distribution to produce a continuous spectrum, whereas radiation from widely-spaced atoms in a gas is in separate bunches that produce discrete "lines" when diffracted by a grating.

17. The many spectral lines from the element hydrogen are the result of the many energy states the single electron can occupy when excited.

19. The "missing" energy may appear as light of other colors or as invisible infrared light. If the atoms are closely packed, as in a solid, some of the "missing" energy may appear as heat. In that case, the illuminated substance warms.

21. Fluorescence is the process in which high-frequency (high energy) ultraviolet radiation converts to low-frequency (lower energy) visible radiation with some energy left over, possibly appearing as heat. If your friend is suggesting that low-energy infrared radiation can be converted to higher-energy visible light, that is clearly a violation of the conservation of energy — a no-no! Now if your friend is suggesting that infrared radiation can cause the fluorescence of still lower-frequency infrared radiation, which is not seen as light, then your friend's reasoning is well founded.

23. Fabrics and other fluorescent materials produce bright colors in sunlight because they both reflect visible light and transform some of the sun's ultraviolet light into visible light. They literally glow when exposed to the combined visible and ultraviolet light of the sun. (Certain fluorescent dyes added to inks are sometimes called "Day-Glo" colors.)

25. Illumination by the lower-frequency light doesn't have sufficiently energetic photons to ionize the atoms in the material, but has photons of enough energy to excite the atoms. In contrast, illumination by ultraviolet light does have sufficient energy for ejecting the electrons, leaving atoms in the material ionized. Imparting different energies produces different results.

27. The photons from the photoflash tube must have at least as much energy as the photons they are intended to produce in the laser. Red photons have less energy than green photons, so wouldn't be energetic enough to stimulate the emission of green photons. Energetic green photons can produce less-energetic red photons, but not the other way around.

29. If it weren't relatively long-lived, there wouldn't be enough accumulation of atoms in this excited state to produce the "population inversion" that is necessary for laser action.

31. Your friend's assertion violates the law of energy conservation. A laser or any device cannot put out more energy than is put into it. Power, on the other hand, is another story, as is treated in the following exercise.

33. $\bar{f}$ is the peak frequency of incandescent radiation — that is, the frequency at which the radiation is most intense. T is the temperature of the emitter. (Scientists also describe the radiation as having a temperature. Incandescent radiation emitted by a body at a certain temperature is said to have that same temperature. Thus we can use radiation to measure temperature, whether it be the radiation emitted by a blast furnace or by the sun or by frigid outer space.)

35. Both radiate energy in accord with $\bar{f} \sim T$. Since the sun's temperature is so much greater than the Earth's, the frequency of radiation emitted by the sun is correspondingly greater than the frequency of radiation emitted by the Earth. Earth's radiation is called terrestrial radiation, treated back in Chapter 15. The amount of energy radiated by the sun is also much greater than the amount radiated by the Earth. (The *amount* varies in proportion to the fourth power of the absolute temperature, so the sun with surface temperature 20 times the Earth's surface temperature radiates $(20)^4$, or 160,000 times as much energy as the Earth.)

37. The metal is glowing at all temperatures, whether we can see the glow or not. As its temperature is increased, the glow reaches the visible part of the spectrum and is visible to human eyes. Light of the lowest energy per photon is red. So the heated metal passes from infra-red (which we can't see) to visible red. It is red hot.

39. Star's relative temperatures — lowish for reddish; midish for whitish; and hotish for bluish.

41. An incandescent source that peaks in the green part of the visible spectrum will also emit reds and blues, which would overlap to appear white. Our sun is a good example. For green light and only green light to be emitted, we would have some other kind of a source, such as a laser, not an incandescent source. So "green-hot" stars are white.

43. Six transitions are possible, as shown. The highest-frequency transition is from quantum level 4 to level 1. The lowest-frequency transition is from quantum level 4 to level 3.

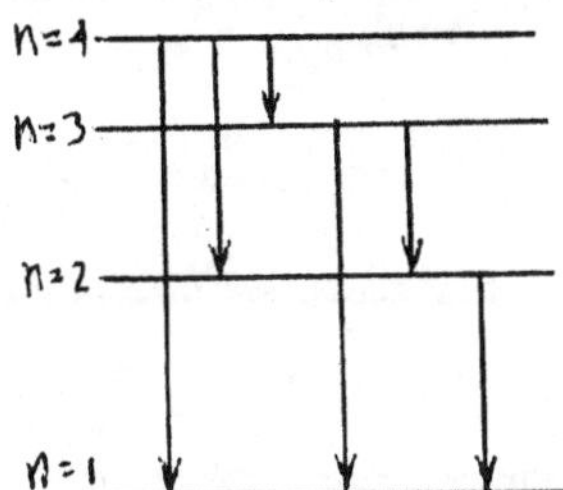

45. Yes, there is a relationship among the wavelengths, but it is not as simple as the relationship among frequencies. Because energies are additive, so are the frequencies. But since wavelength is inversely proportional to frequency, it is the inverses of the wavelengths that are additive. Thus,

$$\frac{1}{\lambda(4 \to 3)} + \frac{1}{\lambda(3 \to 1)} = \frac{1}{\lambda(4 \to 1)}$$

Chapter 30 Problem Solution

(a) The B-to-A transition has twice the energy and twice the frequency of the C-to-B transition. Therefore it will have half the wavelength, or 300 nm. Reasoning: Since $c = f\lambda$, $\lambda = c/f$. Wavelength is inversely proportional to frequency. Twice the frequency means half the wavelength. (b) The C-to-A transition has three times the energy and three times the frequency of the C-to-B transition. Therefore it will have one-third the wavelength, or 200 nm.

Chapter 31: Light Quanta
Answers to Exercises

1. Classical physics is primarily the physics known before 1900 that includes the study of motion in accord with Newton's laws and the study of electromagnetism in accord with the laws of Maxwell. Classical mechanics, often called Newtonian mechanics, is characterized by absolute predictability. After 1900 scientists discovered that Newtonian rules simply don't apply in the domain of the very small — the submicroscopic. This is the domain of quantum physics, where things are "grainy" and where values of energy and momentum (as well as mass and charge) occur in lumps, or quanta. In this domain, particles and waves merge and the basic rules are rules of probability, not certainty. Quantum physics is different and not easy to visualize like classical physics. We nevertheless tend to impress our classical wave and particle models on our findings in an effort to visualize this subatomic world.

3. The equation $E = hf$ illustrates the "weirdness" of the quantum world, for it links a *particle* property, the photon energy E, to a wave property, the frequency f. This link goes under the name "wave-particle duality." The photon has both wave and particle properties. (The appearance of Planck's quantum constant h in the equation is a hint that we are dealing here with a quantum phenomenon. The constant h shows up in every equation of the quantum world. It is the "perpetrator" of quantum weirdness. If h were zero, there would be no quantum phenomena.)

5. It makes no sense to talk of photons of white light, for white light is a mixture of various frequencies and therefore a mixture of many photons. One photon of white light has no physical meaning.

7. Finding materials that would respond photoelectrically to red light was difficult

because photons of red light have less energy than photons of green or blue light.

9. When a photon of ultraviolet light encounters a living cell, it transfers to the cell an amount of energy that can be damaging to the cell. When a photon of visible light encounters a living cell, the amount of energy it transfers to the cell is less, and less likely to be damaging. Hence skin exposure to ultraviolet radiation can be damaging to the skin while exposure to visible light generally is not.

11. It is not the total energy in the light beam that causes electrons to be ejected, but the energy per photon. Hence a few blue photons can dislodge a few electrons, where hoards of low-energy red photons cannot dislodge any. The photons act singly, not in concert.

13. Some automatic doors utilize a beam of light that continuously shines on a photodetector. When you block the beam by walking through it, the generation of current in the photodetector ceases. This change of current then activates the opening of the door.

15. There will be colors toward the red end of the spectrum where the meter will show no reading, since no electrons are ejected. As the color is changed toward the blue and violet, a point will be reached where the meter starts to give a reading. If a color for which the meter reads zero is made more intense, the meter will continue to read zero. If a color for which the meter shows a reading is made more intense, the current recorded by the meter will increase as more electrons are ejected.

17. We can never definitely say what something *is*, only how it behaves. Then we construct models to account for the behavior. The photoelectric effect doesn't prove that light is corpuscular, but supports the corpuscular model of light. Particles best account for photoelectric behavior. Similarly, interference experiments support the wave model of light. Waves best account for interference behavior. We have models to help us conceptualize what something *is*; knowledge of the details of how something behaves helps us to refine the model. It is important that we keep in mind that our models for understanding nature are just that: models. If they work well enough, we tend to think that the model represents what *is*. (More about models in the answer to the next exercise.)

19. An explanation is the following: Light refracting through the lens system is understandable via the wave model of light, and its arrival spot by spot to form the image is understandable via the particle model of light. How can this be? We have had to conclude that even single photos have wave properties. These are waves of probability that determine where a photon is likely or not likely to go. These waves interfere constructively and destructively at different locations on the film, so the points of photon impact are in accord with probability determined by the waves.

21. No. Complementarity isn't a compromise, but suggests that what you see depends on your point of view. What you see when you look at a box, for example, depends on whether you see it from one side, the top, and so on. All measurements of energy and matter show quanta in some experiments and waves in others. For light, we see particle behavior in emission and absorption, and wave behavior in propagation between emission and absorption.

23. By absorbing energy from the impact of a particle or photon.

25. A proton with the same speed as an electron has more momentum than the electron. Therefore, in accord with de Broglie's equation $\lambda = h/p$ (wavelength inversely proportional to momentum), the proton has a shorter wavelength, the electron a longer wavelength. (The formula for wavelength in terms of momentum does not make sense in the domain of classical physics. But why should we suppose that our descriptions of the macroscopic everyday world should carry to the microworld? They don't; hence the invention of quantum physics.)

27. By de Broglie's formula, as velocity increases, momentum increases, so wavelength decreases.

29. The cannonball obviously has more momentum than the BB traveling at the same speed, so in accord with de Broglie's formula the BB has the longer wavelength. (Both wavelengths are too small to measure.)

31. Protons of the same speed as electrons would have more momentum, and therefore smaller wavelengths, and therefore less diffraction. Diffraction is an asset for long-wavelength radio waves, helping them to get around obstructions, but it is a drawback in microscopes, where it makes images fuzzy. Why are there not proton microscopes? There are; we call them atomic accelerators. The high momenta of high-velocity protons make it possible to extract detailed information on nuclear structure, illuminating a domain vastly smaller than the size of a single atom.

33. Planck's constant would be much much larger than its present value.

35. We don't know if an electron *is* a particle or a wave; we know it *behaves* as a wave when it moves from one place to another and behaves as a particle when it is incident upon a detector. The unwarranted assumption is that an electron must *be* either a particle *or* a wave. It is common to hear some people say that something can only be either this or that, as if both were not possible (like those who say we must choose between biological evolution *or* the existence of a supreme being).

37. Heisenberg's uncertainty principle applies *only* to quantum phenomena. However, it serves as a popular metaphor for the macro domain. Just as the way we measure affects what's being measured, the way we phrase a question often influences the answer we get. So to various extents we alter that which we wish to measure in a public opinion survey. Although there are countless examples of altering circumstances by measuring them, the uncertainty principle has meaning only in the submicroscopic world.

39. The question is absurd, with the implication that eradicating butterflies will prevent tornadoes. If a butterfly can, in principle, cause a tornado, so can a billion other things. Eradicating butterflies will leave the other 999,999,999 causes untouched. Besides, a butterfly is as likely to *prevent* a tornado as to cause one.

Chapter 31 Problem Solutions

1. Frequency is speed/wavelength: $f = (3 \times 10^8$ m/s$)/(2.5 \times 10^{-5}$ m$) = 1.2 \times 10^{13}$ Hz. Photon energy is Planck's constant × frequency: $E = hf = (6.6 \times 10^{-34}$ J s$)(1.2 \times 10^{13}$ Hz$)$ $=$ **7.9×10^{-21} J**. (In the electron-volt unit common in atomic and optical physics, this is 0.05 eV, about one-twentieth the energy acquired by an electron in being accelerated through a potential difference of 1 V. 1 eV is equal to 1.6×10^{-19} J.)

3. The ball's momentum is $mv = (0.1$ kg$)$ $(0.001$ m/s$)$ $= 1 \times 10^{-4}$ kg m/s, so its de Broglie wavelength is $h/p = (6.6 \times 10^{-34}$ J s$)/(1 \times 10^{-4}$ kg m/s$)$ $=$ **6.6×10^{-30} m**, incredibly small relative even to the tiny wavelength of the electron. There is no hope of rolling a ball slowly enough to make its wavelength appreciable.

Part 8: Atomic and Nuclear Physics

Chapter 32: The Atom and the Quantum
Answers to the Exercises

1. Photons from the ultraviolet lamp have greater frequency, energy, and momentum. Only wavelength is greater for photons emitted by the TV transmitter.

3. The dense concentration of positive charge and mass in the nucleus of Rutherford's model of the atom accounts for the backscattering of alpha particles as they ricochet off the gold atoms of the thin foil. This backscattering would not occur if the positive charge and mass of the atom were spread throughout the volume of the atom, just as a golf ball would not bounce backward when striking a piece of cake, or even when colliding with a tennis ball or another golf ball. A golf ball will bounce backward if it strikes a massive object such as a bowling ball, and in a similar way, some of the alpha particles bounce backward when encountering the massive atomic nucleus and the enormously strong electric field in its vicinity.

5. Uranium has 92 protons in its nucleus, which is 92 times as much positive charge as hydrogen. This greater charge pulls the surrounding electrons into tighter orbits. The result of this is that the heavier atoms are not appreciably larger in size than the lighter atoms (see Figure 32.6).

7. It would emit a continuous spectrum. Its energy would change gradually and continuously as it spiraled inward and it would radiate at its rotational frequency, which would be continuously increasing.

9. If the energy spacings of the levels were equal, there would be only two spectral lines. Note that a transition between the 3rd and 2nd level would have the same difference in energy as a transition between the 2nd and first level. So both transitions would produce the same frequency of light and produce one line. The other line would be due to the transition from the 3rd to the first level.

11. If we think of electrons as orbiting the nucleus in standing waves, then the circumference of these wave patterns must be a whole number of wavelengths. In this way the circumferences are discrete. This means that the radii of orbits are therefore discrete. Since energy depends upon this radial distance, the energy values are also discrete. (In a more refined wave model, there are standing waves in the radial as well as the orbital direction.)

13. Helium's electrons are in one filled shell. The filled shell means that bonding with other elements is rare. Lithium has two shells, the first filled and the second with only one of eight electrons in it, making it very reactive with other elements.

15. Yes. In atoms, electrons move on the order of 2 million meters per second, and so their wave nature is quite pronounced.

17. Both use Bohr's concept of energy levels in an atom. An orbital is represented by the easier-to-visualize orbit.

19. The amplitude of a matter wave is called its wave function, represented by the symbol ψ (psi). Where ψ is large, the particle (or other material) is more likely to be found. Where ψ is small, the particle is less likely to be found. (The actual probability is proportional to ψ^2.)

21. The laws of probability applied to one or a few atoms give poor predictability, but for hordes of atoms, the situation is entirely different.

Although it is impossible to predict which electron will absorb a photon in the photoelectric effect, it is possible to predict accurately the current produced by a beam of light on photosensitive material. We can't say where a given photon will hit a screen in double-slit diffraction, but we can predict with great accuracy the relative intensities of a wave-interference pattern for a bright beam of light. Predicting the kinetic energy of a particular atom as it bumbles about in an atomic lattice is highly inaccurate, but predicting the average kinetic energy of hordes of atoms in the same atomic lattice, which measures the temperature of the substance, is possible with high precision. The indeterminacy at the quantum level can be discounted when large aggregates of atoms so well lend themselves to extremely accurate macroscopic prediction.

23. Electrons have a definite mass and a definite charge, and can sometimes be detected at specific points—so we say they have particle properties; electrons also produce diffraction and interference effects, so we say they have wave properties. There is a contradiction only if we insist the electron may have only particle OR only wave properties. Investigators find that electrons display both particle and wave properties.

25. Bohr's correspondence principle says that quantum mechanics must overlap and agree with classical mechanics in the domain where classical mechanics has been shown to be valid.

27. The philosopher was speaking of classical physics, the physics of the macroscopic world, where to a high degree of accuracy the same physical conditions do produce the same results. Feynman must have been speaking of the quantum domain where for small numbers of particles and events, the same conditions are not expected to produce the same results.

29. The speed of light is large compared with the ordinary speeds with which we deal in everyday life. Planck's constant is small in that it gives wavelengths of ordinary matter far too small to detect and energies of individual photons too small to detect singly with our eyes.

Problem Solutions

1. When n = 50, the atom is $(50)^2$, or 2500, times larger than when n = 1, so its radius is $(2500)(1 \times 10^{-10}\text{ m}) = 2.5 \times 10^{-7}$ m. The volume of this enlarged atom is $(2500)^3$, or 1.6×10^{10}, times larger than the volume of the atom in its lowest state. Sixteen billion unexcited atoms would fit within this one excited atom! (Atoms this large and larger have been made in recent years by trapping single atoms in an evacuated region. These giant atoms with their barely tethered distant electrons are known as Rydberg atoms.)

Chapter 33: The Atomic Nucleus and Radioactivity
Answers to Exercises

1. X rays are high-frequency electromagnetic waves, and are therefore most similar to even higher-frequency electromagnetic waves—gamma rays. Alpha and beta rays, in contrast, are streams of material particles.

3. It is impossible for a hydrogen atom to eject an alpha particle, for an alpha particle is composed of four nucleons—two protons and two neutrons. It is equally impossible for a 1-kg melon to disintegrate into four 1-kg melons.

5. The alpha particle has twice the charge, but almost 8000 times the inertia (since each of the four nucleons has nearly 2000 times the mass of an electron). Even though the alpha particle is slower than the electron, it has more momentum due to its great mass, and hence deflects less than an electron in a given magnetic field.

7. Alpha radiation decreases the atomic number of the emitting element by 2 and the atomic mass number by 4. Beta radiation increases the atomic number of an element by 1 and does not affect the atomic mass number. Gamma radiation does not affect the atomic number or the atomic mass number. So alpha radiation results in the greatest change in both atomic number and mass number.

9. Gamma predominates inside the enclosed elevator because the structure of the elevator shields against alpha and beta particles better than against gamma-ray photons.

11. An alpha particle undergoes an acceleration due to mutual electric repulsion as soon as it is out of the nucleus and away from the attracting nuclear force. This is because it has the same sign of charge as the nucleus. Like charges repel.

13. They repel by the electric force, and attract each other by the strong nuclear force. The strong force predominates. (If it didn't, there would be no atoms beyond hydrogen!) If the protons are separated to where the longer-range electric force overcomes the shorter-range strong force, they fly apart.

15. The existence of atomic nuclei containing many protons is evidence that something stronger than electric repulsion is occurring in the nucleus. If there were not a stronger attractive nuclear force to keep the repelling electrical force from driving protons apart, the nucleus as we know it wouldn't exist.

17. A positively charged hydrogen atom, an ion, is the nucleus of the atom, since no electron remains. It is usually a proton, but could be a deuteron or triton, one of the nuclei of heavier hydrogen isotopes.

19. Starting from birth, a human population has a certain half life, the time until half have died, but this doesn't mean that half of those still living will die in the next equal interval of time. For radioactive atoms, the chance of "dying" (undergoing decay) is always the same, regardless of the age of the atom. A young atom and an old atom of the same type have exactly the same chance to decay in the next equal interval of time. This is not so for humans, for whom the chance of dying increases with age.

21. The spiral path of charged particles in a bubble chamber is the result of a slowdown of the particles due to collisions with atoms, usually hydrogen, in the chamber. The slower-moving charged particles bend more in the magnetic field of the chamber and their paths become spirals. If the charged particles moved without resistance, their paths would be circles or helixes.

23. The mass of the element is 157 + 104 = 261. Its atomic number is 104, a transuranic element recently named rutherfordium.

25. After the polonium nucleus emits a beta particle, the atomic number increases by 1 to become 85, and the atomic mass is unchanged at 218. However, if an alpha particle is emitted, the atomic number decreases by 2 to become 82, and the atomic mass decreases by 4, becoming 214.

27. An element can decay to an element of greater atomic number by emitting electrons (beta rays). When this happens, a neutron in the nuclues becomes a proton and the atomic number increases by one.

29. If strontium-90 (atomic number 38) emits betas, it should become the element yttrium (atomic number 39); hence the physicist can test a sample of strontium for traces of yttrium by spectrographic means or other techniques. To verify that it is a "pure" beta emitter, the physicist can check to make sure that the sample is emitting no alphas or gammas.

31. The elements below uranium in atomic number with short half-lives exist as the product of the radioactive decay of uranium or another very long-lived element, thorium. For the billions of years that the uranium and thorium last, the lighter elements will be steadily replenished.

33. Your friend will encounter more radioactivity from the granite outcroppings than he or she will living near a nuclear power plant. Furthermore, at high altitude your friend will be treated to increased cosmic radiation. But the radiations encountered in the vicinity of the plant, on the granite outcropping, or at high altitude are not appreciably different than the radiation one encounters in the "safest" of situations. Advise your friend to enjoy life anyway!

35. Although there is significantly more radioactivity in a nuclear power plant than in a coal-fired power plant, almost none of it escapes from the nuclear plant, whereas most of what radioactivity there is in a coal-fired plant does escape, through the stacks. As a result, a typical coal plant injects more radioactivity into the environment than does a typical nuclear plant. (Unfortunately, if you mention this to people you meet at a normal social gathering, you'll be seen as some sort of ally of Darth Vader!)

37. The irradiated food does not become radioactive as a result of being zapped with gamma rays. This is because the gamma rays lack the energy to initiate the nuclear reactions in atoms in the food that could make them radioactive.

39. Stone tablets cannot be dated by the carbon dating technique. Nonliving stone does not ingest carbon and transform that carbon by radioactive decay. Carbon dating works for organic material.

Problem Solutions

1. At the end of the second year 1/4 of the original sample will be left; at the end of the third year, 1/8 will be left; and at the end of the fourth year, 1/16 will be left.

3. 1/16 will remain after 4 half lives, so $4 \times 30 =$ 120 years.

5. The intensity is down by a factor of 16.7 (from 100% to 6%). How many factors of two is this? About 4, since $2^4 = 16$. So the age of the artifact is about 4×5730 years, or about 23,000 years.

Chapter 34: Nuclear Fission and Fusion Answers to Exercises

1. Unenriched uranium—which contains more than 99% of the non-fissionable isotope U-238—undergoes a chain reaction only if it is mixed with a moderator to slow down the neutrons. Uranium in ore is mixed with other substances that impede the reaction and has no foerator to slow down the neutrons, so no chain reaction occurs. (There is evidence, however, that several billion years ago, when the percentage of U-235 in uranium ore was greater, a natural reactor existed in Gabon, West Africa.)

3. A fission reactor has a critical mass. Its minimum size (including moderator, coolant, etc.) is too large to power a small vehicle (although it is practical as a power source for submarines and ships). Indirectly, fission can be used to power automobiles by making electricity that is used to charge electric car batteries.

5. In a large piece of fissionable material a neutron can move farther through the material before reaching a surface. Larger volumes of fissionable material have proportionally less area compared to their greater volumes, and therefore lose less neutrons.

7. The average distance increases. (It's easier to see the opposite process where big pieces broken up into little pieces decreases the distance a neutron can travel and still be within the material. Proportional surface area increases with decreasing size, which is why you break a sugar cube into little pieces to increase the surface area exposed to tea for quick dissolving.) In the case of uranium fuel, the process of assembling small pieces into a single big piece increases average traveling distance, decreases surface area, reduces neutron leakage, and increases the probability of a chain reaction and an explosion.

9. Plutonium has a relatively short half life (24,360 years), so any plutonium initially in the Earth's crust has long since decayed. The same is true for any heavier elements with even shorter half lives from which plutonium might originate. Trace amounts of plutonium can occur naturally in U-238 concentrations, however, as a result of neutron capture, where U-238 becomes U-239 and after beta emission becomes Np-239, which further transforms by beta emission to Pu-239. (There are elements in the Earth's crust with half lives even shorter than plutonium's, but these are the products of uranium decay; between uranium and lead in the periodic table of elements.)

11. The resulting nucleus is $_{92}U^{233}$. The mass number is increased by 1 and the atomic number by 2. U-233, like U-235, is fissionable with slow neutrons. (Notice the similarity to the production of $_{94}Pu^{239}$ from $_{92}U^{238}$.)

13. When a neutron bounces from a carbon nucleus, the nucleus rebounds, taking some energy away from the neutron and slowing it down so it will be more effective in stimulating fission events. A lead nucleus is so massive that it scarcely rebounds at all. The neutron bounces with practically no loss of energy and practically no change of speed (like a marble from a bowling ball).

15. If the difference in mass for changes in the atomic nucleus increased tenfold (from 0.1% to 1.0%), the energy release from such reactions would increase tenfold as well.

17. Both chemical burning and nuclear fusion require a minimum ignition temperature to start and in both the reaction is spread by heat from one region to neighboring regions. There is no critical mass. Any amount of thermonuclear fuel or of combustible fuel can be stored.

19. Each fragment would contain 46 protons (half of 92) and 72 neutrons (half of 144), making it the nucleus of Pd-118, an isotope of palladium, element number 46. If two neutrons were emitted, the identical fragments would still be isotopes of palladium, but Pd-117 instead of Pd-118.

21. Fusing heavy nuclei (which is the way that the heavy transuranic elements are made) costs energy. The total mass of the products is greater than the total mass of the fusing nuclei.

23. Energy would be released by the fissioning of gold and from the fusion of carbon, but by neither fission nor fusion for iron. Neither fission nor fusion will result in a decrease of mass for iron.

25. If the masses of nucleons varied in accord with the shape of the curve of Figure 34.15 instead of the curve of Figure 34.16, then the fissioning of all elements would liberate energy and all fusion processes would absorb rather than liberate energy. This is because all fission reactions (decreasing atomic number) would result in nuclei with less mass per nucleon, and all fusion reactions (increasing atomic number) would result in the opposite; nuclei of more mass per nucleon.

27. Although more energy is released in the fissioning of a single uranium nucleus than in the fusing of a pair of deuterium nuclei, the much greater number of lighter deuterium atoms in a gram of matter compared to the fewer heavier uranium atoms in a gram of matter, results in more energy liberated per gram for the fusion of deuterium.

29. A hydrogen bomb produces a lot of fission energy as well as fusion energy. Some of the fission is in the fission bomb "trigger" used to ignite the thermonuclear reaction and some is in fissionable material that surrounds the thermonuclear fuel. Neutrons produced in fusion cause more fission in this blanket. Fallout results mainly from the fission.

31. Energy from the sun is our chief source of energy, which itself is the energy of fusion. Harnessing that energy on Earth has proven to be a formidable challenge.

33. Minerals which are now being mined can be recycled over and over again with the advent of a fusion-torch operation. This recycling would tend to reduce (but not eliminate) the role of mining in providing raw materials.

35. The lists can be very large. Foremost considerations are these: conventional fossil-fuel power plants consume our natural resources and convert them into greenhouse gases and poisonous contaminants that are discharged into the atmosphere, producing among other things, global climate change and acid rain. A lesser environmental problem exists with nuclear power plants, which do not pollute the

atmosphere. Pollution from nukes is concentrated in the radioactive waste products from the reactor core. Any rational discussion about the drawbacks of either of these power sources must acknowledge that *both* are polluters—so the argument is about which form of pollution we are more willing to accept in return for electrical power. (Before you say "No Nukes!", rational thinking suggests that you first be able to say that you "Know Nukes!")

37. The nuclei will be positively charged and will move toward the negative plate (and away from the positive one). The negative electrons will move in the opposite direction, toward the positive plate (and away from the negative one.)

39. The lighter nuclei with less mass deflect the most, while the more massive one are less deflected due to greater inertia. The mass spectrometer deflects ions in the same way, with less massive ions sweeping into circular paths of small radii and more massive ions sweeping in wider circular paths. In this way ions are separated according to their mass.

Problem Solutions

1. The energy released by the explosion in kilocalories is (20 kilotons)(4.2×10^{12} J/kiloton)/(4,184 J/kilocalorie) = 2.0×10^{10} kilocalories. This is enough energy to heat 2.0×10^{10} kg of water by 1 °C. Dividing by 50, we conclude that this energy could heat 4.0×10^{8} kilograms of water by 50 °C. This is nearly half a million tons.

3. The neutron and the alpha particle fly apart with equal and opposite momentum. But since the neutron has one-fourth the mass of the alpha particle, it has four times the speed. Then consider the kinetic-energy equation, $KE = (1/2)mv^2$. For the neutron, $KE = (1/2)m(4v)^2 = 8mv^2$, and for the alpha particle, $KE = (1/2)(4m)v^2 = 2mv^2$. The KEs are in the ratio of 8/2, or 80/20. So we see that the neutron gets 80% of the energy, and the alpha particle 20%. (Alternative method: The formulas for momentum and KE can be combined to give $KE = p^2/2m$. This equation tells us that for particles with the same momentum, KE is inversely proportional to mass.)

Part 8: Relativity

Chapter 35: Special Theory of Relativity
Answers to Exercises

1. The effects of relativity become pronounced only at speeds near the speed of light or when energies change by amounts comparable to mc^2. In our "non-relativistic" world, we don't directly perceive such things, whereas we do perceive events governed by classical mechanics. So the mechanics of Newton is consistent with our common sense, based on everyday experience, but the relativity of Einstein is not consistent with common sense. Its effects are outside our everyday experience.

3. (a) The bullet is moving faster relative to the ground when the train is moving (forward). (b) The bullet moves at the same speed relative to the freight car whether the train is moving or not.

5. Michelson and Morley considered their experiment a failure in the sense that it did not confirm the result that was expected. What was expected, that differences in the velocity of light would be encountered and measured, turned out not to be true. The experiment was successful in that it widened the doors to new insights in physics.

7. The *average* speed of light in a transparent medium is less than c, but in the model of light discussed in Chapter 25, the photons that make up the beam travel at c in the void that lies between the atoms of the material. Hence the speed of individual photons is always c. In any event, Einstein's postulate is that the speed of light in *free* space is invariant.

9. It's all a matter of relative velocity. If two frames of reference are in relative motion, events can occur in the order AB in one frame and in the order BA in the other frame. (See the next exercise.)

11. More and more energy must be put into an object that is accelerated to higher and higher speeds. This energy is evidenced by increased momentum. As the speed of light is approached, the momentum of the object approaches infinity. In this view there is infinite resistance to any further increase in momentum, and hence speed. Hence c is the speed limit for material particles. (Kinetic energy likewise approaches infinity as the speed of light is approached.)

13. As explained in the answer to Exercise 12, the moving points are not material things. No mass or no information can travel faster than c, and the points so described are neither mass nor information. Hence, their faster motion doesn't contradict special relativity.

15. When we say that light travels a certain distance in 20,000 years we are talking about distance in our frame of reference. From the frame of reference of a traveling astronaut, this distance may well be far shorter, perhaps even short enough that she could cover it in 20 years of her time (traveling, to be sure, at a speed close to the speed of light). Someday, astronauts may travel to destinations many light years away in a matter of months in their frame of reference.

17. A twin who makes a long trip at relativistic speeds returns younger than his stay-at-home twin sister, but both of them are older than

when they separated. If they could watch each other during the trip, there would be no time where either would see a reversal of aging, only a slowing or speeding of aging. A hypothetical reversal would result only for speeds greater than the speed of light.

19. If you were in a high-speed (or no speed!) rocket ship, you would note no changes in your pulse or in your volume. This is because the velocity between the observer, that is, yourself, and the observed is zero. No relativistic effect occurs for the observer and the observed when both are in the same reference frame.

21. Making such an appointment would not be a good idea because if you and your dentist moved about differently between now and next Thursday, you would not agree on what time it is. If your dentist zipped off to a different galaxy for the weekend, you might not even agree on what day it is!

23. Elongated like an ellipse, longer in the direction of motion than perpendicular to that direction. The Lorentz contraction shortens the long axis of the elliptical shape to make it no longer than the short axis. (A circle in motion, such as the wheel of the relativistic bicycle (grandson Manuel with his dog Grey) is contracted to appear shorter in the direction of motion than perpendicular to that direction. If it appeared circular to the observer on the ground, it would have to be longer in the direction of motion for the person moving with it.)

25. Both the frequency and the wavelength of the light change (and, of course, its direction of motion changes). Its speed stays the same.

27. The stick will appear to be one-half meter long when it moves with its length along the direction of motion. Why one half its length? Because for it to have a momentum equal to $2mv$, its speed must be $0.87c$.

29. For the moving electron, length contraction reduces the apparent length of the 2-mile long tube. Because its speed is nearly the speed of light, the contraction is great.

31. The acid bath that dissolved the latched pin will be a little warmer, and a little more massive (in principle). The extra potential energy of the latched pin is transformed into a little more mass.

33. To make the electrons hit the screen with a certain speed, they have to be given more momentum and more energy than if they were non-relativistic particles. The extra energy is supplied by your power utility. You pay the bill!

35. $E = mc^2$ means that energy and mass are two sides of the same coin, mass-energy. The c^2 is the proportionality constant that links the units of energy and mass. In a practical sense, energy and mass are one and the same. When something gains energy, it gains mass. When something loses energy, it loses mass. Mass is simply congealed energy.

37. Just as time is required for knowledge of distant events to reach our eyes, a lesser yet finite time is required for information on nearby things to reach our eyes. So the answer is yes, there is always a finite interval between an event and our perception of that event. If the back of your hand is 30 cm from your eyes, you see it as it was one-billionth of a second ago.

39. Kierkegaard's statement, "Life can only be understood backwards; but it must be lived forwards.", is consistent with special relativity. No matter how much time might be dilated as a result of high speeds, a space traveler can only effectively slow the passage of time relative to various frames of reference, but can never reverse it—the theory does not provide for traveling backward in time. Time at whatever rate, flows only forward.

Problem Solutions

1. Frequency and period are reciprocals of one another (Chapter 18). If the frequency is doubled, the period is halved. For uniform motion, one senses only half as much time between flashes that are doubled in frequency. For accelerated motion, the situation is different. If the source gains speed in approaching, then each successive flash has even less distance to travel and the frequency increases more, and the period decreases more as well with time.

3. $V = \dfrac{c + c}{1 + \dfrac{c^2}{c^2}} = \dfrac{2c}{1 + 1} = c$

5. In the previous problem we see that for $v = 0.99c$, γ is 7.1. The momentum of the bus is more than seven times greater than would be calculated if classical mechanics were valid. The same is true of electrons, or anything traveling at this speed.

7. Gamma at $v = 0.10\,c$ is $1/\sqrt{[1 - (v^2/c^2)]}$ $= 1/\sqrt{[1 - (0.10)^2]} = 1/\sqrt{[1 - 0.01]} = 1/\sqrt{0.99} =$ 1.005. You would measure the passenger's catnap to last 1.005 (5 m) = 5.03 min.

9. Gamma at $v = 0.5\,c$ is $1/[\sqrt{1 + (v^2/c^2)}]$ $= 1/[\sqrt{1 - 0.5^2}] = 1/[\sqrt{1 - 0.25}] = 1/\sqrt{0.75} = 1.15$. Multiplying 1 h of taxi time by γ gives 1.15 h of Earth time. The drivers' new pay will be 11.5 stellars for this trip.

Chapter 36: General Theory of Relativity
Answers to Exercises

1. In accord with the principle of equivalence, one cannot discern between accelerated motion and gravitation. The effects of each are identical. So unless she has other clues, she will not be able to tell the difference.

3. An astronaut, when in orbit, is in an accelerated frame of reference and is experiencing gravity. But the astronaut is weightless because the effect of gravity and the effect of the frame's acceleration cancel. The same is true in any example of free fall.

5. Ole Jules called his shot wrong on this one. In a space ship that drifts through space, whether under the influence of moon, Earth, or whatever gravitational field, the ship and its occupants are in a state of free fall — hence there is no sensation of up or down. Occupants of a spaceship would feel weight, or sense an up or down, only if the ship were made to accelerate — say, against their feet. Then they could stand and sense that down is toward their feet, and up away from their feet.

7. We don't notice the bending of light by gravity in our everyday environment because the gravity we experience is too weak for a noticeable effect. If there were stellar black holes in our vicinity, the bending of light near them would be quite noticeable.

9. A beam of light traveling horizontally for one second in a uniform gravitational field of strength 1 *g* will fall a vertical distance of 4.9 meters, just as a baseball would. This is providing it remains in a 1-*g* field for one second, for it would travel 300,000 kilometers during this second also, nearly 25 Earth diameters away, and well away from the 1-*g* field strength of the Earth's surface (unless it were confined to the 1-*g* region as shown in Figure 36.8, with mirrors). If light were to travel in a 1-*g* region for two seconds, then like a baseball, it would fall $1/2\ g\ 2^2 = 19.6$ meters.

11. The clock will run slower at the bottom of a deep well than at the surface, because in going down the well, we are moving in the direction that the gravitational force acts, and this results in a slowing of clocks.

13. The light will be red-shifted. The accelerating car is equivalent to a stationary car standing vertically with its rear end down. The light going from the back to the front of the accelerating car behaves just like light going upward away from the surface of a planet. It is gravitationally red-shifted. (If your friend were moving toward you, it would be a different story. Then the Doppler effect arising from your relative motion would produce a blue-shift. But in this example, with no relative motion, there is no Doppler effect. Only the effect of gravity—or equivalent acceleration—remains.)

15. It will run slightly slower. For observers on Earth, this is because moving a clock from a pole to the equator is moving it in the direction of the centrifugal force, which slows the rate at which clocks run (the same as if it were moving in the direction of a gravitational force). For observers outside in an inertial frame, the slowing of the clock at the equator is an example of time dilation, an effect of special relativity caused by the motion of the clock. (The situation is much like that shown in Figure 36.9.)

17. Prudence is older. Charity's time runs slower during the time she is at the edge of the rotating kingdom (see Figure 36.9).

19. Light emitted from the star is red shifted. This can be understood as the result of gravity slowing down time on the surface of the star, or as gravity taking energy away from the photons as they propagate away from the star.

21. The astronaut falling into the black hole would see the rest of the universe blue-shifted. The astronaut's time scale is being slowed, which makes time scales elsewhere look fast to the astronaut. The blue shift can also be understood as the result of the black hole's gravity adding energy to photons that "fall" toward the black hole. The added energy means greater frequency.

23. Yes. If the star is massive enough and concentrated enough, its gravity could be strong enough to make light follow a circular path. This is what light does at the "horizon" of a black hole.

25. Yes. For example, place the sun just outside one of the legs in Figure 36.14.

27. Oscillating mass (or, more generally, accelerated mass) is the mechanism for emission of gravitational waves, just as oscillating or accelerated charge is the mechanism for emission of electromagnetic waves. When it is absorbed, a gravitational wave can set mass into oscillation, just as an absorbed electromagnetic wave can set charge into oscillation. (Scientists seeking to detect gravitational waves try to detect tiny oscillations of matter caused by the absorption of the waves. See Figure 36.17.)

29. Open.

Appendix E

Answers to Questions to Ponder

1. A dollar loses 1/2 its value in 1 doubling time of the inflationary economy; this is 70/7% = 10 years. It the dollar is loaned at 7% compound interest, it loses nothing.

3. For a 5% growth rate, 42 years is three doubling times (70/5% = 14 years; 42/14 = 3). Three doubling times is an eightfold increase. So in 42 years the city would have to have 8 sewerage treatment plants to remain as presently loaded; more than 8 if load per plant is to be reduced while servicing 8 times as many people.

5. Doubling one penny for 30 days yields a total of $10,737,418.23!

7. It is generally acknowledged that if the human race is to survive while alleviating even part of the misery that afflicts so much of humankind, the present rates of population growth and energy consumption must be reduced. The chances of achieving reduced growth rates are greater in a climate of scarce energy than in a climate of abundant energy. We can hope that by the time fusion is a viable power source, that we will have learned to optimize our numbers and to use energy more wisely.

Sample MECHANICS Exam

Write the BEST answer to each of the following only in the small box. No penalty for guessing. Good Energy!

1. The acceleration of a bowling ball rolling along a smooth horizontal bowling alley is
   a. zero. b. about 10 m/s^2. c. constant.

2. You stand at the top of a cliff and throw a rock downward, and another rock horizontally at the same speed. The rock that stays in the air for the longest time is the one thrown
   a. downward. b. horizontally. c. both take the same time.

3. If an object moves along a straight-line path, then it *must* be
   a. accelerating. b. acted on by a force. c. both of these. d. none of these.

4. A heavy rock and a light rock in free fall have the same acceleration. The *reason* the heavy rock does not have more acceleration is because
   a. the force of gravity on each is the same.
   b. there is no air resistance.
   c. the inertia of both rocks is the same.
   d. all of these. e. none of these.

5. A ball rolls down a curved ramp as shown. As its speed increases, its rate of gaining speed
   a. increases. b. decreases. c. remains unchanged.

6. Apply the equation $Ft = \Delta\ mv$ to the case of a person falling on a wooden floor. If v is the speed of the person as she strikes the floor, then m is the
   a. mass of the person. b. mass of the floor. c. both. d. none of these.

7. Looking at the seesaw you can see that compared to the weight of the boy, the weight of the board is
   a. more b. less c. the same d. there's no way to tell

8. A pair of tennis balls fall through the air from a tall building. One is regular and the other is filled with lead pellets. The ball to reach the ground first is the
   a. regular ball b. lead-filled ball c. is the same for both

9. The same pair of tennis balls (regular and lead filled) fall from a tall building. Air resistance just before they hit is actually greater for the
   a. regular ball b. lead-filled ball c. is the same for both

10. When a bullet is fired from a rifle, the force that accelerates the bullet is equal in magnitude to the force that makes the rifle recoil. But compared to the rifle, the bullet has a greater
    a. inertia b. potential energy. c. kinetic energy. d. momentum.

11. The reason for the answer to the preceding question has to do with the fact that the force on the bullet acts over
    a. the same time. b. a longer time. c. a longer distance. d. none of these.

12. Which pulls with the greater force on the earth's oceans?
    a. moon. b. sun. c. both the same.

1a, 2b, 3d, 4e, 5b, 6a, 7a, 8b, 9b, 10c, 11c, 12b.

Sample PROPERTIES OF MATTER Exam

Write the BEST answer to each of the following only in the small box. No penalty for guessing. Good Energy!

1. What makes one element distinct from another is the number of
   a. protons in its nucleus.
   b. neutrons in its nucleus.
   c. electrons in its nucleus.
   d. total particles in its nucleus.

2. In the atomic nucleus of a certain element are 26 protons and 28 neutrons. The ATOMIC NUMBER of the element is
   a. 26. b. 27. c. 28. d. 54. e. none of these.

3. Which is bigger in size? A kilogram of aluminum or a kilogram of lead.
   a. aluminum b. lead c. the same.

4. Consider two oranges, one with twice the diameter of the other. How much heavier is the larger orange?
   a. twice. b. four times. c. eight times. d. sixteen times e. none of these.

5. Consider the same two oranges, one with twice the diameter of the other. How much more skin does the larger orange have?
   a. twice. b. four times. c. eight times. d. sixteen times. e. none of these.

6. Three bowling balls are suspended at various depths in the water as shown. Buoyant force is greatest on ball
   a. A. b. B. c. C. d. same on each.

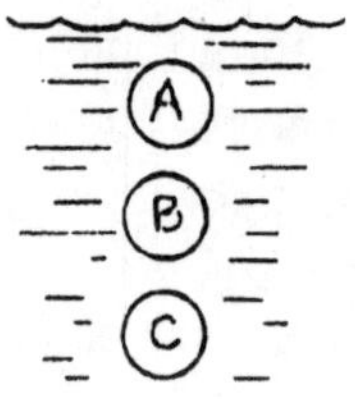

7. Compared to an empty ship, a ship with a cargo of styrofoam floats
   a. deeper in the water.
   b. higher in the water.
   c. with no change in level.

8. The greatest amount of water is displaced by a rock when it
   a. floats in a light pie pan. b. is submerged. c. ...same either way.

9. Two life preservers have identical volumes, but one is filled with styrofoam while the other is filled with sand. When the two life preservers are fully submerged, the buoyant force is greater on the one filled with
   a. styrofoam.
   b. sand.
   c. same on each as long as their volumes are the same.

10. As a weighted air-filled balloon sinks deeper and deeper into water, the buoyant force on it
    a. increases. b. decreases. c. remains essentially the same.

11. Squeeze an air-filled balloon to half size and the pressure inside
    a. remains the same. b. halves. c. doubles. d. none of these.

12. As a helium-filled balloon rises in the air, it becomes
    a. lighter. b. less dense. c. non buoyant. d. all of these e. none of these.

1a, 2a, 3a, 4c, 5b, 6d, 7a, 8a, 9c, 10b, 11c, 12b.

Sample HEAT Exam

Write the BEST answer to each of the following only in the small box. No penalty for guessing. Good Energy!

1. In a mixture of hydrogen gas, oxygen gas, and nitrogen gas, the molecules with the greatest average speed are those of
   a. hydrogen.
   b. oxygen.
   c. nitrogen.
   d. ...all will have the same average speed at the same temperature.

2. The reason that the white-hot sparks that strike your face from a 4th-of-July-type sparkler don't harm you is because
   a. they have a low temperature.
   b. the energy per molecule is very low.
   c. the energy per molecule is high, but little energy is transferred because of relatively few molecules in the spark.

3. As a piece of metal with a hole in it cools, the diameter of the hole
   a. increases. b. decreases. c. remains the same size.

4. If water had a higher specific heat, in cold weather, ponds would be
   a. less likely to freeze.
   b. more likely to freeze.
   c. neither more nor less likely to freeze.

5. Consider a sample of water at 2°C. It the temperature is increased slightly, say by one degree, the volume of water
   a. increases. b. decreases. c. remains unchanged.

6. The temperature of water at the bottom of a deep ice-covered lake is
   a. slightly below freezing temperature.
   b. itself at the temperature of freezing.
   c. somewhat above freezing temperature.

7. The principle reason one can walk barefoot on red-hot wooden coals without burning the feet has to do with
   a. low temperature of the coals.
   b. low thermal conductivity of the coals.
   c. mind-over-matter techniques.

8. If the slower-moving molecules in a liquid were more likely to undergo evaporation, then evaporation would make the remaining liquid
   a. warmer. b. cooler. c. no warmer or cooler than without evaporation.

9. Melting ice actually
   a. tends to warm the surroundings.
   b. tends to cool the surroundings.
   c. has no effect on the temperature of the surroundings.

10. Consider a piece of metal that has a temperature of 5°C. If it is heated until it has twice the internal energy, its temperature will be
    a. 10°C. b. 273°C. c. 278°C. d. 283°C. e. 556°C.

1a, 2c, 3b, 4a, 5b, 6c, 7b, 8a, 9b, 10d.

Sample SOUND Exam

Write the BEST answer to each of the following only in the small box. No penalty for guessing. Good Energy!

1. A portion of water oscillates up and down two complete cycles in one second as a water wave passes by. The wave's wavelength is 5 meters. What is the wave's speed?
   a. 2m/s. b. 5 m/s. c. 10 m/s. d. 15 m/s. e. none of these.

2. A 60-vibration per second wave travels 30 meters in one second. Its frequency is
   a. 30 hertz and it travels at 60 m/s.
   b. 60 hertz and it travels at 30 m/s.
   c. neither of these.

3. A mass on the end of a spring bobs up and down one complete cycle every two seconds. Its frequency is
   a. 0.5 Hz. b. 2 Hz. c. neither of these.

4. When a source of sound approaches you, you detect an increase in its
   a. speed. b. wavelength. c. frequency. d. all of these.

5. True or false: A sonic boom is typically produced when an aircraft goes from subsonic to the supersonic speed.
   a. true. b. false.

6. The speed of sound in air depends on
   a. its frequency.
   b. its wavelength.
   c. air temperature.
   d. all of these.
   e. none of these.

7. A singer holds a high note and shatters a distant crystal wine glass. This phenomenon best demonstrates
   a. forced vibrations.
   b. the Doppler Effect.
   c. interference
   d. resonance.

8. To set a tuning fork of 400 Hz into resonance, it is best to use another of
   a. 200 Hz. b. 400 Hz. c. 800 Hz. d. any of these three.

9. About how many octaves are present between 100 Hz and 1600 Hz?
   a. 4. b. 5. c. 6. d. 7. e. 8.

10. True or false: Any radio wave travels faster under all conditions than any sound wave.
    a. true. b. false.

1c, 2b, 3a, 4c, 5b, 6c, 7d, 8b, 9a, 10a.

Sample ELECTRICITY & MAGNETISM Exam

Write the BEST answer to each of the following only in the small box. No penalty for guessing. Good Energy!

1. Protons and electrons
   a. repel each other. b. attract each other. c. do not interact.

2. Particle A interacts with particle B, which has twice the charge of particle A. Compared to the force on particle A, the force on particle B is
   a. four times as much.
   b. two times as much
   c. the same.
   d. half as much.
   e. none of these.

3. When you touch a negative Van de Graaff generator, your standing hair is
   a. negative also. b. positive.

4. Two charged particles held close to each other are released. As the particles move, the velocity of each increases. Therefore the particles have
   a. the same sign of charge. b. opposite signs of charge. c. not enough information is given.

5. You can touch and discharge a 10,000-volt Van de Graaff generator with little harm because although the voltage is high, there is relatively little
   a. resistance. b. energy. c. grounding d. all of these. e. none of these.

6. The current through a 12-ohm hairdryer connected to 120-V is
   a. 1 A. b. 10 A. c. 12 A d. 120 A. e. none of these.

7. Double the voltage that operates a hair dryer and the current within tends to
   a. halve. b. remain the same. c. double. d. quadruple.

8. A woman experiences an electrical shock with a faulty hairdryer. The electrons making the shock come from the
   a. woman's body.
   b. ground
   c. power plant
   d. hairdryer.
   e. electric field in the air.

9. As more lamps are connected to a series circuit, the overall current in the power source
   a. increases. b. decreases. c. stays the same.

10. As more lamps are connected to a parallel circuit, the overall current in the power source
    a. increases. b. decreases. c. stays the same.

11. Change the magnetic field in a closed loop of wire and you induce in the loop a
    a. voltage. b. current. c. electric field. d. all these. e. none of these.

12. A step-up transformer increases
    a. power. b. energy. c. both of these. d. neither of these.

1b, 2c, 3a, 4c, 5b, 6b, 7c, 8a, 9b, 10a, 11d, 12d.

Sample LIGHT Exam

Write the BEST answer to each of the following only in the small box. No penalty for guessing. Good Energy!

1. Which of the following does not fit in the same family?
   a. light wave. b. radio wave. c. sound wave. d.microwave. e. X-ray.

2. If the resonant frequency of the outer electron shells in atoms in a particular material match the frequency of green light, the material will be
   a. transparent to green light. b. opaque to green light.

3. If water naturally absorbed blue and violet light rather than infrared, water would appear
   a. green-blue, as it presently appears.
   b. a more intense green-blue.
   c. orange-yellowish.
   d. black.
   e. to have no color at all.

4. The sky is blue because air molecules in the sky act as tiny
   a. mirrors that reflect primarily blue light.
   b. oscillators that scatter high-frequency light.
   c. incandescant blue-hot sources.
   d. prisms.
   e. none of these.

5. The average speed of light is greatest in
   a. red glass. b. yellow glass. c. green glass. d. blue glass. e. all the same.

6. If different colors of light had the same speed in matter, there would be no
   a. rainbows. b. dispersion by prisms. c. colors from diamonds. d. all of these.

7. When light is refracted there is a change in its
   a. frequency. b. wavelength. c. both of these. d. none of these.

8. Lenses work because in different materials light has different
   a. wavelengths. b. frequencies. c. speeds. d. energies. e. none of these.

9. A fish outside water will see better if it has goggles that are
   a. tinted green-blue. b. flat. c. filled with water. d. none of these.

10. Waves diffract most when their wavelengths are
    a. long. b. short. c. same each way.

11. A hologram best illustrates
    a. resonance. b. interference. c. laser light. d. a new photography.

12. Which photons have the most energy of those listed below?
    a. red. b. white. c. blue. d. all the same.

1c, 2b, 3c, 4b, 5a, 6d, 7b, 8c, 9c, 10a, 11b, 12c.

Sample LIGHT Exam

Write the BEST answer to each of the following only in the small box. No penalty for guessing. Good Energy!

1. Which of the following does not fit in the same family?
   a. light wave. b. radio wave. c. sound wave. d.microwave. e. X-ray.

2. If the resonant frequency of the outer electron shells in atoms in a particular material match the frequency of green light, the material will be
   a. transparent to green light. b. opaque to green light.

3. If water naturally absorbed blue and violet light rather than infrared, water would appear
   a. green-blue, as it presently appears.
   b. a more intense green-blue.
   c. orange-yellowish.
   d. black.
   e. to have no color at all.

4. The sky is blue because air molecules in the sky act as tiny
   a. mirrors that reflect primarily blue light.
   b. oscillators that scatter high-frequency light.
   c. incandescant blue-hot sources.
   d. prisms.
   e. none of these.

5. The average speed of light is greatest in
   a. red glass. b. yellow glass. c. green glass. d. blue glass. e. all the same.

6. If different colors of light had the same speed in matter, there would be no
   a. rainbows. b. dispersion by prisms. c. colors from diamonds. d. all of these.

7. When light is refracted there is a change in its
   a. frequency. b. wavelength. c. both of these. d. none of these.

8. Lenses work because in different materials light has different
   a. wavelengths. b. frequencies. c. speeds. d. energies. e. none of these.

9. A fish outside water will see better if it has goggles that are
   a. tinted green-blue. b. flat. c. filled with water. d. none of these.

10. Waves diffract most when their wavelengths are
    a. long. b. short. c. same each way.

11. A hologram best illustrates
    a. resonance. b. interference. c. laser light. d. a new photography.

12. Which photons have the most energy of those listed below?
    a. red. b. white. c. blue. d. all the same.

1c, 2b, 3c, 4b, 5a, 6d, 7b, 8c, 9c, 10a, 11b, 12c.

Sample ATOMIC & NUCLEAR PHYSICS Exam

Write the BEST answer to each of the following only in the small box. No penalty for guessing. Good Energy!

1. Which of the following forms an interference pattern when directed toward two suitably-spaced thin slits?
   a. light. b. sound. c. electrons. d. all of these. e. none of these.

2. An electron and a baseball move at the same speed. Which has the shorter de-Broglie wavelength?
   a. electron. b. baseball. c. both the same.

3. Electrical forces within the atomic nucleus tend to
   a. hold it together. b. push it apart. c. neither of these.

4. Which of these experiences the greatest electrical force in an electric field?
   a. alpha particle. b. beta particle. c. gamma ray. d. none of these.

5. The radioactive half life of a certain isotope is 1 day. At the end of 3 days the amount remaining is
   a. 1/2. b. 1/4. c. 1/8. d. 1/16. e. none of these.

6. When U-238 emits an alpha particle, the nucleus left behind has
   a. 90 protons. b. 91 protons. c. 92 protons. d. 93 protons. e. 94 protons.

7. When U-239 emits a beta particle, the nucleus left behind has
   a. 90 protons. b. 91 protons. c. 92 protons. d. 93 protons. e. 94 protons.

8. When U-235 undergoes fission, the pair of nuclei that result have a total of
   a. less than 92 protons. b. 92 protons. c. more than 92 protons.

9. A nuclear proton has a greater mass in
   a. helium. b. iron. c. uranium. d. same in each.

10. If an iron nucleus undergoes fission, the masses of its nucleons
    a. increase. b. decrease. c. neither increase or decrease.

11. If an iron nucleus undergoes fusion, the masses of its nucleons
    a. increase. b. decrease. c. neither increase or decrease.

12. The type of ray that originates in the cosmos is the
    a. alpha ray. b. beta ray. c. cosmic ray. d. hoo ray!

1d, 2b, 3b, 4a, 5c, 6a, 7d, 8b, 9a, 10a, 11a, 12c.

Sample RELATIVITY Exam

Write the BEST answer to each of the following only in the small box. No penalty for guessing. Good Energy!

1. Which equation is the triumph of the theory of Special Relativity?
   a. $E = ma^2$. b. $E = mb^2$. c. $E = mc^2$. d. $E = md^2$. e. $E = me^2$.

2. Relativistic equations for time, length, and mass hold true for
   a. speeds near the speed of light.
   b. everyday low speeds.
   c. both of these.
   d. none of these.

3. According to special relativity, if you measure your pulse while traveling a very high speeds, you would notice that your pulse rate would be
   a. increased. b. decreased. c. no different.

4. When a light source approaches you, your measurements of it will show an increase in its
   a. speed. b. wavelength. c. frequency. d. all of these. e. none of these.

5. Because of relativistic effects, the masses of the electrons that are fired against the inside surface of a TV tube are
   a. slightly greater. b. slightly less. c. unchanged.

6. A spear has a rest mass of 1 kilogram. When properly thrown past you your instruments show it to have a mass of 2 kilograms. Your instruments also show the speed of the spear to be
   a. 0.5*c*. b. 0.75*c*. c. 0.87*c*. d. 0.99*c*. e. none of these.

7. Since there is an upper limit on the speed of a particle, there is also an upper limit on its
   a. momentum. b. kinetic energy. c. temperature. d. all of these.
   e. none of these.

8. Compared to special relativity, general relativity is more concerned with
   a. acceleration. b. gravitation. c. space-time geometry. d. all of these.
   e. none of these.

9. From a general relativistic point of view, compared to a watch at the earth's poles, a watch at the earth's equator should run
   a. faster. b. slower. c. at the same rate.

10. From a general relativistic point of view, a person on the ground floor of a skyscraper ages
    a. faster than a person on the top floor.
    b. slower than a person on the top floor.
    c. at the same rate as a person on the top floor.

11. If the orbit of Mercury were perfectly circular, its rate of precession would be
    a. larger. b. smaller. c. no different that it is now. d. zero.

12. An astronaut falling into a black hole would see the universe
    a. red shifted. b. blue shifted.

1c, 2c, 3c, 4c, 5a, 6c, 7e, 8d, 9a, 10b, 11d, 12b.

Sample Final Exam

Answer in detail, using examples and diagrams to make your points clear. Keep your answers to less than a page for each question.

1. "Inertia is that property of all matter that causes it to resist being moved." Is there anything wrong with this statement, and if not, why not, and if so, why so?
2. Distinguish between mass and weight. Why, for example, does a heavy object accelerate no more than a light object in free fall?
3. Give at least two examples of an object or collection of objects that have some (not zero) kinetic energy, but a net momentum of zero.
4. The force of gravity pulls apples off trees and they fall. The same force extends to the moon—so why doesn't the moon fall also?
5. What accounts for the fact that there are <u>two</u> ocean tides per day?
6. What evidence can you cite to support the atomic theory of matter?
7. Discuss the principles that account for the flight of both lighter-than-air and heavier-than-air aircraft?
8. How does a suction cup work?
9. Why does warm air rise?
10. Why is it that deep bodies of water, such as Lake Tahoe in California, do not freeze over even in the coldest of winters?
11. What is a sonic boom and how is it produced?
12. What does it mean to say the electrical outlets in your home are rated at 110 volts?
13. Explain how an electrical transformer operates.
14. Distinguish between light waves and sound waves.
15. What is the evidence for the claim that the stars are composed of the same elements found on earth?
16. Why is the sky blue and the sunsets orange?
17. What produces the spectrum of colors of gasoline on a wet street?
18. Distinguish between nuclear fission and nuclear fusion.
19. How is radioactivity used to determine the ages of ancient organic and inorganic objects?
20. What would be unusual about your observations of occupants in a space ship traveling at relativistic speeds. Assume you can clearly see them make any measurements you like.

A Do-It-Yourself Recipe for a Simple Electric Motor

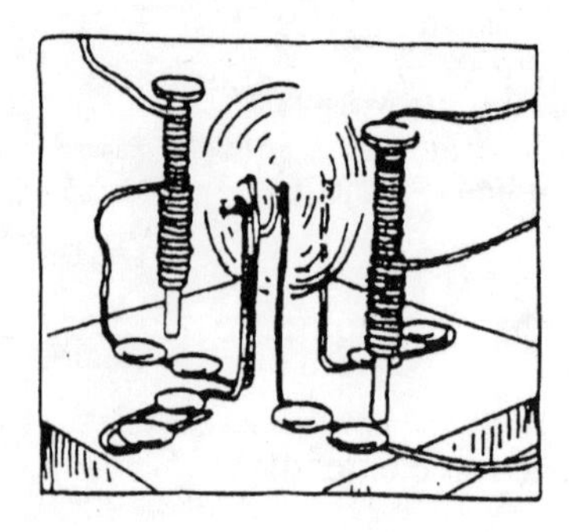

The finished motor shown to the left can be built with the following commonplace tools and materials: eight thumbtacks, three 2-inch paper clips, two 3½-inch nails, needle-nosed pliers, electrical or adhesive tape, a wooden board about five inches square, about seven feet of No. 20 insulated copper wire, and a knife to scrape the ends with. Two 1½-volt dry cells provide an adequate power supply.

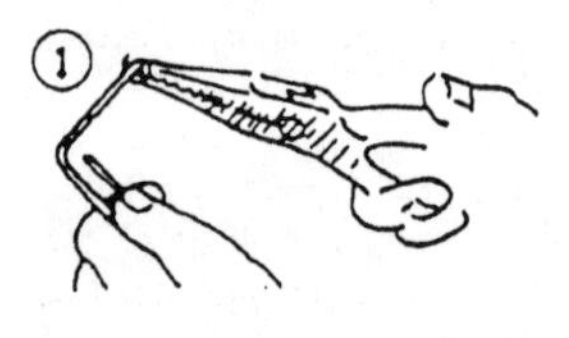

Step 1. The first step in making the motor is straightening the smaller loop of one of the paper clips, and then twisting it so that it stands upright at right angles to the larger loop. Then use the pliers to bend a tiny loop in the upright end. Do the same with a second paper clip.

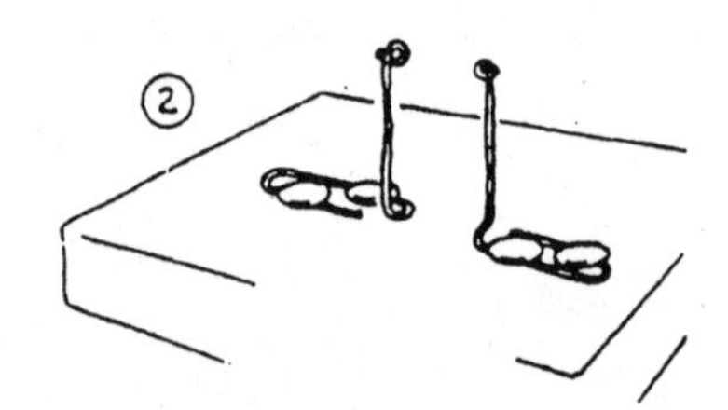

Step 2. Next, attach the paper clips to the board with tacks as shown. The upright ends of the clips should be about an inch apart. The tacks should be loose enough for final adjustment later. These clips are the supports for the axle of the motor's rotor.

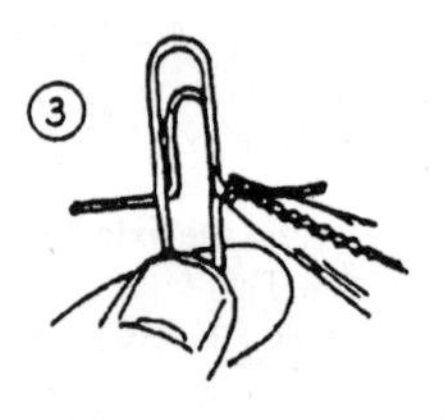

Step 3. Next make the rotor. With pliers, bend the ends of the third paper clip perpendicular to the clip's midpoint as shown. The ends, which will serve as the rotor's axle, should each be about a half-inch long.

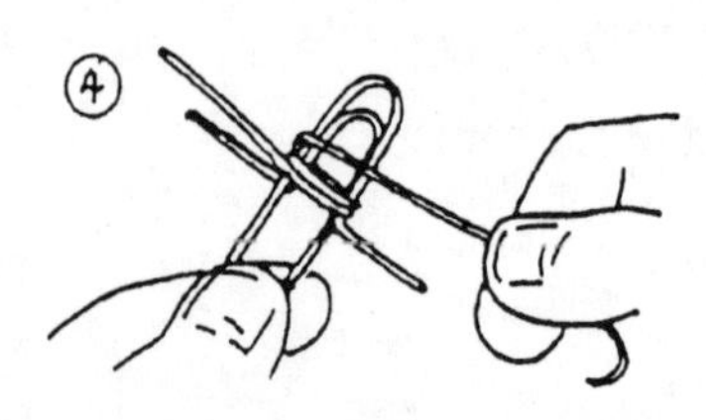

Step 4. Leaving one inch free, wrap the copper wire tightly around the rotor clip, working out from the middle. Wind the turns of wire closely together, but not so tightly that the clip is bent out of shape.

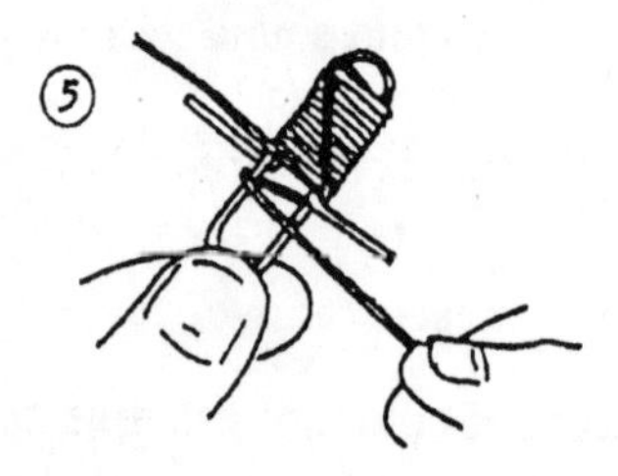

Step 5. Wrap about 20 coils out toward the end of the rotor clip. Then take the wire back to the center and wrap—in the same direction—an equal number of turns around the other half. These coils will make the clip an electromagnet.

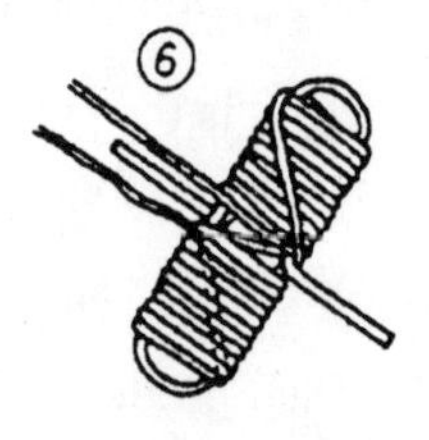

Step 6. When the copper wire has been wound around the second half of the rotor clip as shown, it is brought back to the center of the clip. The ends of the wire will serve as the rotor's *commutator*, which reverses its current with each rotation.

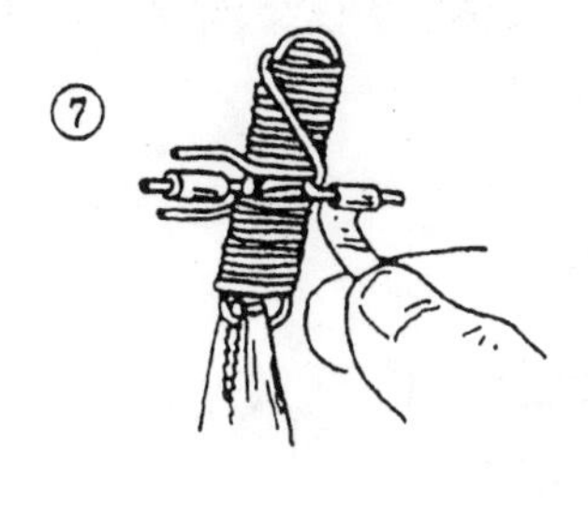

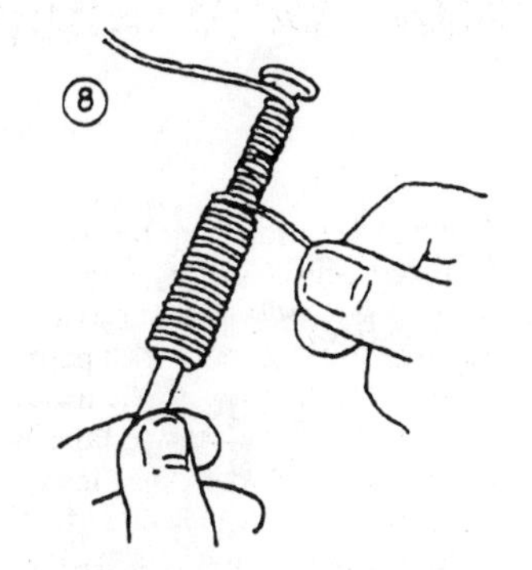

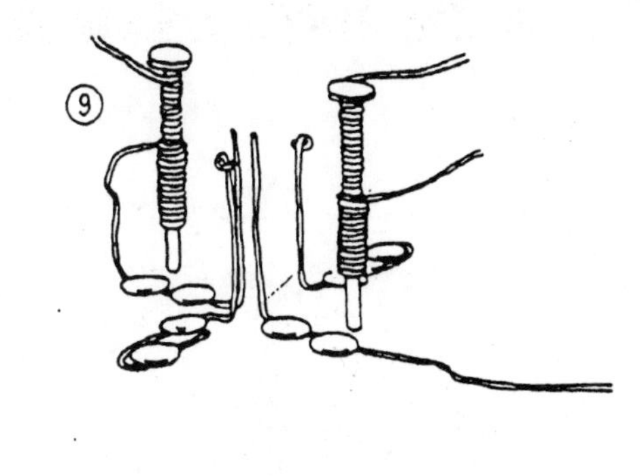

Step 7. The next step is to cut the ends of the wire so that they are slightly shorter than the projecting end of the clip. Then scrape the coating off the ends of the wires making sure to expose the bare copper.

Then take two strips of electrical or adhesive tape—each about $\frac{1}{4}$ inch wide and 2 inches long—and wrap them around the axle next to the clip as shown. This tape keeps the rotor-clip axle in the paper-clip supports. The center of gravity of the finished rotor should be along the axle so that it will twirl without wobbling.

Step 8. Make two stationary electromagnets by wrapping each nail with wire, leaving about 9 inches of wire free close to the head. Wind the wire evenly for about $2\frac{1}{2}$ inches down from the top, then about half-way back up again. Both nails should be wound in the same direction.

Leave about 6 inches of wire sticking out from the middle of each nail and cut it. Each nail should now have a 9-inch and a 6-inch tail. Scrape about $\frac{3}{4}$ inch of insulation from the end of each tail, exposing the bare copper.

Step 9. Hammer the nails into the board just far enough apart to make room for the rotor. Tack the 6-inch tail from one nail to the board. Lead it to within $\frac{1}{4}$ inch of either support and bend it up so its tip is slightly higher than the support. Do the same with an unattached 12-inch length of wire. These form the *brushes*. About $\frac{3}{4}$ inch of insulation should be thoroughly scraped from each end of the 12-inch length of wire. Now all loose ends of wire are scraped free of insulation.

Step 10. Fit the axle of the rotor into the loops of each support so that the rotor's commutators, when twirling, will make contact with the brushes. Twist the end of the 6-inch tail from the second nail around the 9-inch wire from the first nail. The 9-inch wire from the second nail will connect with the dry-cell terminal. Link the free end of the 12-inch wire to the opposite dry-cell terminal and the circuit is complete.

It is important to make final adjustments so that the rotor will spin freely. As the rotor spins, both commutators should touch the brushes simultaneously. Only then will current be established in the entire circuit, making the rotor and nails electromagnets. Each time the rotor makes a half-turn, the direction of current in the rotor alternates, changing its magnetic-field polarity. It may be necessary to give the rotor a gentle nudge for the motor to operate, just as you sometimes have to do with some types of electric shavers.

Utilizing the fact that like magnetic poles repel each other and unlike poles attract, can you explain the operation of this motor?

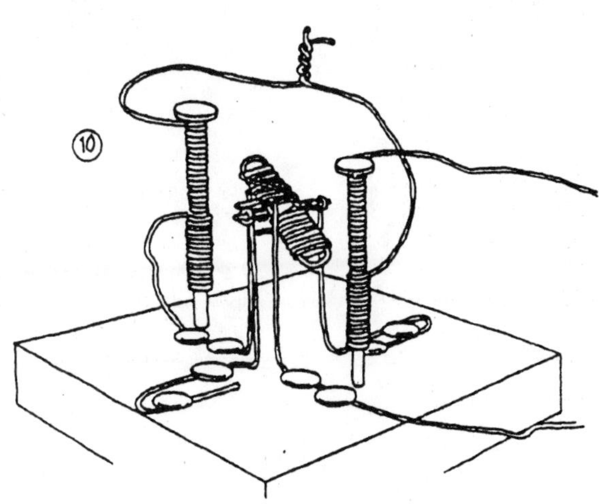

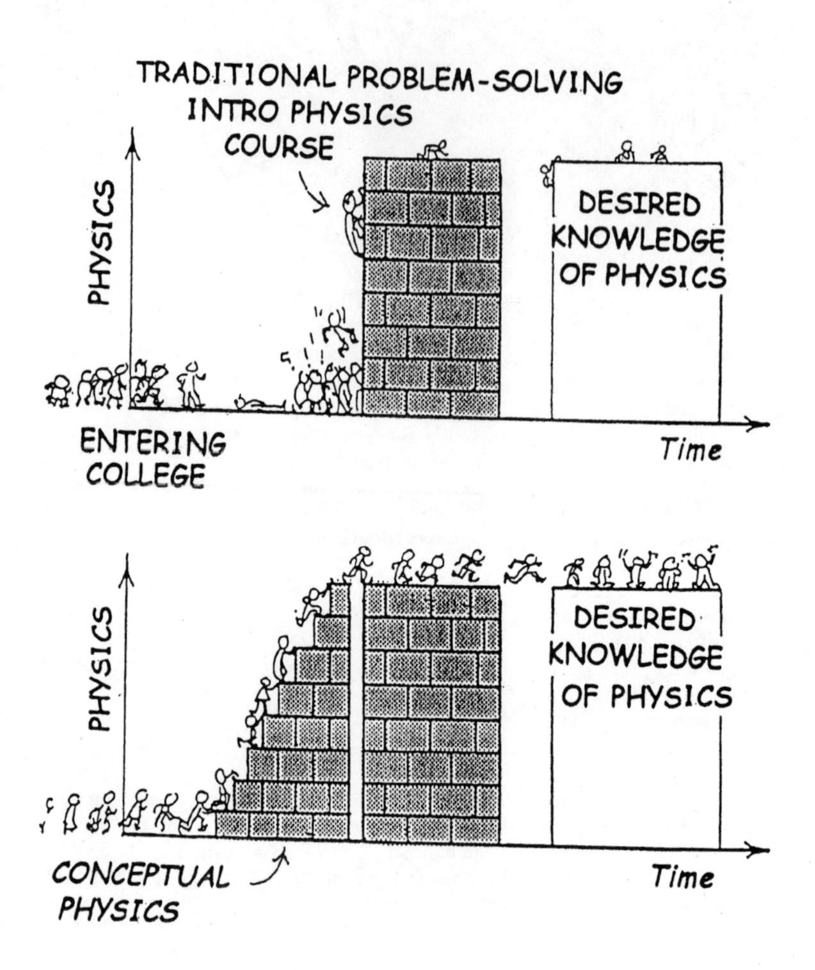
TRADITIONAL PROBLEM-SOLVING
INTRO PHYSICS
COURSE
PHYSICS
DESIRED
KNOWLEDGE
OF PHYSICS
ENTERING
COLLEGE
Time
PHYSICS
DESIRED
KNOWLEDGE
OF PHYSICS
CONCEPTUAL
PHYSICS
Time